ADVANCES IN CHEMICAL PHYSICS

VOLUME IX

ADVANCES IN CHEMICAL PHYSICS

Edited by I. PRIGOGINE
University of Brussels, Brussels, Belgium

VOLUME IX

INTERSCIENCE PUBLISHERS
a division of John Wiley & Sons Ltd., London - New York - Sydney

First Published 1965

Library of Congress Catalog Card Number 58 - 9935

PRINTED IN GREAT BRITAIN AT THE PITMAN PRESS, BATH

INTRODUCTION

In the last decades, chemical physics has attracted an ever increasing amount of interest. The variety of problems, such as those of chemical kinetics, molecular physics, molecular spectroscopy, transport processes, thermodynamics, the study of the state of matter, and the variety of experimental methods used, makes the great development of this field understandable. But the consequence of this breadth of subject matter has been the scattering of the relevant literature in a great number of publications.

Despite this variety and the implicit difficulty of exactly defining the topic of chemical physics, there are a certain number of basic problems that concern the properties of individual molecules and atoms as well as the behavior of statistical ensembles of molecules and atoms. This new series is devoted to this group of problems which are characteristic of modern chemical physics.

As a consequence of the enormous growth in the amount of information to be transmitted, the original papers, as published in the leading scientific journals, have of necessity been made as short as is compatible with a minimum of scientific clarity. They have, therefore, become increasingly difficult to follow for anyone who is not an expert in this specific field. In order to alleviate this situation, numerous publications have recently appeared which are devoted to review articles and which contain a more or less critical survey of the literature in a specific field.

An alternative way to improve the situation, however, is to ask an expert to write a comprehensive article in which he explains his view on a subject freely and without limitation of space. The emphasis in this case would be on the personal ideas of the author. This is the approach that has been attempted in this new series. We hope that as a consequence of this approach, the series may become especially stimulating for new research.

Finally, we hope that the style of this series will develop into something more personal and less academic than what has become the standard scientific style. Such a hope, however, is not likely to be completely realized until a certain degree of maturity has been attained—a process which normally requires a few years.

At present, we intend to publish one volume a year, but this schedule may be revised in the future.

In order to proceed to a more effective coverage of the different aspects of chemical physics, it has seemed appropriate to form an editorial board. I want to express to them my thanks for their cooperation.

I. PRIGOGINE

CONTRIBUTORS TO VOLUME IX

GORDON L. GOODMAN, Chemistry Division, Argonne National Laboratory, Argonne, Illinois, U.S.A.

JAROSLAV KOUTECKÝ, Institute of Physical Chemistry, Czechoslovak Academy of Sciences, Prague, Czechoslovakia

R. K. NESBET, I.B.M. San Jose Research Laboratory, San Jose, California, U.S.A.

HOWARD REISS, Director, North American Aviation Science Center, Thousand Oaks, California, U.S.A.

BERNARD WEINSTOCK, Scientific Laboratory, Ford Motor Company, Dearborn, Michigan, U.S.A.

ROBERT H. WENTORF, JR., General Electric Research Laboratory, General Electric Company, Schenectady, New York, U.S.A.

CONTENTS

SCALED PARTICLE METHODS IN THE STATISTICAL THERMODYNAMICS OF FLUIDS*

HOWARD REISS, *North American Aviation Science Center, Thousand Oaks, California*

CONTENTS

I. INTRODUCTION

During the past 25 years theories of the statistical mechanics of fluids have either been very formal and based upon first principles or they have been extremely heuristic, depending largely upon models. Examples of the formal theories are the Mayer cluster theory,[1] the radial distribution function theory,[2] and the Percus–Yevick theory, which develops a collective coordinate approach.[3] Examples of the more heuristic theories are free volume theories,[4] quasi-lattice theories,[5] and the significant structures theory due to Eyring.[6] The formal theories, based upon first principles, have not been too successful in providing numerical results for comparison with experiment. The heuristic theories have been more successful in this respect, but only at the expense of some real detachment from fundamentals.

On the other hand, the formal theories have strengthened our

* See Note Added in Proof, p. 84.

insight into the general properties of fluids; for example, into the natures of phase transitions and the internal structures of fluids, e.g., the radial distribution function. Cluster theory has shown us how to calculate the virial coefficients, and in the case of the first few coefficients such calculations have actually been performed. In fact, for the hard sphere fluid we are in possession of exact values for the first four virial coefficients,[7] and approximate but accurate values for some higher ones.[8]

The theory of liquids has also been attacked by machine computation techniques. These include calculations on systems of hard spheres[9–11] and even on systems in which the molecules attract one another with a variety of intermolecular potentials.[12] Machine calculations have been of two types, either involving Monte Carlo techniques[9–10] or based on the direct integration of the dynamical equations of motion.[11–12] Only systems containing small numbers of molecules have been considered, and for the purpose of simulating a macroscopic assembly periodic boundary conditions have been applied. With such boundary conditions a molecule leaving the container at a certain point re-enters through an image point. Just how much approximation is introduced by this device is not clear, but it has been observed that changing the number of molecules in the unit cell of this periodic system does not seem to change the result.

During the last five years a new theory has appeared. This has come to be known as the scaled particle theory of fluids. The first paper on the subject was published in 1959 by Reiss, Frisch, and Lebowitz[13] and is concerned primarily with systems of hard spheres.* Since then a number of papers have been published by Reiss and his co-workers, and others. These have concerned themselves with various idealized systems such as hard spheres, hard discs, and one-dimensional systems, as well as real fluids, and particularly with attempts to provide numerical results which can be compared with experiment. We shall draw attention to these papers and the relevant authors in the body of this text. For the moment, however, we limit our comments to what must be considered the remarkable success of the theory.

* In this connection the reader is referred to the excellent discussion of hard sphere systems presented by H. L. Frisch in Volume VI in this series. Frisch also reviews some aspects of scaled particle theory.

In the first place, the theory is formal and based on first principles to the same extent as the cluster theory, the radial distribution function theory, and the Percus–Yevick theory. At the same time, however, it permits one to derive numbers which agree very well with experiment. *A particularly dramatic feature of the theory is the fact that it contains no adjustable parameters and that the numbers which agree with experiment are obtained from formulas containing quantities which are derived from independent measurements.* Since the applicable theories, either formal or heuristic, have been useful in separate ways, it represents a distinct departure to discover a single theory which can by itself accomplish the kinds of things that the previous theories have accomplished separately. In fact, its degree of correlation with experiment is as good if not better than that found in connection with most of the heuristic theories.

The treatment of hard spheres leads among other things to an analytical expression for the equation of state which agrees almost precisely with the equations of state obtained through the medium of machine computation. Furthermore, this simple analytical expression when expanded as a power series in the density yields virial coefficients for the system of hard spheres which are in excellent agreement with those computed by cluster theory. For example, it turns out that the first three virial coefficients are exact, the fourth virial coefficient is in error by some 3%, and the fifth in error by 4 or 5%. Another remarkable development is the fact that an exact solution of the Percus–Yevick equation for hard spheres has been obtained recently by Wertheim,[14] and that this solution leads to an equation of state for hard spheres identical with that obtained by means of the scaled particle theory. This is the first time that two theories of fluids, both professing to be based on first principles (although containing approximations), have led to identical results. For example, the radial distribution functions obtained on the one hand by the solution of the Kirkwood integral equation and on the other hand from the Born–Green and Yvon integral equations[2] are different. The equations themselves are exact, but when the superposition approximation[15] is introduced in order to effect closure different approximate results emerge.

Attempts to extend the scaled particle theory to real fluids in an effort to derive their equations of state have not met with the success obtained in connection with hard spheres. On the other hand, many exact relations can be derived, and this aspect of the theory is still under development.

Much success has been achieved with real fluids by relinquishing the ideal of deriving the entire equation of state. It proves possible to compute such things as compressibilities, expansivities, heats of vaporization, and heats of fusion, as well as surface tensions of real liquids. Even fused salts can be treated in this manner. Simple formulas become available which specify these quantities in terms of temperature, a parameter defining the hard core diameter of the molecule,* and the measured density. Since the density must be measured and is not determined by theory, the result falls short of the ideal of supplying the equation of state. Nevertheless, the formulas are extremely valuable. Numbers which agree with experiment are obtained without the introduction of models such as the quasi-crystalline fluid, the free volume or cell model, or the Eyring significant structures concept. That so much can be achieved once the density of the fluid is known suggests that we have solved the more crucial problem in the theory of dense fluids (fluids having densities common to most liquids).

A careful examination of scaled particle theory reveals that it is geometric in nature and deals in a rigorous manner with the problem of the packing in a sufficiently dense fluid of molecular hard cores. Apparently what has been suspected by several authors for some time, that the most important problem in the theory of liquids is concerned with the packing of hard cores, is actually so. One is thus driven to a model of the liquid state which is a direct outgrowth of fundamental considerations. In this model the soft intermolecular potential (or the non-hard-core part of the potential) acts primarily to establish the overall density of the fluid, while the internal structure is determined by the packing of the hard cores. Thus it might be said that the

* This hard core diameter *is not* used as an adjustable parameter when the theory is applied to experiment. It has a well defined physical meaning and can usually be determined by some sort of independent measurement (see Sections VII–X, and especially Section X).

soft potential determines the *volume* of a container which in turn is filled with a hard sphere fluid.

Another aspect of scaled particle theory is this. Although it is formal and based on first principles, each step in its mathematical development bears a close relation to the associated physical picture. Thus, it is possible to discuss those features of the theory which bear directly on experiment in a highly conceptual manner, and the experimentalist who is not interested in working through the bulk of the mathematical formalism can acquire insight and understanding. Because the theory is of interest to *both* theorists and experimentalists we have chosen to present it in a manner which should please the tastes of both. Thus Sections II, III, IV, V, and VI are formal and systematic in nature. Exact relations are derived from the basic principles of statistical mechanics and are used to establish equations of state. The remaining sections are devoted to a less ambitious but nevertheless fruitful application of scaled particle theory to real fluids, and to a comparison of the results derived from it with those obtained from experiment. These last sections are written so that an experimentalist can begin with them (provided that he accepts certain relations which have been derived in the earlier sections). Many of these relations can be made plausible intuitively. For their precise justification, however, it will be necessary to refer to the earlier sections.

II. SOME EXACT RELATIONS

The scaled particle theory of fluids is so named because it makes use of a coupling parameter which measures the breadth of a molecule and its potential field rather than the amplitude of this potential. A molecule is coupled formally to a fluid by letting it "grow" in size until it achieves the scale of its neighbors.

We shall restrict attention to fluids composed of spherically symmetric molecules which may exert attractive forces upon one another, but which possess hard cores of diameter a. No real fluid exactly satisfies this requirement, but at low enough temperatures the repulsive part of the intermolecular potential rises so steeply that its increment for a small fractional change of the intermolecular distance is many times kT (where k is the Boltzmann constant and T is the temperature). Thus, it is possible to supply a reasonably unambiguous value for a.* However, it must be expected that the best choice for a is likely to depend upon the temperature, smaller values being more suitable at higher temperatures.†

Consider a fluid consisting of N molecules contained within a volume V such that the concentration is $\rho = N/V$. Within this fluid consider a spherical region of radius r. We denote by $p_0(r)$ the probability that no molecule has its center within this spherical cell. In passing we note that a spherical cavity would be created within the original fluid of N molecules with diameters a if a hard molecule of diameter b were inserted into the fluid such that

$$r = \frac{a + b}{2} \tag{1}$$

We shall endeavor to calculate $p_0(r)$.

To illuminate the situation it is worthwhile to examine the probability $p_0(r + dr)$ that the centers of all molecules are excluded from the augmented spherical cell of radius $r + dr$. Now the probability that the spherical shell of thickness dr and volume $4\pi r^2 dr$ contains the center of a molecule is $4\pi r^2 \rho G(r) dr$, where $G(r)$ is so defined that $\rho G(r)$ measures the concentration of molecular centers just outside the spherical exclusion region. Thus $G(r)$ measures the conditional probability that the center of a molecule will be found within the spherical shell at r when the region enclosed by the shell is *known* to be free of molecular centers. The chance that the spherical shell is itself free of centers is

$$1 - 4\pi r^2 \rho G(r) dr \tag{2}$$

For the volume of radius $r + dr$ to be devoid of centers, it is necessary that the volume of radius r and the shell of thickness dr be *simultaneously* free of centers. Thus the chance $p_0(r + dr)$

* Although some of the relations of the scaled particle theory can be derived without the requirement of this hard core, without it the general usefulness of the method is severely limited, and in what follows we shall frequently demand that the molecules do possess hard cores.

† Methods are available (see Section X) for determining a by independent measurements at each temperature. Thus a is not an adjustable parameter.

that the volume of radius $r + dr$ is empty is given by the product of $p_0(r)$ and Eq. (2)

$$p_0(r + dr) = p_0(r)[1 - 4\pi r^2 \rho G(r)dr] = p_0 + \frac{\partial p_0}{\partial r}\,dr \tag{3}$$

or

$$\frac{\partial \ln p_0}{\partial r} = -4\pi r^2 \rho G(r) \tag{4}$$

Upon integration we obtain

$$p_0(r) = e^{-\int_0^r 4\pi\lambda^2 \rho G(\lambda)d\lambda} \tag{5}$$

where the initial condition

$$p_0(0) = 1 \tag{6}$$

has been applied. This asserts that there is unit chance for a molecule not to have its center in a region of zero radius.

Thus we have been able to express $p_0(r)$ in terms of $G(r)$, but this only shifts the problem to the determination of $G(r)$. Nevertheless, Eq. (5) will prove useful as an intermediate result.

We shall now express $p_0(r)$ in another way. Suppose $p_1(r)$, $p_2(r)$, etc. represent the probabilities that the centers of *exactly* one particle, *exactly* two particles, etc. lie in the spherical region of radius r. Then

$$p_0 + p_1 + p_2 + p_3 + \cdots = 1 \tag{7}$$

since either no particles or *some* combination of them have centers in the region. Equation (7) may be manipulated algebraically as follows:

$$\begin{aligned} p_0 + \sum_{m=1}^{\infty} p_m &= p_0 - \sum_{m=1}^{\infty} \{[1 + (-1)]^m - 1\}p_m \\ &= p_0 - \sum_{m=1}^{\infty} p_m \sum_{n=1}^{\infty} (-1)C_n^m \\ &= p_0 - \sum_{n=1}^{\infty} (-1)^n \sum_{m=n}^{\infty} C_n^m p_m = 1 \end{aligned} \tag{8}$$

where the C_n^m are the binomial coefficients.

Solving for p_0 we have

$$p_0 = 1 - (p_1 + 2p_2 + 3p_3 + \dots) + (p_2 + 3p_3 + 6p_4 + \dots) - (p_3 + 4p_4 + \dots) + \dots \tag{9}$$

Now the first sum in brackets represents the average number of individual particles to be found in the region of radius r, while the second sum is the average number of pairs, the third the average number of triplets, and so on. This is obvious because the average number of n-tuplets I_n is given by

$$I_n = \sum_{m=n}^{\infty} C_n^m p_m \tag{10}$$

where C_n^m is the number of distinct n-tuplets which can be fashioned from m particles. Thus, in the second bracketed expression in Eq. (9) (which refers to pairs, i.e., $n = 2$) $C_2^2 = 1$, $C_2^3 = 3$, $C_2^4 = 6$, etc., because only one distinct pair can be formed from two particles, three pairs can be formed from three particles, six from four, and so on.

We introduce the n-particle correlation function $g^{(n)}(1, 2, \dots, n)$ (ref. 16), where the numbers in the brackets refer to the coordinates of position of the volume elements $d\tau_1, d\tau_2, \dots, d\tau_n$, and $\rho^n g^{(n)}(1, 2, \dots, n) d\tau_1 d\tau_2 \dots d\tau_n$ is the probability that the n volume elements simultaneously contain the centers of particles, i.e., that an n-tuplet occupies the set of volume elements. Therefore, if we integrate $\rho^n g^{(n)} d\tau_1 d\tau_2 \dots d\tau_n$ throughout a spherical cell of radius r, we should obtain the sum of the average numbers of n-tuplets found in each distinct set of volume elements and hence the average number of n-tuplets in the cell. This is true except for the fact that the integration contains $n!$ indistinguishable permutations of volume elements so that the final result should be divided by $n!$; thus we arrive at the result

$$I_n = \sum_{m=n}^{\infty} C_n^m p_m = \frac{\rho^n}{n!} \int_{\left[\substack{\text{cell of}\\ \text{radius } r}\right]} g^{(n)} d\tau_1 d\tau_2 \dots d\tau_n \tag{11}$$

which upon substitution into (8) or (9) yields

$$p_0(r) = 1 + \sum_{n=1}^{\infty} \frac{(-1)^n \rho^n}{n!} \int_{\left[\substack{\text{cell of}\\ \text{radius } r}\right]} g^{(n)} d\tau_1 d\tau_2 \dots d\tau_n \tag{12}$$

For molecules possessing hard cores with diameter a, $g^{(n)}$ will be zero (for all values of $n \geqslant n^*$, where n^* depends upon r) when $d\tau_1, d\tau_2, \ldots, d\tau_n$ all lie within the cell of radius r. Thus if $r \leqslant a/2$, it is clear that only the center of one molecule at a time can lie within the cell, otherwise hard cores would be forced to overlap. Similarly, for $a/2 < r \leqslant a/\sqrt{3}$ only two molecules may simultaneously occupy the cell. With $r = a$, twelve molecules can be accommodated. As a result of this exclusion feature the series in Eq. (12) contains a finite number of terms as long as r is finite. This feature of Eq. (12) has many valuable consequences, some of which have still to be explored.

With $r \leqslant a/2$ there is only room for one molecule in the cell and Eq. (12) becomes

$$p_0(r) = 1 - \rho\int_0^r 4\pi\lambda^2 g^{(1)} d\lambda = 1 - \tfrac{4}{3}\pi r^3\rho, \quad r \leqslant a/2 \tag{13}$$

since in a fluid $g^{(1)}$ is always unity.[17] Thus, in this case, an exact expression for $p_0(r)$ can be derived. Substituting (13) into (4) then yields

$$G(r) = [1 - \tfrac{4}{3}\pi r^3\rho]^{-1}, \quad r \leqslant a/2 \tag{14}$$

so that an exact expression for $G(r)$ is also available when $r \leqslant a/2$.

In the region ($a/2 < r \leqslant a/\sqrt{3}$) two molecules can be accommodated and Eq. (12) becomes (using Eq. (13))

$$p_0 = 1 - \tfrac{4}{3}\pi r^3\rho + \frac{\rho^2}{2}\int_{\left[\substack{\text{Cell of}\\ \text{radius } r}\right]} g^{(2)} d\tau_1 d\tau_2, \quad \frac{a}{2} < r \leqslant \frac{a}{\sqrt{3}} \tag{15}$$

Now $g^{(2)}(1, 2)$ depends only upon the relative coordinates of molecules 1 and 2, and because of this it is possible in Eq. (15) to integrate by parts.[18] The result is

$$p_0(r) = 1 - \tfrac{4}{3}\pi r^3\rho + \frac{\rho^2}{2}\int_0^{2r} 4\pi s^2 g(s)\left[\frac{\pi r^3}{3}\left\{4 - \frac{3s}{r} + \frac{s^3}{4r^3}\right\}\right] ds, \quad r \leqslant \frac{a}{\sqrt{3}} \tag{16}$$

where g (the radial distribution function) is $g^{(2)}$, and

$$s = |\mathbf{r}_2 - \mathbf{r}_1| \tag{17}$$

where $\mathbf{r}_2$ and $\mathbf{r}_1$ are the position vectors of particles 2 and 1, respectively. Equating the right-hand sides of Eqs. (5) and (16) and differentiating with respect to r then yields the following relations for $r \geqslant a/2$:

$$\left(\frac{\partial G}{\partial r}\right)_{r=a/2} = \frac{\pi a^2 \rho}{(1 - \frac{1}{6}\pi a^3 \rho)^2} \tag{18}$$

$$\left(\frac{\partial^2 G}{\partial r^2}\right)_{r=a/2} = \frac{4\pi a \rho}{(1 - \frac{1}{6}\pi a^3 \rho)^2} + \frac{2\pi^2 a^4 \rho^2}{(1 - \frac{1}{6}\pi a^3 \rho)^3} - \frac{8\pi a \rho}{(1 - \frac{1}{6}\pi a^3 \rho)} g(a) \tag{19}$$

The right-hand side of Eq. (18) is the derivative at $r = a/2$ of (14). Since (18) represents the derivative at $r = a/2$ of the form of G which applies in the interval $(a/2, a/\sqrt{3})$, this means that the first derivative of G is continuous at $r = a/2$, the first two terms on the right of Eq. (19) are the second derivative of Eq. (14), but the third term is something new. Consequently, the second derivative of G at $a/2$ possesses a discontinuity of magnitude

$$-\frac{8\pi a \rho}{(1 - \frac{1}{6}\pi a^3 \rho)} g(a) \tag{20}$$

It is evident that this discontinuity is related to the appearance of a new physical process at $r = a/2$, namely, the process in which the centers of two molecules are accommodated simultaneously in the spherical cell for the first time. The possibility of this new configuration activates an additional term in Eq. (12). Equation (20) shows that the discontinuity is proportional to $g(a)$, which measures the probability that two molecules are in contact. Thus, the discontinuity is proportional to the probability of the new event, for when the radius of the spherical cell is exactly $a/2$ any two molecules whose centers simultaneously occupy it must be in contact.

In ref. 13 a more detailed study of discontinuities of this sort is presented. Discontinuities appear whenever it becomes possible to increase the number of particles which can simultaneously occupy the cell. Thus at $r = a/\sqrt{3}$, when it becomes possible to include *three* particles, there is a discontinuity proportional to the triplet correlation function $g^{(3)}(a, a, a)$ in the *fourth* derivative of G. At this point the first three derivatives of G are continuous.

It appears as though the order of the derivative in which the discontinuity first appears increases by two at each succeeding singular point. Thus $G(r)$ is non-analytic but quite smooth because the singularities occur in the second or higher order derivatives. One might expect that G will be approximated rather closely by a suitably chosen analytic function.

In addition to (5) and (12) there is another useful way to express $p_0(r)$. We appeal to the well known theorem of statistical mechanics[19] which states that the probability of observing some fluctuated configuration of a statistical system is given by $\exp(-W/kT)$, where W is the reversible work required to produce the fluctuated configuration through the application of a constraint. Thus

$$p_0(r) = e^{-W(r)/kT} \tag{21}$$

where $W(r)$ is the reversible work required to produce a spherical cavity of radius r in the fluid. Comparing this expression with Eq. (5) we see that

$$W(r) = kT\rho \int_0^r 4\pi\lambda^2 G(\lambda)\,d\lambda \tag{22}$$

This expression might have been arrived at simply in another manner. Suppose the cavity is produced (as we have suggested earlier) by expanding a hard sphere immersed in the fluid. The diameter b of this hard sphere is specified by Eq. (1). Then the force exerted by the fluid on the surface of the sphere will be the kinetic force due to encounters with the surrounding molecules. The force per unit area will be kT times the local concentration, and for the entire spherical surface it is

$$f(r) = 4\pi r^2 kT\rho G(r) \tag{23}$$

Consequently, in expanding through the interval dr the sphere must perform the work

$$dW = f(r)dr = 4\pi r^2 kT\rho G(r)dr \tag{24}$$

Integrating from $r = 0$ we obtain Eq. (22).

Substituting (14) into (22) we have

$$W(r) = -kT \ln\left(1 - \tfrac{4}{3}\pi r^3\rho\right), \quad r \leqslant a/2 \tag{25}$$

so that for $r \leqslant a/2$, an exact expression is available for $W(r)$. In general, the ability to arrive at these exact expressions for $r \leqslant a/2$ is a direct consequence of the limited occupancy of the cell in this range. Only the center of one molecule can be accommodated, and therefore molecular correlation need not be considered.

III. SPECIALIZATION TO HARD SPHERES; PARTICLE SCALING

The equation of state for a system of hard spheres can be expressed in the following form:

$$p/kT = \rho + \tfrac{2}{3}\pi a^3 \rho^2 g(a, \rho, T) \tag{26}$$

where a represents the diameter of a typical sphere and ρ is the concentration. In Eq. (26) we have indicated the temperature and concentration dependence of the pair correlation function g explicitly. In general, g and G depend upon both temperature and concentration, but in the interest of notational simplicity we shall not always indicate this dependence. No confusion should result from this simplification. Equation (26) can be derived from the virial theorem[20] or by another method due to H. S. Green.[21] It is instructive to note that values of g for all values of its distance argument are not required for the equation of state of hard spheres. Only the value at a appears in Eq. (26), i.e., the value of the pair correlation function for two molecules in contact. Thus, for the equation of state it appears wasteful to attempt the evaluation of the pair correlation function for all values of its arguments. This is one of the factors which motivated the initial development of scaled particle theory.

Let us return for the moment to normal (non-hard-sphere) fluids. The quantity $\rho G(r)$ has been defined as the local concentration of molecular centers adjacent to a spherical cavity of radius r from which all molecular centers have been excluded. As we know, it is also possible to regard the cavity as having been produced by the addition of a hard sphere of diameter b such that b satisfies Eq. (1). This hard sphere may be regarded as (and is) a *solute* molecule whose intermolecular potential with the solvent molecules (molecules of diameter a) involves their hard cores and not the soft potentials which these molecules also possess.

From this point of view $W(r)$ in Eq. (22) is the reversible work which must be expended in adding a solute molecule to a fixed point in a pure solvent. Furthermore, $\rho G(r)$ measures the concentration of solvent molecules in contact with the solute, so that if $g_{ab}^{(2)}(R)$ is the pair correlation function between a solvent and a solute molecule whose centers are separated by the distance R

$$G(r) = g_{ab}^{(2)}(R = r = (a + b)/2) \tag{27}$$

The reversible work which must be performed in adding a solute molecule to a fixed point is part of the chemical potential μ_b of that solute in the solution. When the solute is free to wander throughout the entire fluid an additional term corresponding to a free energy of mixing must be included. This quantity is rigorously[13]

$$kT \ln \Lambda_b^3 \rho_b \tag{28}$$

where Λ_b^3 is the reciprocal of the momentum partition function and is given by

$$\Lambda_b^3 = h^3/(2\pi m_b kT)^{3/2} \tag{29}$$

where h is Planck's constant and m_b is the mass of the solute. ρ_b is the solute concentration; in this case $1/V$ since only one solute molecule has been added. Equation (28), which is based upon random mixing, applies in dilute solution when more than one solute molecule is present, provided that the solution is sufficiently attenuated so that the solute molecules do not interact with one another. In this case ρ_b is larger than $1/V$.

The chemical potential of the solute is then the sum of Eqs. (22) and (28)

$$\mu_b = kT \ln \Lambda_b^3 \rho_b + kT\rho \int_0^r 4\pi\lambda^2 \rho G(\lambda)\, d\lambda \tag{30}$$

For this special case the origin of the term "scaled particle" is now revealed, since the molecule can be added by scaling up from an initial radius of $-a/2$ to a final radius $b/2$ and the reversible work can be computed during this process. The fact that the initial radius is negative stems from Eq. (1) and the requirement $r = 0$ in order that the solute be completely absent. If one wishes, one may interpret the negative radius as corresponding to a situation involving a *point* molecule having the ability to penetrate the solvent molecules to the depth $a/2$. Such a point molecule permits the assembly of solvent molecules to assume all of the configurations which were accessible before the addition of the solute. In this circumstance the solute brings nothing with it except its free energy of mixing. In any event, the added solute is a formal device and the interpretation of the negative radius should cause no difficulty.

It should also be noted, however, that the addition of a point molecule which cannot penetrate the cores of the solvent molecules requires the expenditure of a considerable amount of work. Such a molecule has $b = 0$ so that according to Eq. (1) it is equivalent to a cavity of radius $r = a/2$; thus $W(a/2)$ can be computed using Eq. (25), and we have

$$\mu_b = kT \ln \frac{\Lambda_b^3}{V} - kT \ln \left(1 - \frac{\pi a^3 \rho}{6}\right) \tag{31}$$

This is an exact expression.

The fact that λ is used in (31) to denote the changing size of the solute molecule has led Helfand[22] to call the solute molecule a λ-cule. However, Helfand defines a more general sort of λ-cule than the hard sphere employed here. This generalization will be discussed in Section V.

Let us return to the consideration of a fluid of hard spheres, each sphere having a diameter a. If the λ-cule to which Eq. (30) refers is permitted to grow until its radius is $a/2$, it now becomes a solvent molecule since all the molecules in the fluid, both solute and solvent, are now hard spheres. According to (30) the chemical potential of the solvent is then

$$\mu = kT \ln \Lambda^3 \rho + kT\rho \int_0^a 4\pi\lambda^2 G(\lambda) d\lambda \tag{32}$$

where Λ^3 now contains m, the mass of the solvent molecule. Furthermore, in this case Eq. (27) requires

$$G(a) = g_{aa}^{(2)}(R = a) = g(a) \tag{33}$$

so that for the particular value of its argument, $R = a$, $g(a)$ has the same numerical value as $G(a)$. This represents a key relation.

It was mentioned earlier in connection with Eq. (27) that the determination of the hard sphere equation of state did not require full information concerning the dependence of the pair correlation function upon its distance argument R, but only the numerical value $g(a)$ of g for molecules in contact. For this purpose it is sufficient to know the dependence of G upon its argument r, for then as a special case the numerical value of G at $r = a$, namely $G(a)$, will be known. But according to Eq. (33) this is also the numerical value of $g(a)$.

It turns out that it is easier to gather detailed information concerning G than g. Thus it is expedient to derive the equation of state by studying G. For the remainder of this section and also in the next we shall employ $G(a)$ in place of $g(a)$ wherever the latter occurs. As a first step we can replace $g(a)$ in Eq. (27) by $G(a)$, so that we obtain

$$\frac{p}{kT} = \rho + \tfrac{2}{3}\pi a^3 \rho^2 G(a) \tag{34}$$

Furthermore, under this condition Eq. (19) becomes

$$\left(\frac{\partial^2 G}{\partial r^2}\right)_{r=a/2} = \frac{4\pi a\rho}{(1 - \frac{1}{6}\pi a^3\rho)^2} + \frac{2\pi^2 a^4 \rho^2}{(1 - \frac{1}{6}\pi a^3\rho)^3} - \frac{8\pi a\rho}{(1 - \frac{1}{6}\pi a^3\rho)}\, G(a) \tag{35}$$

This interesting relation connects the properties of G at $r = a/2$ and at $r = a$. In doing so it bridges over the twelve singular points between $a/2$ and a.

We can derive another expression for the pressure p besides the one contained in (34). This time the pressure is related to the chemical potential μ by a straightforward thermodynamic formula which is essentially the Gibbs–Duhem relation.[23] Thus

$$p = \int_0^\rho \rho' \left(\frac{\partial \mu}{\partial \rho'}\right)_T d\rho' \tag{36}$$

Substitution of (32) in (36) together with an integration by parts yields

$$\frac{p}{kT} = \rho + \rho^2 \int_0^a 4\pi r^2 G(r, \rho)\,dr - \int_0^\rho \rho' d\rho' \int_0^a 4\pi r^2 G(r, \rho')\,dr \tag{37}$$

where G is now written as $G(r, \rho)$ to emphasize the explicit dependence of G upon ρ. Equating the right-hand sides of (34) and (37) gives

$$\tfrac{2}{3}\pi a^3\rho^2 G(a, \rho) = \rho^2 \int_0^a 4\pi r^2 G(r, \rho)dr - \int_0^\rho \rho' d\rho' \int_0^a 4\pi r^2 G(r, \rho')dr \quad (38)$$

which is the desired relation.

Equation (38) influences the form of G in the interval $r = 0$ to $r = a$. Actually G is known exactly (Eq. (14)) in the interval $r = 0$ to $r = a/2$, so that the range over which (38) must be applied is even more limited, extending only from $r = a/2$ to $r = a$. In this sense Eq. (38) may be regarded as still another connection formula, associating the properties of G at $r = a/2$ and $r = a$. However, (38) is more than a connection formula, for it influences the behavior of G in the entire interval between $a/2$ and a.

The discussion of Section II indicated that singularities in G occurred in high order derivatives so that G was at least smooth if not analytic. It might be expected therefore that a simple form for G in which the r-dependence is specified and which contains arbitrary ρ-dependent parameters will, when made to conform to Eqs. (14), (18), (35), and (38), provide physically reasonable forms for the ρ-dependence of these parameters. Thus G might be represented by a power series in r with ρ-dependent coefficients. Determination of the forms of these coefficients should then provide a simple approximation for G.

This method has been utilized in ref. 13, and it is demonstrated there that the approximation is fairly good considering its simplicity. With G known in this way, it may be substituted into Eq. (34) to yield the equation of state.

Insight into the accuracy of this result is obtained by expanding the resulting equation of state as a power series in the density in order to generate the virial coefficients. These may then be compared with the exact values.[7,8] As we have indicated, the agreement is fairly good. The reader is referred to ref. 13 for details.

Equation (38) together with the appropriate boundary conditions fails to determine G uniquely because the integrals in it possess definite rather than indefinite limits on r. However, the limits on ρ are indefinite so one may expect it to furnish a powerful tool for the exploration of the dependence of G upon ρ, once its dependence upon r is known. Below, another integral equation will be derived whose limits upon r are indefinite. Although this equation will prove useful, it contains a term whose functional dependence on G is not specified in a manner which permits the unique determination of G.

Before deriving this equation, however, it is possible to demonstrate still another exact condition which limits the form of G, this time in the interval $r = a$ to $r = \infty$. To accomplish this we note that $\rho G(\infty)$ measures the concentration of molecules next to a rigid plane wall; for as the radius of the spherical cavity becomes infinite the curvature vanishes. As in Eq. (23) the force per unit area (on the plane wall, the pressure) is given by kT times the local concentration. Thus the pressure of the fluid is

$$p = kT\rho G(\infty) \tag{39}$$

Combining this with (34) yields

$$G(\infty) = 1 + \tfrac{2}{3}\pi a^3 \rho G(a) \tag{40}$$

which connects the properties of G at $r = a$ with those at $r = \infty$.

We now turn to the task of deriving the more general integral equation promised above. Consider a fluid containing solutes of diameter b and solvent molecules of diameter a. In particular, assume that the solution is dilute so that the concentration of solute ρ_b is very much less than the concentration of solvent ρ. The generalization of Eq. (26) for the pressure of this solution may be shown to be[24]

$$\frac{p}{kT} = \rho + \rho_b + \tfrac{2}{3}\pi a^3 \rho^2 g_{aa}(a) + \tfrac{4}{3}\pi \left(\frac{a+b}{2}\right)^3 \rho\rho_b g_{ab}\left(\frac{a+b}{2}\right) + \tfrac{2}{3}\pi b^3 \rho_b^2 g_{bb}(b) \tag{41}$$

where $g_{aa}(a)$ is the radial distribution function on contact between two solvent molecules, $g_{bb}(b)$ the radial distribution function on contact between two solute molecules, and $g_{ab}[(a + b)/2]$ the similar quantity between solute and solvent molecules.

The chemical potential μ_b of the solute is given by Eq. (31). Introducing (5) into (31) yields

$$\mu_b = kT \ln \left\{\rho_b \Lambda_b^3 / p_0 \left(\frac{a+b}{2}\right)\right\} \tag{42}$$

By straightforward thermodynamic methods involving the Gibbs–Duhem relation[23] it can be shown that

$$\left(\frac{\partial p}{\partial \rho_b}\right)_\rho = \rho_b \left(\frac{\partial \mu_b}{\partial \rho_b}\right)_\rho + \rho \left(\frac{\partial \mu_b}{\partial \rho}\right)_{\rho_b} \tag{43}$$

This is an extension of (36) to binary systems. Substitution of (41) and (42) into (43) yields

$$\left(\frac{\partial \ln p_0}{\partial \rho}\right)_{\rho_b} = -\frac{1}{\rho}\left\{1 + \tfrac{2}{3}\pi a^3\rho^2 \left(\frac{\partial g_{aa}}{\partial \rho_b}\right)_\rho + \tfrac{4}{3}\pi \left(\frac{a+b}{2}\right)^3 \rho g_{aa}\left(\frac{a+b}{2}\right) + \tfrac{4}{3}\pi\left(\frac{a+b}{2}\right)^3 \rho\rho_b \left(\frac{\partial g_{ab}}{\partial \rho_b}\right)_\rho - \frac{\rho_b}{kT}\left(\frac{\partial \mu_b}{\partial \rho_b}\right)_\rho\right\} \tag{44}$$

where the term in g_{bb} has been omitted because we shall be interested in the case $\rho_b \to 0$ and it is of higher order in ρ_b. In the limit $\rho_b \to 0$

$$g_{ab}\left(\frac{a+b}{2}\right) \to G\left(\frac{a+b}{2}, \rho\right), \quad g_{aa}(a) \to G(a, \rho) \tag{45}$$

and

$$\left(\frac{\partial \mu_b}{\partial \rho_b}\right)_\rho \to \frac{kT}{\rho_b} \tag{46}$$

since μ_b is given by (42) and in this limit $p_0[(a+b)/2]$ is independent of ρ_b. Substituting (45) and (46) into (44) we then have in the limit $\rho_b \to 0$

$$\begin{aligned}\left(\frac{\partial \ln p_0}{\partial \rho}\right)_{\rho_b=0} &= -\tfrac{2}{3}\pi a^3\rho\left(\frac{\partial G(a,\rho)}{\partial \rho_b}\right)_{\rho_b=0} - \tfrac{4}{3}\pi\left(\frac{a+b}{2}\right)^3 G\left(\frac{a+b}{2}, \rho\right) \\ &= \frac{Q(b,a,\rho)}{\rho} - \tfrac{4}{3}\pi\left(\frac{a+b}{2}\right)^3 G\left(\frac{a+b}{2}, \rho\right)\end{aligned} \tag{47}$$

It is evident that

$$Q(b, a, \rho) = -\tfrac{2}{3}\pi a^3\rho^2 \left(\frac{\partial G(a,\rho)}{\partial \rho_b}\right)_{\rho_b=0} \tag{48}$$

At this stage it is convenient to introduce the transformation

$$\begin{aligned}\xi &= \frac{a+b}{2a} \\ x &= a^3\rho\end{aligned} \tag{49}$$

From Eqs. (47) and (4) one can then derive

$$Q(\xi, x) = x \frac{\partial \ln p_0}{\partial x} - \frac{\xi}{3} \frac{\partial \ln p_0}{\partial \xi}$$

$$= -x \frac{\partial}{\partial x} \left\{ x \int_0^{\xi} 4\pi\xi^2 G(\xi, x) d\xi \right\} + \tfrac{4}{3}\pi x \xi^3 G(\xi, x) \qquad (50)$$

This is an integral equation in $G(\xi, x)$ containing indefinite integrals. Unfortunately, the function $Q(\xi, x)$ defined by (48) is not available in a useful form and the utility of (50) will depend upon how explicitly $Q(\xi, x)$ can be determined.

IV. THE EQUATION OF STATE OF HARD SPHERES

In Sections II and III several exact relations have been derived. Among these are the expressions for $p_0(r)$ contained in Eqs. (5), (12), and (22). Furthermore, exact expressions are available for $p_0(r)$, $G(r)$, and $W(r)$ for $r = a/2$. These are contained in Eqs. (13), (14), and (25). In addition, there are the conditions on G and its derivatives at $r = a/2$. The value of G at $r = a/2$ is obtained by setting $r = a/2$ in (14). The result is

$$G(a/2) = (1 - \tfrac{1}{6}\pi a^3 \rho)^{-1} \qquad (51)$$

The values of the first two derivatives of G at $r = a/2$ are given by (18) and (19), respectively. It should be borne in mind that these derivatives belong to G to the right of $r = a/2$, whereas the appropriate derivatives for G to the left of $a/2$ are obtained through the straightforward differentiation of (14) followed by substitution of $a/2$ for r. When this is done it turns out that Eq. (18) is the derivative of G to the left of $a/2$ at $r = a/2$, and that the first two terms on the right of (19) are the derivative of G to the left of $a/2$ at $r = a/2$. Thus, as we have mentioned before, the first derivative of G is continuous at $r = a/2$, but the second derivative is discontinuous possessing a jump given by the third term on the right of (19) and proportional to $g(a)$, the radial distribution function on contact.

At $r = a/\sqrt{3}$ there is a singularity also, but the discontinuity appears first in the fourth derivative of G with respect to r, and is proportional to the contact triplet correlation function $g^{(3)}(a, a, a)$. In general, singularities occur whenever r reaches a magnitude such that an additional particle can be packed into the spherical cell of radius r. As indicated in Section II, these singularities occur in derivatives of increasingly higher order as r increases.

The relations (5), (12), and (22) can be inverted to furnish several means of expressing $G(r)$. Thus

$$G(r) = \frac{1}{4\pi r^2 \rho kT}\frac{\partial W}{\partial r} = -\frac{1}{4\pi r^2 \rho}\frac{\partial \ln p_0}{\partial r} \tag{52}$$

From this relation we note that the singularities in the derivatives of W with respect to r occur in the $(n + 1)$th derivative if they occur in the nth derivative of G. Thus, at $r = a/2$ the singularity occurs first in the third derivative of W with respect to r.

All the relations derived thus far require that the fluid molecules possess hard cores, although they may also possess soft intermolecular potentials and exert attractive forces upon one another. If we dispense with the soft potential and deal only with hard spheres, then according to Eq. (33), $g(a)$ and $G(a)$ are identical and (19) becomes (35), which is a connection formula relating the properties of G at $a/2$ to the properties of G at a. When we deal strictly with a hard sphere system several additional conditions exist on $G(r)$. These are (38), (40), and (50). Now Q in Eq. (50) is expressed as a function of ξ and x; whereas in (48) it is defined in terms of the untransformed variables r and ρ. If we introduce the transformation (49) together with the additional transformation

$$x_b = a^3 \rho_b \tag{53}$$

we obtain for $Q(\xi, x)$ the following relation:

$$Q(\xi, x) = -\tfrac{2}{3}\pi x^2 \left(\frac{\partial G(1, x)}{\partial x_b}\right)_{x_b=0} \tag{54}$$

In view of (34) the equation of state for hard spheres is determined when $G(a, \rho)$ is specified. We therefore turn our attention to the application of these exact relations to the task of evaluating $G(a, \rho)$.

We begin with a more careful inspection of the properties of $Q(\xi, x)$. In the first place, $Q(\xi, x)$ like $G(\xi, x)$ must be non-analytic in ξ. Substituting N_b/V for ρ_b in (53) (where N_b is the number of solute molecules in the solution) and the result into (54) we have

$$Q(\xi, x) = -\frac{2\pi x^2 V}{3a^3}\left(\frac{\partial G(1, x)}{\partial N_b}\right)_{N_b=0} \tag{55}$$

The derivative in this expression represents the change in the radial distribution function on contact between two solvent molecules in pure solvent. When a single solute molecule is added to the solvent the volume of the system remains constant. Q depends upon the properties of the solute and in particular upon the diameter b and therefore upon ξ. The physical significance of Q revealed in this manner may be helpful in arriving at a more convenient form for it.

Some insight into the behavior of Q can be developed by examining it in the limits $\xi = \frac{1}{2}$ and $\xi = \infty$, where the exact relations which we have derived throw some light on the subject; thus substitution of (14) into (50) yields

$$Q = \frac{\partial Q}{\partial \xi} = 0, \quad \xi \leqslant \tfrac{1}{2} \tag{56}$$

From the established continuity of G and $\partial G/\partial \xi$ at $\xi = \frac{1}{2}$, the continuity of both Q and $\partial Q/\partial \xi$ can be demonstrated. Thus for Q to the right of $\xi = \frac{1}{2}$

$$Q(\tfrac{1}{2}, x) = \left(\frac{\partial Q}{\partial \xi}\right)_{\xi=\frac{1}{2}} = 0 \tag{57}$$

Furthermore, the use of Eq. (35), the continuity conditions on G, and Eq. (50) (twice differentiated) yields

$$\left(\frac{\partial^2 Q}{\partial \xi^2}\right)_{\xi=\frac{1}{2}} = -\frac{4\pi^2 x^2}{3(1 - \frac{1}{6}\pi x)}\, G(1, x) \tag{58}$$

Thus $Q = 0$ for $\xi = \frac{1}{2}$, and to the right of $\xi = \frac{1}{2}$ an expansion in powers of $(\xi - \frac{1}{2})$ must begin with a quadratic term since both Q and $(\partial Q/\partial \xi)_{\xi=\frac{1}{2}}$ are 0. The coefficient of this term is given by Eq. (58), so that

$$\lim_{\xi \to \frac{1}{2}} Q(\xi, x) = -\frac{2\pi^2 x^2}{3(1 - \frac{1}{6}\pi x)}\, G(1, x)\, (\xi - \tfrac{1}{2})^2 \tag{59}$$

Equation (50) may also be used to investigate Q in the limit $\xi = \infty$. If $G(\xi, x)$ in the integrand is assumed to be $G(\infty, x)$, the integral can be evaluated immediately. This substitution is suggested by the fact that over the infinite range extending from some large value of ξ, ξ^*, to $\xi = \infty$, $G(\xi, x)$ is sensibly equal to $G(\infty, x)$. The finite range between $\xi = 0$ and $\xi = \xi^*$ contributes a negligible term. Therefore, the dominant part of Q at $\xi = \infty$ goes as ξ^3 and in fact is

$$Q_{\text{dom.}} = -\tfrac{4}{3}\pi\xi^3 x^2 \frac{\partial G(\infty, x)}{\partial x} \tag{60}$$

We have indicated that Q and G are smooth functions of ξ even though they are non-analytic. This suggests that they may be approximated by analytic functions. If we concentrate upon an approximation for Q it seems reasonable to choose a function which has the proper dependence on ξ in the two limits $\xi = \frac{1}{2}$, $\xi = \infty$. The simplest function of this type is a simple sum of the forms exhibited in Eqs. (59) and (60)

$$Q(\xi, x) = w_2(x)(\xi - \tfrac{1}{2})^2 + w_3(x)(\xi - \tfrac{1}{2})^3, \quad \xi > \tfrac{1}{2} \tag{61}$$

In essence this is an interpolation function. At $\xi = \infty$, the dominant term of (61) goes as ξ^3, while as $\xi \to \frac{1}{2}$ the limiting term goes as $(\xi - \frac{1}{2})^2$.

Substituting (61) into (50) we derive as the most general analytic solution (subject to the restriction that G remains finite as $\xi \to \infty$) for $\xi > \frac{1}{2}$

$$G(\xi, x) = \varphi(\xi^3 x) + \alpha_0'(x) + \frac{\alpha_1(x)}{\xi} + \frac{\alpha_2(x)}{\xi^2} \tag{62}$$

where φ is a function of the single variable $\xi^3 x$. As $\xi \to \infty$ we require that $G(\xi, x) \to G(\infty, x)$ so that $\varphi(\xi^3 x)$ must be a function of x alone as $\xi \to \infty$. This violates the assumption that φ depends on $\xi^3 x$ unless

$$\varphi(\xi^3 x) = \text{constant}, \quad \xi \to \infty \tag{63}$$

Thus Eq. (63) implies that

$$\varphi = \text{constant} \quad \text{(all values of } \xi, x) \tag{64}$$

This constant may be incorporated into $\alpha_0'(x)$ in (62). Thus we have

$$G(\xi, x) = \alpha_0(x) + \frac{\alpha_1(x)}{\xi} + \frac{\alpha_2(x)}{\xi^2} \tag{65}$$

The only approximation thus far has been the assumption of the form (61) *for* Q. This in itself is not severe. The chosen form has the proper behavior in the limits $\xi = \frac{1}{2}$ and $\xi = \infty$. Furthermore, we know that Q although it is non-analytic is smooth, so that interpolation by a simple analytic function of the form (61) is reasonable.

The solution (65) receives support from another quarter. Up until this point our arguments have been founded upon statistical methods and therefore Eq. (65) should apply to the range of ξ corresponding to molecular dimensions. On the other hand, it must also hold for values of ξ corresponding to macroscopic dimensions. It is therefore satisfying to know that in the macroscopic range arguments which are based on thermodynamic considerations alone demand the same relation.

For example, it is known from thermodynamic reasoning that the increment dW of reversible work which must be expended in increasing the reduced radius ξ of a macroscopic spherical cavity by $d\xi$ is

$$dW = p(4\pi a^3\xi^2 d\xi) + \sigma(x)\left\{1 - \frac{2\delta(x)}{\xi}\right\}(8\pi a^2 \xi d\xi) \tag{66}$$

Here the first term on the right is the volume work associated with the volume increase $4\pi a^3\xi^2 d\xi$, while the second term is the surface work. The quantity

$$\sigma\left\{1 - \frac{2\delta}{\xi}\right\} \tag{67}$$

is the surface tension. $\delta(x)$ is the distance (divided by a) between that dividing surface for which the superficial density of matter vanishes and the Gibbs surface of tension,[25–27] while $\sigma(x)$ is the surface tension at infinite ξ, i.e., infinite radius of curvature.

Substitution of Eq. (66) into Eq. (52) (expressing (52) in terms of ξ and x) then yields

$$G(\xi, x) = \frac{a^3 p}{xkT} + \left\{\frac{2a^2\sigma(x)}{xkT}\right\}\frac{1}{\xi} - \left\{\frac{4a^2\sigma(x)\delta(x)}{xkT}\right\}\frac{1}{\xi^2} \tag{68}$$

This is identical in form with Eq. (65) *and is exact in the limit of large* ξ.

On the other hand, Eq. (65) has been derived for *small* ξ. This lends support to the idea that Eq. (66) is a good approximation over the entire range between $\xi = \frac{1}{2}$ and $\xi = \infty$. Attempts to extrapolate surface tension considerations down to molecular dimensions have been common in the past and not without some degree of success.[28,29] Extrapolations of this sort have been used most effectively in nucleation theory.[30] Equation (39) shows that the first term on the right of (68) is $G(\infty, x)$, so that G given by (68) approaches the correct limit for $\xi \to \infty$.

If it is assumed that Eq. (68) is valid right down to $\xi = \frac{1}{2}$, we can substitute it into both (51) and (18), suitably transformed so that they depend upon ξ and x. This provides two equations in the three unknowns $G(1, x)$, $\sigma(x)$, and $\delta(x)$. A third equation is obtained by replacing a^3p/xkT in (68) by $G(\infty, x)$, making use of (40) so as to write $G(\infty, x)$ in terms of $G(1, x)$, and then setting ξ in (68) equal to unity. The three equations are:

$$\frac{1}{1 - \frac{1}{6}\pi x} = 1 + \tfrac{2}{3}\pi x G(1, x) + \frac{4a^2\sigma(x)}{kTx} - \frac{16a^2\sigma(x)\delta(x)}{kTx} \tag{69}$$

$$\frac{\pi x}{(1 - \frac{1}{6}\pi x)^2} = -\frac{8a^2\sigma(x)}{kTx} + \frac{64a^2\sigma(x)\delta(x)}{kTx} \tag{70}$$

$$(1 - \tfrac{2}{3}\pi x)G(1, x) = 1 + \frac{2a^2\sigma(x)}{kTx} - \frac{4a^2\sigma(x)\delta(x)}{kTx} \tag{71}$$

These can be solved simultaneously to yield

$$G(1, x) = \frac{1}{4(1 - \frac{1}{6}\pi x)} + \frac{3}{4(1 - \frac{1}{6}\pi x)^2} + \frac{\pi x}{8(1 - \frac{1}{6}\pi x)^3} \tag{72}$$

$$\sigma(x) = -\frac{kT\pi x^2}{8a^2}\frac{[1 + \frac{1}{6}\pi x]}{[1 - \frac{1}{6}\pi x]^3} \tag{73}$$

$$\delta(x) = \tfrac{1}{8} - \tfrac{1}{8}\left\{\frac{1 - \frac{1}{6}\pi x}{1 + \frac{1}{6}\pi x}\right\} \tag{74}$$

Thus not only has the equation of state of the fluid been determined insofar as it depends on $G(1, x)$, but the surface tension and its dependence upon the curvature have been determined as well.

Up to this point we have not employed all of the exact relations derived in Sections II and III. Thus neither Eq. (35) nor Eq. (38) has been used. It seems therefore that more relations are available than can be accommodated by the form (65) with its three undetermined parameters. We shall return to this question below.

As we mentioned earlier, the equation of state for hard spheres has been investigated by machine computation techniques. Wainwright and Alder[11] performed such computations by integrating the dynamical equations of motion for an assembly of hard sphere molecules. Combining their data with Eq. (26) it is possible to specify $G(1, x)$ in its dependence on the reduced concentration x. Wainwright and Alder discovered, as have several authors, that the system of hard spheres undergoes what appears to be a phase transition to a crystalline configuration. This transition takes place at concentrations less than that characteristic of close packing. Our result for $G(1, x)$ gives no indication of this singular behavior, and this is not surprising since pains have been taken to avoid incorporating obvious singularities into our theory. Thus we have approximated the non-analytic function G with an analytic one.

Fortunately, Wainwright and Alder also observed what appeared to be a metastable equation of state for hard spheres, i.e., a continuation of the fluid equation of state into the range of concentrations characteristic of the crystalline state. The points which appear in Fig. 1 are taken from their data while the curve is a plot of Eq. (72). Agreement is very good. The broken vertical line denotes the value of x at which Wainwright and Alder observed the phase transition. The points plotted for values of x beyond this line are the metastable ones. For orientation it should be pointed out that the value of the reduced density x which corresponds to face-centered-cubic close packing is

$$\sqrt{2} = 1.412$$

Because the phase transition may be associated with the singularities, it may be valuable to utilize Eq. (34), which deals directly with a singularity. In this case, however, G would have to be specified by a form other than that provided by (65).

An expansion of $G(1, x)$ as a power series in x can be made. The coefficients of the expansion are related to the virial coefficients.

In the expansion of G the coefficient for the nth power is related to the $(n+2)$th virial coefficient (obtained by expanding p as a power series in x). This follows from Eq. (26) where it is seen that p is related to G in a manner which multiplies the latter by

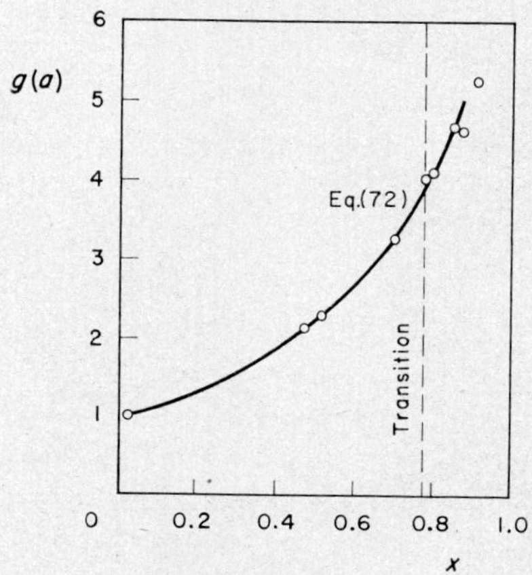

Fig. 1. Comparison of Eq. (72) with the machine data of Wainwright and Alder.

the square of the concentration. From our knowledge of the first five virial coefficients[7,8] the series representation of $G(1, x)$ is

$$G(1, x) = 1 + 0.417(\pi x) + 0.1275(\pi x)^2 + (0.0341 \pm 0.0015)(\pi x)^3 \tag{75}$$

whereas the expansion of (72) yields

$$G(1, x) = 1 + 0.417(\pi x) + 0.1319(\pi x)^2 + 0.0359(\pi x)^3 + \dots \tag{76}$$

Equation (76) compares quite favorably with (75). In fact, up to and including the third virial coefficient (the first two terms of (75) and (76)) the agreement is exact, while the fourth virial coefficient is in error by 3% and the fifth only by 5%.

Application of Eq. (26) to Eq. (72) yields the equation of state

$$\frac{\pi a^3 p}{6kT} = \frac{y(1 + y + y^2)}{(1 - y)^3} \tag{77}$$

where the substitution

$$y = \tfrac{1}{6}\pi x \tag{78}$$

has been used.

Recently there has been an interesting development. Percus and Yevick[3,31] have derived an integral equation approach to the theory of fluids based upon a collective coordinate analysis. The solution of their integral equation yields an approximation to the pair correlation function $g^{(2)}$ in much the same way that this function is obtained from the Kirkwood or the Born–Green and Yvon integral equations.[2] The possession of $g^{(2)}$ then permits the derivation of the equation of state. At first, the Percus–Yevick equation was not solved exactly, but Stell[32] employed it to evaluate the first five virial coefficients. It emerged that up to and *including* the fifth virial coefficient the Percus–Yevick equation (applied to hard spheres) yielded the same results as the scaled particle theory. Thus the Percus–Yevick equation gives Eq. (76) *exactly* for $G(1, x)$!

On the basis of this evidence, Wertheim was moved to seek an exact solution of the Percus–Yevick equation. (It must be remembered, however, that the Percus–Yevick equation is itself approximate.) In an elegant paper[14] he accomplished this solution and found that the equation of state derived from it was *precisely* (77). Thus the Percus–Yevick theory and the scaled particle theory yield identical results, at least for the case of hard spheres. As we mentioned in Section I, this is the first time that two separate non-heuristic theories of fluids have yielded identical analytical results. Furthermore, with the exception of the cluster theory,[1] which is designed specifically to yield the virial coefficients, both theories yield the best values of these coefficients which have been provided by any treatment of dense fluids. It is almost unnecessary to emphasize the importance of exploring the relation between these two harmonious developments.

Recently Lebowitz[33] has generalized the Percus–Yevick equation to include a scaled particle solute, and has shown that the generalization requires $W(r)$ to be a third-order polynomial in r, for $r = a$. This is consistent with Eq. (66) and further substantiates the correspondence between the Percus–Yevick and the scaled particle theories.

We have mentioned that more exact conditions are available

than can be accommodated by the form (65). One's first thought is to add to Eq. (65) an additional function possessing additional parameters. For example, with an additional parameter one might employ (34). Thus far all attempts to accomplish this have led to poorer results for $G(1, x)$, i.e., to poorer results in terms of the agreement of an expansion in powers of x with (75). A very interesting note is sounded, however, when Eq. (68) (into which Eqs. (69), (70), and (71) have been substituted with the assistance of Eq. (26)) is introduced into Eq. (38). It emerges that (43) satisfies (38)! Since (38) is an *independent* exact condition, this increases our reliance in the form (65). In effect, α_0, α_1, and α_2 are the same functions of x whether we determine them by use of (51) and (18) together with (39) or alternatively together with (38).

An examination of Eq. (77) reveals that the pressure depends linearly upon temperature, and that the only dependence appears in the factor T which multiplies the more complicated factor depending upon the concentration y. This feature is characteristic of all the relations with which we have dealt in connection with the hard sphere fluid. This emphasizes the fact that the hard sphere problem is geometric and is involved essentially with the packing of the hard spheres within the available volume V. This is why all of our relations have been concerned with concentration rather than temperature. To the extent that the problem of real fluids depends upon the packing of the real molecules so that the solution can be advanced by an appeal to the results of these first four sections, the problem of real fluids is also geometric. We shall see that many of the results of these first four sections can be successfully adapted to the treatment of real fluids, especially when they are as dense as most liquids. Thus, the major problems in the theory of liquids (at least for simple liquids containing spherical molecules) appear to be geometric.

V. SYSTEMATIC EXTENSION OF THE THEORY TO REAL FLUIDS

As we mentioned earlier, the attempt to extend the theory to real fluids so as to make possible the derivation of their equations of state has not met with the same success as the treatment of hard spheres. Much progress can be made by relinquishing the ideal of attaining the equation of state and focusing attention upon more limited goals. In Section VII and beyond we shall do this. On the other hand, *some* results have been obtained in the attempt to extend the theory systematically, and we shall report on these in the present section.

In the case of hard spheres, progress was made by examining the properties of a function G which was simply related to the equation of state. This suggests that a similar function might exist when the molecules are not hard spheres but possess intermolecular potentials which are partly soft. Such a function does indeed exist and Helfand, Reiss, Frisch, and Lebowitz[22] have discussed its properties. Helfand *et al.* have used θ to symbolize this function. In what follows we shall do the same; and the remainder of this section will be devoted to a discussion of the properties of θ.

The algebraic manipulations surrounding θ can become quite involved. In view of this, and the fact that the θ formalism has not yet permitted an advance to the equation of state, we shall list and discuss formulas without derivation. The reader is referred to ref. 22 for details. This approach should be effective because many of the formulas are intuitively reasonable, especially in view of their resemblance to the corresponding formulas for G. Although the θ formalism is still incomplete, the relations to be discussed are exact and consequently important, and one may look forward to further progress in this field. At the moment, progress is limited because θ is a more complicated function than G, and a larger number of exact conditions are required to specify it than were required in the case of G. Unfortunately, the number of exact conditions available is the same as in the G case. Thus θ cannot be specified in as detailed a manner as G.

Suppose that the intermolecular potential in the real fluid is denoted by $u(r)$. We do not insist (at least for the present) that $u(r)$ contain a hard core part. As in the previous sections, we consider a fluid of N molecules confined to a volume V such that the concentration is once again $\rho = N/V$. Into this fluid we introduce a *solute* molecule characterized by the fact that the intermolecular potential between it and any of the solvent (original fluid) molecules is $u(r/\lambda)$. Thus for $\lambda = 1$ the solute becomes another solvent molecule. Because of the dependence of

$u(r/\lambda)$ upon λ we call the solute molecule a λ-cule. If $u(r)$ has a conventional form such as the form illustrated in Fig. 2 it follows that

$$\begin{aligned} u(0) &= \infty \\ u(\infty) &= 0 \end{aligned} \tag{79}$$

As a result when

$$\lambda \to 0, \quad u(r/\lambda) \to u(\infty) = 0 \tag{80}$$

$$\lambda \to 1, \quad u(r/\lambda) \to u(r) \tag{81}$$

so that the λ-cule is decoupled entirely from the fluid by having $\lambda \to 0$. On the other hand, (81) indicates that coupling is complete

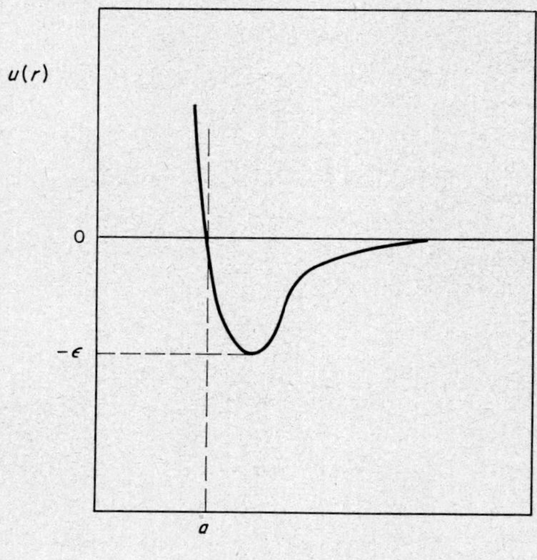

Fig. 2. Conventional intermolecular potential.

when $\lambda \to 1$ in the sense that the λ-cule becomes an ordinary solvent molecule.

This coupling of the λ-cule to the rest of the fluid by scaling the breadth of the intermolecular potential (and in effect the size or breadth of the molecule) has led to the nomenclature "scaled particle theory". The growing hard sphere which produces the spherical cavity of the preceding sections is a special case of such particle scaling.

In many cases the intermolecular potential may be expressed in the form

$$u(r) = \varepsilon\varphi(r/a) \tag{82}$$

where ε measures the depth of the potential well and a is a distance parameter usually proportional to the value of r at the position of the potential minimum (see Fig. 2). Frequently we choose for a the value of r at this minimum. Another choice, and one which we shall adopt in view of our interest in the hard core, is the value of r at the point where $u(r)$ is 0 as it rises steeply to assure repulsion. The intermolecular potential for hard spheres is a special case of Eq. (82) in which ε may be chosen equal to unity and

$$\begin{aligned} \varphi(r/a) &= 0, \quad r/a > 1 \\ \varphi(r/a) &= \infty, \quad r/a \leqslant 1 \end{aligned} \tag{83}$$

Equation (82) serves to define a for our present purposes.

The function θ can now be defined

$$\theta(\lambda, \rho, T) = -\int_0^\infty \left(\frac{r}{\lambda}\right) \frac{u'(r/\lambda)}{kT}\, g(r, \lambda)\, \frac{r^2 dr}{\lambda^3 a^3} \tag{84}$$

In this relation u' represents the derivative of $u(r/\lambda)$ with respect to r/λ, not merely with respect to r, and $g(r, \lambda)$ is the pair correlation function between the λ-cule and the solvent molecule whose center is a distance r from its center. In the case where the intermolecular potential u reduces to that for hard spheres (Eq. (83)) θ reduces to G. This becomes clear when it is recognized that u' then becomes a delta function with its peak at $r = \lambda a$, so that the integrand in Eq. (84) may be moved in front of the integral sign with only $g(\lambda a)$ appearing. In the parlance of the preceding sections this is $G(\lambda a)$.

Helfand *et al.*[22] have established the counterparts in the present formalism of Eqs. (34), (30), (52), (14), (40), (48), (38), and (50).

We list all but the last two of these, employing a notation suitable for the θ formalism:

$$p/kT = 1 + \tfrac{2}{3}\pi a^3\rho^2\theta(1, \rho) \tag{85}$$

$$\mu_\lambda = kT \ln \Lambda_\lambda^3\rho_\lambda - kT\rho a^3 \int_0^\lambda 4\pi\lambda'^2\theta(\lambda', \rho)d\lambda' \tag{86}$$

$$\theta(\lambda, \rho) = \frac{1}{4\pi\rho a^3 kT\lambda^2}\frac{\partial W}{\partial\lambda} \tag{87}$$

$$\theta = \frac{B^*}{1 - \frac{4}{3}\pi\rho a^3\lambda^3 B^*}, \quad \lambda < \frac{1}{2\gamma} \tag{88}$$

$$\rho\theta(\infty, \rho) = \gamma^3[\rho + \tfrac{2}{3}\pi a^3\rho^2\theta(1, \rho)] = (\gamma^3 - 1)[\rho_* + \tfrac{2}{3}\pi a^3\rho_*^2\theta(1, \rho_*)] \tag{89}$$

$$Q(\lambda, \rho) = -\tfrac{2}{3}\pi a^3\rho^2\left[\frac{\partial\theta(1, \rho)}{\partial\rho_\lambda}\right]_{\rho_\lambda=0} \tag{90}$$

In Eq. (86) ρ_λ is the concentration of λ-cules and Λ_λ^3 is given by Eq. (29) with λ substituted for b. The chemical potential μ of the solvent is obtained from (86) by setting λ equal to unity, $\Lambda_\lambda = \Lambda$, and $\rho_\lambda = \rho$. In Eq. (87) $W(\lambda, \rho)$ is the reversible work which must be performed to expand the λ-cule to λ. In Eq. (88) B^* has the following meaning:

$$B^* = B/\tfrac{2}{3}\pi a^3 \tag{91}$$

$$B = -\tfrac{1}{2}\int_0^\infty (e^{-u(r)/kT} - 1)4\pi r^2 dr \tag{92}$$

From (92) it appears that B is the second virial coefficient. We have still to define γ. γ appears in Eq. (88) because the relation (unlike (85) through (87)) assumes an intermolecular potential with a hard core at $r = a$ and a long-range cutoff at γa as illustrated in Fig. 3. Thus γa measures the position of the long-range cutoff. The intermolecular potential may be rendered more general by allowing γ to pass to infinity, but then $\frac{1}{2}\gamma = 0$ and Eq. (88) will have no range of validity.

The question now arises concerning the analogs of Eqs. (18) and (19). Are there continuity and connection relations at $\lambda = \frac{1}{2}\gamma$ as there were in the case of hard spheres at $r = a/2$? As a matter of fact, Helfand *et al.* demonstrate that θ and its first derivative at $\lambda = \frac{1}{2}\gamma$ *are* continuous and, as in the case of hard spheres, the second derivative is discontinuous.

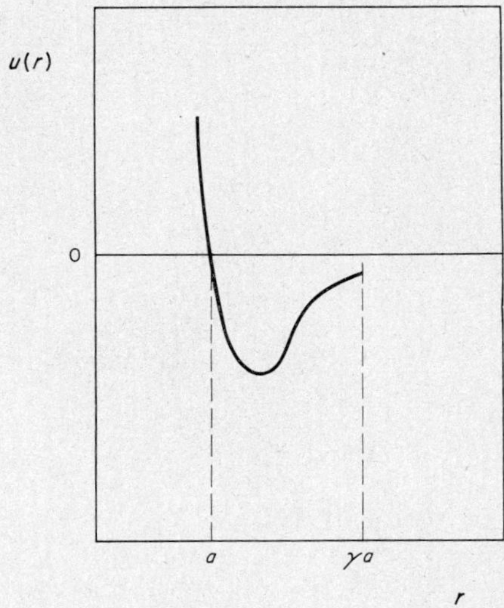

Fig. 3. Conventional intermolecular potential with upper cutoff.

Equation (90) is based upon an even more restricted potential, namely the square-well potential depicted in Fig. 4. Here the boundaries of the square well occur at $r = a$ and $r = \gamma a$. There is a natural long-range cutoff. The quantity ρ_* which appears in Eq. (90) must still be defined. It is determined by the relation

$$\frac{\varepsilon}{kT} = \ln \frac{\rho_*}{\rho} + 4\pi a^3 \int_0^1 \lambda^2[\rho_* \theta(\lambda, \rho_*) - \rho\theta(\lambda, \rho)]d\lambda \tag{93}$$

The almost complete formal similarity between the exact relations which have been derived for θ and those which have been derived in the preceding sections for G might lead one to hope that θ itself would have the same simple form exhibited by G in Eq. (65), i.e.,

$$\theta = A + \frac{B}{\lambda} + \frac{C}{\lambda^2} \tag{94}$$

Unfortunately, this simple choice for the form of θ does not appear to be satisfactory since when real or almost real potentials are employed it leads to results which are at variance with experiment. Some indication of the difficulty may be obtained through

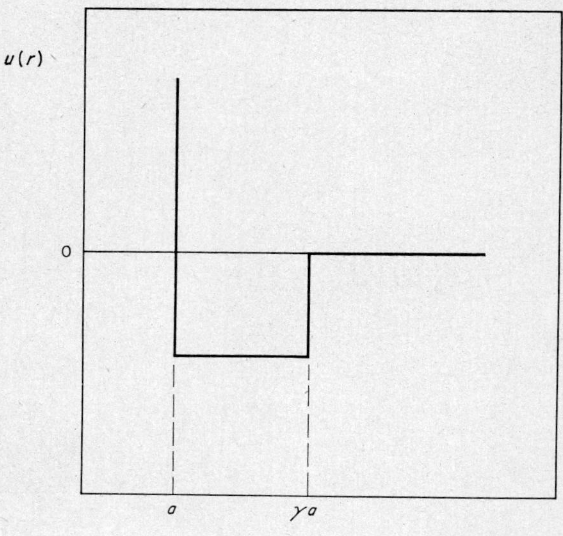

Fig. 4. Square-well potential.

reference to the square-well potential shown in Fig. 4. This potential is specified by the following conditions:

$$\begin{aligned} u(r) &= \infty, \quad r < a \\ &= -\varepsilon, \quad a < r < \gamma a \\ &= 0, \quad \gamma a < r \end{aligned}$$

It is clear that the derivative of the square-well potential contains two delta functions, one at $r = a$ and the other at $r = \gamma a$. As a result, (84) requires

$$\theta(\lambda, \rho) = g(\lambda a, \lambda) - \gamma^3 g(\gamma\lambda a, \lambda)[e^{\varepsilon/kT} - 1] \qquad (96)$$

We see that θ is specified by *two* values of g, one at $r = \lambda a$ and the other at $r = \gamma\lambda a$. In the case of hard spheres only one value of g, $g(\lambda a)$ (or for solvent molecules $g(a) = G(a)$), is required. In essence, more functions must be known when potential wells, and therefore attractive forces, are present; but we are still limited to the number of exact conditions available in the case of hard spheres.

Helfand *et al.* were able to express $Q(\lambda, \rho)$, specified by Eq. (89), in terms of certain integrals involving pair, triplet, and fourth-order correlation functions. For the limiting case of hard spheres $Q(\lambda, \rho)$ assumes the properties appropriate to hard spheres. These properties were discussed in Section IV. In particular, this limiting form of Q and its first derivative with respect to size vanish for $\lambda \leqslant \frac{1}{2}$.

As we shall see in the following sections, the equation of state of a real fluid is closely related to the equation of state for hard spheres when the concentration ρ is high enough. Thus θ should be closely related to G with increasing ρ. An examination of the properties of θ in this limit may be effective in stimulating further developments. For example, it would be interesting to prove that θ converges on G as the value of ρ corresponding to close packing is approached.

VI. MISCELLANEOUS

Insofar as the systematic application of scaled particle theory is concerned, there are still some formal things which we have not discussed. These are more or less disconnected items stemming from the work of several authors. This section will be devoted to a brief discussion of these aspects of the theory.

Item 1 concerns the relation between G and g. In their earliest paper, Reiss *et al.*[13] drew attention to the fact that in the interval lying between $r = a/2$ and $r = a/\sqrt{3}$, $g(2r)$ could be expressed as a function of $G(r)$. This relation follows:

$$g(2r) = \left[\frac{F'''(r)}{96}\right] + \frac{1}{96r}\left[\frac{8F'(r)}{r} + 7F''(r)\right], \quad \frac{a}{2} < r \leqslant \frac{a}{\sqrt{3}} \qquad (97)$$

in which the primes refer to differentiation with respect to r and

$$F(r) = \frac{\exp\left\{-\rho\int_0^r 4\pi s^2 G(s, \rho)ds\right\} - 1 + \frac{4}{3}\pi r^3\rho}{\frac{2}{3}\pi^2 r^3\rho^2} \qquad (98)$$

This relation (Eq. (97)), which of course is only valid for systems of hard spheres, is limited to that range of r corresponding to the activation of the third term in Eq. (12), but only up until the point at which the fourth term becomes activated. That is, it corresponds to values of r such that an associated spherical cell may contain the centers of two but not three molecules simultaneously. Equation (97) is obtained through the combined use of Eqs. (16), (18), and (35). For the details the reader is referred to ref. 13.

Another interesting relation (which has not been published before) expresses, as a function of Q and G, the chemical potential of a hard sphere solvent molecule in a solution to which one hard sphere solute molecule has been added.

This relation, which will now be derived, may prove useful in the further analysis of the properties of Q. It should be clear at this juncture that additional progress in the theory of hard spheres will depend upon our ability to express Q in a more effective manner than we have done in Eq. (61). Thus, all relations which provide additional insight into the physical significance of Q are important.

Suppose we have a dilute solution consisting of N_b hard sphere solute molecules of diameter b dissolved in N hard sphere solvent molecules of diameter a, all contained within the volume V. Then at constant temperature the Gibbs–Duhem relation[23] assumes the form

$$-Vdp + Nd\mu + N_b d\mu_b = 0 \tag{99}$$

which can be rearranged to give

$$\left(\frac{\partial \mu}{\partial \mu_b}\right)_V = \frac{1}{\rho}\left(\frac{\partial p}{\partial \mu_b}\right)_V - \frac{\rho_b}{\rho} \tag{100}$$

This may be further transformed to

$$\left(\frac{\partial \mu}{\partial \mu_b}\right)_V = \frac{\left(\frac{\partial p}{\partial \rho_b}\right)_V}{\rho\left(\frac{\partial \mu_b}{\partial \rho_b}\right)_V} - \frac{\rho_b}{\rho} \tag{101}$$

Now since the solution is dilute, ρ_b, the concentration of solute molecules, is much less than ρ, the concentration of solvent molecules. Ignoring terms in ρ_b^2 we may differentiate Eq. (41) with respect to ρ_b to yield

$$\begin{aligned}\left(\frac{\partial p}{\partial \rho_b}\right)_V &= kT\left\{1 + \tfrac{2}{3}\pi a^3\rho^2\left(\frac{\partial G(a)}{\partial \rho_b}\right)_V + \tfrac{4}{3}\pi\left(\frac{a+b}{2}\right)^3 \rho G\left(\frac{a+b}{2}\right)\right\} \\ &= kT\left\{1 - Q + \tfrac{4}{3}\pi\left(\frac{a+b}{2}\right)^3 \rho G\left(\frac{a+b}{2}\right)\right\} \qquad (102)\end{aligned}$$

where in the final form we have used the definition of Q offered in Eq. (48). Utilizing Eqs. (102) and (46) in (101) we obtain

$$\left(\frac{\partial \mu}{\partial \mu_b}\right)_V = \frac{\rho_b}{\rho}\left[\tfrac{4}{3}\pi\left(\frac{a+b}{2}\right)^3 \rho G\left(\frac{a+b}{2}\right) - Q\right] \qquad (103)$$

Now we have in a straightforward manner

$$d\mu_b = \left(\frac{\partial \mu_b}{\partial N_b}\right)_V dN_b = \frac{1}{V}\left(\frac{\partial \mu_b}{\partial \rho_b}\right)_V dN_b = \frac{kT}{V\rho_b}\,dN_b \qquad (104)$$

Substituting this into (103) and setting dN_b equal to unity (corresponding to the addition of a single solute molecule) we obtain for the change in the chemical potential of the solvent attending this process

$$\Delta\mu = \frac{kT}{\rho V}\left[\tfrac{4}{3}\pi\left(\frac{a+b}{2}\right)^3 \rho G\left(\frac{a+b}{2}\right) - Q\right] \qquad (105)$$

This is the desired relation. Thus we see that Q is simply related to G and $\Delta\mu$, the change in chemical potential of the solvent attending the addition at constant volume of a single solute molecule.

Helfand, Frisch, and Lebowitz[34] have applied the methods of Sections II, III, and IV to two- and one-dimensional hard sphere fluids. The results which they obtain are much the same as those obtained in connection with three-dimensional fluids. In the one-dimensional case an exact solution of the problem is possible, but this is nothing new since exact solutions for the one-dimensional system have been obtained by other methods. In the two-dimensional case, however, an exact solution is no longer possible, and it turns out that the approximate solution obtained by scaled particle techniques, although very good, is not quite as good as the solution obtained for the three-dimensional case. The accuracy of the solution was measured by comparing it with the results for virial coefficients and machine computations.

In the two-dimensional case the analog of Eq. (34) is

$$\frac{p}{kT} = \rho + \tfrac{1}{2}\pi\rho^2 a^2 G(a, \rho) \tag{106}$$

while the analog of Eq. (32) is

$$\frac{\mu}{kT} = \ln \Lambda^2\rho + 2\pi\rho \int_0^a \lambda G(\lambda)\,d\lambda \tag{107}$$

The two-dimensional equation corresponding to Eq. (52) is

$$G(r, \rho) = \frac{1}{2\pi\rho r kT}\frac{\partial W}{\partial r} \tag{108}$$

For $r \leqslant a/2$, where a is now the diameter of a hard circle, the exact solution for G in two dimensions is

$$G(r, \rho) = (1 - \pi r^2\rho)^{-1} \tag{109}$$

The analogs of Eqs. (38), (40), (48), (50), (57), and (60) in two dimensions are

$$\tfrac{1}{2}\pi a^2\rho^2 G(a, \rho) = \rho^2 \int_0^a 2\pi r G(r, \rho)\,dr - \int_0^\rho \rho' d\rho' \int_0^a 2\pi r G(r, \rho')\,dr \tag{110}$$

$$G(\infty) = 1 + \tfrac{1}{2}\pi\rho a^2 G(a) \tag{111}$$

$$Q(r, \rho) = -\tfrac{1}{2}\pi\rho^2 a^2 \left\{\frac{\partial G(a, \rho, \rho_b)}{\partial \rho_b}\right\}_{\rho_b = 0}, \quad r = \frac{a + b}{2} \tag{112}$$

$$Q(r, \rho) = -\rho \int_0^r 2\pi r' G(r', \rho)\,dr' - \rho^2 \int_0^r 2\pi r' \frac{\partial G}{\partial \rho}\,dr' + \pi\rho r^2 G(r, \rho) \tag{113}$$

$$Q = \frac{\partial Q}{\partial r} = 0, \quad r \leqslant \frac{a}{2} \tag{114}$$

$$Q_{\text{dom.}} = -\pi\rho^2 r^2 \,\frac{\partial G(\infty, \rho)}{\partial \rho} \tag{115}$$

Once again it is found that G and $\partial G/\partial r$ are continuous at $r = a/2$, but that $\partial^2 G/\partial r^2$ possesses a discontinuity. Just as in the case of three dimensions, Q and its first derivative vanish for $r \leqslant a/2$ (see Eq. (114)) so that the expansion of Q about $r = a/2$ begins with a quadratic term. At infinite values of r, however, the dominant term in Q now goes as r^2 rather than as r^3. Thus, the appropriate interpolation formula for Q assumes the following form:

$$Q(r, \rho) = \omega_2(\rho)[r - a/2]^2 \tag{116}$$

This is the analog of Eq. (61). Based upon this formula the suggested form for G is

$$G = A(\rho) + \frac{B(\rho)}{r} \tag{117}$$

This is the analog in two dimensions of Eq. (65). Substituting this into Eq. (108) yields a form for dW which agrees with that derived from thermodynamic considerations when r is large enough to be macroscopic. The form required by thermodynamics in the two-dimensional case which corresponds to Eq. (66) is

$$dW = 2\pi prdr + 2\pi\sigma dr \tag{118}$$

Making use of Eq. (32) and the continuity conditions at $r = a/2$ as in Section IV, we can determine the coefficients in Eq. (117) and thus, through (106), the equation of state. This turns out to be

$$\frac{p}{\rho kT} = \frac{1}{(1 - z)^2}, \quad z = \frac{\pi a^2 \rho}{4} \tag{119}$$

and is the two-dimensional analog of Eq. (77).

Just as in the three-dimensional case, we can use Eq. (110) in place of (111) in our arguments and obtain the same result. This increases our confidence in the form (116). Equation (118) does not include a correction for curvature. However, (118) is consistent with (117) and therefore with (116). Employment of a correction for the dependence of surface tension upon curvature leads to a corresponding form of G which gives poor results.

Incidentally, the formula for the surface tension in two dimensions analogous to Eq. (73) in three dimensions is

$$\sigma = \frac{a\rho kT}{2}\left[\frac{1}{1-z} - \frac{1}{(1-z)^2}\right] \tag{120}$$

We close this section by mentioning that Helfand and Stillinger[35] have blended the scaled particle theory with the radial distribution function theory to obtain the pair correlation function for rigid spheres. They show that a considerable improvement over the usual superposition approach results upon retention of the superficial form of the superposition approximation but with a suitably defined effective local density. These authors employ particle scaling as a coupling procedure and as usual their equations become exact over the half range of the coupling parameter (i.e., for $r \leqslant a/2$). The most striking result of their theory is that it gives the exact fourth virial coefficient, whereas the ordinary superposition approximation yields a value of the fourth virial coefficient which is 61% too low. The theory is too detailed to present here, and the reader is referred to the original paper for further details.

VII. COMPARISON WITH EXPERIMENTAL DATA

The plan outlined in Section I called for a discussion of the application of scaled particle theory to real fluids which could be read without recourse to Sections II through VI of this review. We begin such a discussion in this section.

The scaled particle theory developed in the preceding sections is concerned primarily with systems of molecules possessing spherical symmetry. In particular, these spherical molecules are imagined to have hard billiard ball-like cores which repel one another with infinite force. Now it is known that molecules do not really possess such hard cores, but at low enough temperatures (for example, room temperature) most molecules behave, at least approximately, as though they do have hard cores. Thus, if one considers a conventional intermolecular potential such as the one illustrated in Fig. 2 at the point $r = a$ where the potential is rising very steeply, it rises by many times kT as r decreases by some small fraction of a. Therefore, it may be expected that molecules subscribing to the intermolecular potential diagrammed in Fig. 2 will behave approximately as though they possess hard cores with diameters approximately equal to a.

On the other hand, it is clear that the best choice of hard core diameter will depend slightly upon temperature because the more energetic particles available at higher temperatures can penetrate to smaller distances of separation. This effect is slight, and when not endeavoring to account for it we shall sometimes choose a in the manner indicated in Fig. 2, that is, we shall choose for a the value of r corresponding to the point at which the intermolecular potential passes through zero as it rises on the repulsive side. Experimental values of a defined in this manner can usually be obtained from data on second virial coefficients or viscosity, once a suitably defined form for the intermolecular potential (e.g., the Lennard-Jones potential) has been chosen.[36] We shall, however, engage in considerable discussion concerning the temperature dependence of a. Notice that any difficulty in measuring a precisely does not make a an adjustable parameter. It possesses a well defined physical significance and can be determined from independent measurements such as viscosity or second virial coefficients, or preferably by the technique discussed in Section X. One of the outstanding features of the present theory is its ability to provide numerical values for thermodynamic quantities without making use of even a single adjustable parameter.

An important result of the statistical considerations of the preceding sections had to do with the work of producing spherical cavities in fluids consisting of molecules of the kind we have just described. These cavities are so defined that the centers of all fluid molecules are excluded from them. If the radius of such a cavity is r, an alternative method of defining it involves the assumption that it is produced by the introduction into the fluid of a hard spherical solute molecule of diameter b such that

$$r = (a + b)/2 \tag{121}$$

where a represents the diameter of the hard cores of the original fluid molecules which we may now look upon as solvent molecules. The solute molecule with diameter b is mostly a mathematical device useful for focusing attention upon the properties of the cavity which has been defined, but in certain circumstances it may be regarded as a real solute. Under such circumstances the reversible work, $W(r)$, expended in creating the cavity is related to the chemical potential of the solute.

It should be borne in mind that the solvent molecules are real molecules and therefore they interact with one another with soft intermolecular potentials as well as repelling each other with their hard cores. On the other hand, the solute molecule is a real billiard ball in the sense that its interaction with solvent molecules involves only their hard cores and not the soft parts of the intermolecular potentials.

An important aspect of the formal discussions of Sections II–VI was the development of an expression for $W(r)$. For $r = a/2$ an exact expression was derived. We shall reproduce the argument here. Before doing so, however, it is worth noting that Eq. (121) demands that b be negative for values of $r \leqslant a/2$. Rather than interpret this as corresponding to a solute having a negative diameter, it is easier to assume that for the range $r \leqslant a/2$ the intermolecular potential has been redefined so that the solute is a point molecule which can penetrate the cores of the surrounding solvent molecules. No real physical difficulty is involved in this redefinition because the solute is a mathematical device in the first place. As a matter of fact, it is just as easy to forget about the solute entirely and focus attention upon the cavity with its radius r.

Notice that for $r = a/2$ the diameter b of the solute molecule is zero. Thus a cavity with $r = a/2$ corresponds to a point solute molecule. Even the insertion of a point molecule into the fluid creates a considerable cavity and therefore requires the expenditure of an appreciable amount of work. To remove the solute effectively from the fluid it is necessary to permit the point molecule to penetrate to the centers of the surrounding solvent molecules. Under this circumstance $b = -a/2$, and $r = 0$. Thus, all the configurations accessible to the solvent molecules before the introduction of the solute are once again accessible, and the solute will have been in effect removed.

When we look upon the cavity as having been produced by a solute, the point $r = a/2$ is unique. It corresponds to a point solute and any further reduction in the size of the corresponding cavity requires us to redefine the intermolecular potential. On the other hand, looked upon from the point of view of the cavity itself, $r = a/2$ is unique because any cavity with a radius smaller than this can accommodate the center of but one solvent molecule at a time. Occupancy by two solvent molecules would require them to penetrate one another's hard cores. This situation is diagrammed in Fig. 6.

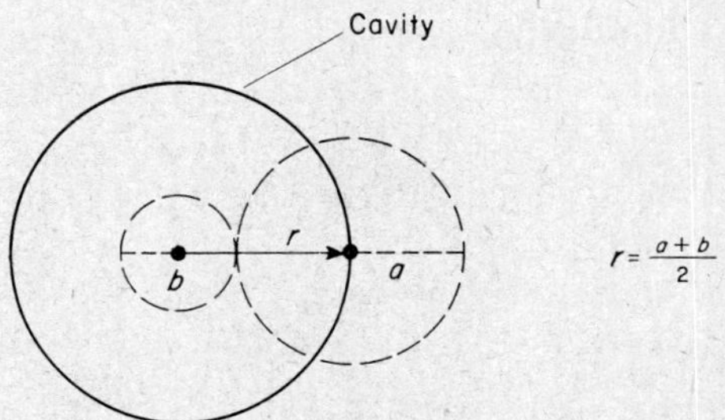

Fig. 5. Illustration of the cavity created by a solute molecule of diameter b.

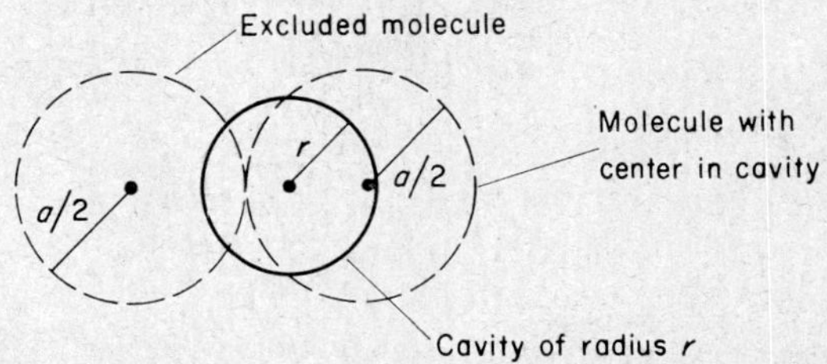

Fig. 6. Illustration demonstrating the exclusion of the center of a second solvent molecule when the cavity radius is less than $a/2$.

Because only one particle at a time can occupy the spherical cell with $r \leqslant a/2$, the probability of finding the center of a molecule in the cell is given by

$$p_1 = \tfrac{4}{3}\pi r^3 \rho \tag{122}$$

where N is the number of fluid molecules occupying the volume V and $\rho = N/V$ is the concentration. The chance p_0 that the spherical cell of radius r is devoid of solvent molecule centers is given by

$$p_0 = 1 - p_1 = 1 - \tfrac{4}{3}\pi r^3 \rho \tag{123}$$

If more than one solvent molecule could occupy the spherical cell at a time, i.e., if $r > a/2$, the calculation of the chance that the cell is occupied would not be so simple. This follows from the fact that then a solvent could occupy the cell if either one molecule were present or two molecules were present, etc. Thus, one would have to compute the sum $p_1 + p_2 + \cdots$, where p_2 denotes the probability that two molecules are simultaneously present in the cell. To compute p_2 and the higher-order terms, higher-order correlation functions are necessary and we do not have these at our disposal. On the other hand, the computation of p_1 involves only singlet distribution functions (namely ρ) and so we are led to the result (123).

Now there is a general theorem of statistical mechanics which states that the probability of observing a fluctuation is equal to the exponential of the reversible work required to produce the fluctuation divided by kT.[19] The occurrence of a spherical cavity of radius r in the fluid may be regarded as a fluctuation whose probability is p_0. Since the reversible work required to produce the cavity is $W(r)$, this theorem demands that

$$p_0(r) = e^{-W(r)/kT} \tag{124}$$

Substituting (123) into (124) and clearing of exponentials, we arrive at the desired result for the reversible work

$$W(r) = -kT \ln\left(1 - \tfrac{4}{3}\pi r^3 \rho\right), \quad r \leqslant a/2 \tag{125}$$

Thus we have an exact expression for $W(r)$ for $r \leqslant a/2$.

For $r > a/2$ an exact expression for $W(r)$ is not available. However, the considerations of the preceding sections indicate that an excellent approximation is the following:

$$W(r) = K_0 + K_1 r + K_2 r^2 + K_3 r^3 \tag{126}$$

where K_0, K_1, K_2, and K_3 are functions of ρ and T. One might consider this to be an obvious approximation, at least over a range of r, because it resembles the beginning of an expansion of W in a power series in r. However, the discussion of the preceding sections shows that W for $r > a/2$ is non-analytic in r. There are discontinuities in the derivatives of W with respect to r. Fortunately, these discontinuities appear in high-order derivatives and the function $W(r)$ is at least *smooth* if non-analytic. The first singularity occurs at $r = a/2$, where the discontinuity occurs in the third derivative of W with respect to r. The next discontinuity occurs at $r = a/\sqrt{3}$, where the singularity occurs first in the fifth derivative of W with respect to r. Singularities at larger values of r occur in still higher-order derivatives. For a detailed discussion of this phenomenon the reader is referred to Section II and in particular to Eqs. (18) and (19).

The points $r = a/2$ and $r = a/\sqrt{3}$ are values of r at which it first becomes possible to introduce the centers of at first two and then three molecules into the spherical cell simultaneously. Thus the singularities are associated with the appearance of new physical processes, namely the ability to pack additional molecules into the cell. For our purposes, the important result is that W and its first two derivatives are continuous at $r = a/2$. Thus, if we denote W, specified by Eq. (125), for $r \leqslant a/2$ by W', we have at $a/2$ the following three conditions:

$$\begin{aligned} W &= W' \\ \frac{\partial W}{\partial r} &= \frac{\partial W'}{\partial r}, \qquad r = a/2 \\ \frac{\partial^2 W}{\partial r^2} &= \frac{\partial^2 W'}{\partial r^2} \end{aligned} \tag{127}$$

These conditions are sufficient to eliminate three of the coefficients in Eq. (126). In fact, the possession of one more condition would permit us to calculate the dependence of all four coefficients upon ρ and T.

Now the form (126) has been suggested by purely statistical considerations and is therefore valid down to very small values of r, e.g., down to $r = a/2$. If we leave the domain of molecular dimensions, however, and deal with a macroscopic cavity, then it should be possible to specify W by thermodynamic rather than statistical methods. One of the most pleasing aspects of scaled particle theory is the fact that thermodynamic considerations demand the same form (126) as do statistical considerations.

Thus thermodynamics requires

$$W(r) = \tfrac{4}{3}\pi r^3 p + 4\pi r^2 \sigma \left(1 - \frac{4a\delta}{r}\right) + K_0 \tag{128}$$

In this equation the term $\frac{4}{3}\pi r^3 p$ is the volume work which must be expended in producing the cavity, while the term $4\pi r^2\sigma$ is the surface work, with σ playing the role of a surface tension. The factor in parentheses involving the quantity δ represents the asymptotic dependence of the surface tension upon curvature.[25–27] δ is a quantity of the order of unity so that the effect of curvature upon the surface tension does not become pronounced until one deals with cavities of molecular dimensions. The quantity K_0 does not usually appear in the thermodynamic expression. It is the same K_0 which appears in Eq. (126) and turns out to be so small that it may be neglected as r becomes of macroscopic size. Thus, in effect, Eq. (128) is the proper macroscopic expression. It is clear that p, σ, δ, and K_0 all depend upon ρ and T, and therefore (128) like (126) is cubic in r with coefficients which depend upon ρ and T. Comparing Eqs. (126) and (128) K_3 is identified immediately:

$$K_3 = \tfrac{4}{3}\pi p \tag{129}$$

Sometimes it is convenient to rewrite Eq. (126) in the form of an expansion about $r = a/2$

$$W(r) = k_0 + k_1(r - a/2) + k_2(r - a/2)^2 + k_3(r - a/3)^3 \tag{130}$$

This is actually a valid asymptotic expansion since beyond $r = a/2$ the first singularity occurs in the fifth derivative of W with respect to r at the point $r = a/\sqrt{3}$.

Introducing Eqs. (125) and (126) into Eq. (127) and making use of Eq. (129) we can evaluate the remaining coefficients K_0, K_1, and K_2. The results for all the coefficients are

$$\begin{aligned}
K_0 &= kT\left[-\ln(1-y) + \frac{9}{2}\left(\frac{y}{1-y}\right)^2\right] - \frac{\pi p a^3}{6} \\
K_1 &= -\left(\frac{kT}{a}\right)\left\{\frac{6y}{1-y} + 18\left(\frac{y}{1-y}\right)^2\right\} + \pi p a^2 \\
K_2 &= \left(\frac{kT}{a^2}\right)\left\{\frac{12y}{1-y} + 18\left(\frac{y}{1-y}\right)^2\right\} - 2\pi p a \\
K_3 &= \tfrac{4}{3}\pi p
\end{aligned} \tag{131}$$

where we have used the reduced concentration

$$y = \tfrac{1}{6}\pi a^3 \rho \tag{132}$$

Employing these coefficients in Eq. (126) and comparing it with Eq. (128) we obtain

$$\sigma = \frac{kT}{4\pi a^2}\left\{\frac{12y}{1-y} + 18\left(\frac{y}{1-y}\right)^2\right\} - \frac{pa}{2} \tag{133}$$

$$\delta = \tfrac{1}{8}\left\{\frac{[6y/(1-y)] + 18[y/(1-y)]^2 - \pi p a^3/kT}{[6y/(1-y)] + 9[y/(1-y)]^2 - \pi p a^3/kT}\right\} \tag{134}$$

The quantity K_0 is associated with the work of introducing the point molecule. This work, which is obtained by replacing r in (125) by $a/2$, has no counterpart in macroscopic considerations, and therefore K_0 has not been discussed in that context.

Equation (133) is an expression for the surface tension. However, it cannot represent the interfacial tension between a liquid and its vapor. This is especially clear if the cavity is regarded as having been produced by the introduction of a hard sphere solute with diameter b so that r is given by Eq. (121). The cavity is not a bubble filled with vapor but is entirely occupied by the large hard sphere. When b becomes infinite the fluid adjacent to the cavity is adjacent to a perfectly repelling plain hard wall. (Such a wall could not be employed experimentally. Not only must it be hard, but the wall cannot exert attractive forces upon the adjacent molecules since by definition it is part of the surface of the hard sphere solute.) Therefore, the surface tension specified by Eq. (133) must refer to the tangential stress minus the hydrostatic pressure adjacent to this hard wall.

Now, although the change in density which occurs on passing through the interface between a liquid and its vapor does not occur discontinuously, it does occur very abruptly over the space of a few angstroms. As a result, a set of forces must exist at this interface which resembles the forces produced by a plane hard wall. In view of this, one might expect the surface tension given by Eq. (133) to approximate the interfacial tension between a liquid and its vapor. The approximation ought to improve as the difference between the densities of the liquid and vapor increases so that the change in density at the interface occurs more abruptly. Thus the approximation ought to be poor near the critical point, but should improve as the temperature is reduced below this point.

The analysis developed thus far in this section was first presented by Reiss, Frisch, Helfand, and Lebowitz[37] in 1959. It can be seen from Eq. (134) that the dependence of the surface tension upon curvature is also specified by this theory. Reiss *et al.* even went so far as to calculate the chemical potential of the hard sphere solute from the reversible work expended in creating the cavity, and from this they calculated the partial pressure of the solute and compared the result with experiment wherever it was reasonable to do so.

The results in this case were acceptable but not dramatic. This was to be expected since there are no solute molecules which can really imitate hard spheres. For further details the reader is referred to the original article. Much more dramatic results have been obtained with other thermodynamic quantities, and we shall turn our attention to these.

We return to Eq. (133) and note that the surface tension is determined by the temperature T, the hard core diameter a, the reduced concentration y characteristic of the temperature T, and the pressure p. For real fluids it turns out that the last term in Eq. (133) involving the pressure is negligible compared to the first. In most cases it is smaller than the first by a factor of a thousand or more. Thus it can be ignored. When this is done, σ is determined by the temperature, the concentration, and a. One might inquire as to how the effect of the soft part of the intermolecular potential enters the theory. It does not appear explicitly in Eq. (133). Only the parameter a associated with the hard core seems to be involved. Actually the soft part of the potential does appear, but implicitly. It acts to determine y. It must be remembered that y is the concentration *measured* at T.

This points up a characteristic which will permeate all of our applications of scaled particle theory to experimental data. The theory as employed will fail to provide the equation of state. Thus y is not specified as a function of T by the theory but must be measured. Nevertheless, relations such as Eq. (133) are valuable and do yield much information. They are definitely statistical rather than thermodynamic since, like (133), they contain molecular parameters such as a.

A method of application of the theory which should lead to the equation of state for real fluids is discussed in Section V. This form of the theory has not yet been carried to completion. On the other hand, in the special case of pure hard sphere fluids, scaled particle theory has enabled us to compute the equation of state. This is discussed in Sections II, III, and IV, and we shall use the hard sphere equation of state in Section VIII.

It will be demonstrated shortly that Eq. (133) can be applied to real fluids with success, and we might pause to consider why such a simple development can be fruitful. In the first place, it is only successful when applied to *dense* fluids (fluids whose densities lie in the range common to most liquids). An examination of the formalism leading to scaled particle theory (Sections II, III, IV, V, and VI) reveals that it is geometric in nature, concerning itself primarily with the manner in which hard spheres can be packed into a given volume V. In fact, in the case of simple hard spheres the entire problem revolves about the volume dependence and not the temperature dependence. The equation of state for hard spheres has the form (135) in which the pressure equals the temperature multiplied by a function of the volume alone (assuming that N remains constant). It is the derivation of this function of the volume to which the early sections of this article address themselves, i.e., to the solution of a geometric problem. Scaled particle theory provides an accurate solution to the packing problem.

Many authors have suggested that the packing of hard cores presents the most crucial problem in the theory of dense real fluids. The success of scaled particle theory when applied to real fluids is a reaffirmation of this point of view. The model of a liquid to which we are driven is the following. The soft part of the intermolecular potential acts primarily to establish the density, i.e., the volume V of a pseudo-container to which the molecules are confined. The molecules in this pseudo-container constitute a hard sphere fluid which is susceptible to treatment by scaled particle theory.

That the soft part of the intermolecular potential acts primarily to establish the volume of the container is suggested by several other sources. For example, at an early date Lennard-Jones and Devonshire,[38] employing a rudimentary form of the cell method, calculated the average field to which a molecule in a cell was subjected by its nearest neighbors. At liquid densities this average potential resembles that of a square well within which a molecule can move freely, experiencing forces only at the boundaries. Increasing the density merely reduces the breadth of the square well, but still fails to establish an *appreciable* potential gradient within the cell.

Reiss *et al.* applied Eq. (133) to some simple liquids. Values of a were taken from Table I-A in ref. 36 and values of the concentration y were obtained from the critical tables.[39] Experimental values for the surface tension were taken from the *Handbook of Chemistry*.[40] The term $pa/2$ in (133) is ignored. The results of Reiss *et al.*[37] are reproduced in Table I.

TABLE I. Comparison of Calculated and Experimental Surface Tensions of Some Simple Liquids

Substance	T, °K	$\sigma_{\text{exp.}}$, dynes/cm	$\sigma_{\text{calc.}}$, dynes/cm
N	27.2	4.8	6.09
Ar	85.1	13.2	16.40
He	4.2	0.098	0.223
H_2	20.4	1.80	1.91
N_2	70.1	10.5	14.9
O_2	70.1	18.3	23.6
C_6H_6	273.1	29.02	34.3

The agreement in Table I is satisfactory when the simplicity of Eq. (133) is considered together with the fact that it specifies the surface tension at a hard wall rather than the interfacial tension between a liquid and its vapor. In the case of the diatomic molecules and benzene the strict requirement of spherical symmetry is not met, and furthermore the meaning of a as obtained from ref. 36 is somewhat obscure. Presumably some rotational averaging has occurred so that the molecules behave more as spheres than they actually are. It is of interest to note that the range of surface tensions involved is quite large (10^{-1} dynes/cm–10^2 dynes/cm).

Reiss and Mayer[41] have applied Eq. (133) to fused salts. How may this be done! At the outset we are confronted with the fact that the simplest fused salts contain ions having rigid cores with at least *two* different diameters. Thus, (133) cannot be applied without further discussion. The situation is saved by a theorem due to Reiss, Mayer, and Katz[42] which states that the thermodynamic properties of a simple fused salt depend (to a high degree of approximation) only upon the sum of the radii of ions of opposite sign and not upon the individual diameters. Thus two salts, one with anions and cations of the same size and another with ions of different sizes, but whose ionic radii add to the same value, will have identical properties. This theorem represents a law of corresponding states. One may therefore apply Eq. (133) to a hypothetical salt in which the anions and cations have the same diameter, and the sum of whose radii equals the sum of the corresponding radii for the real salt. The surface tension of the real salt should be identical (within the limits of the Reiss, Mayer, and Katz theory) with that computed for the hypothetical salt. In dealing with the hypothetical salt, however, Eq. (133) can be applied since only one diameter a is involved. This diameter is the sum of the anion and cation radii.

Since the values of a should represent distances of closest approach and not bond distances, reliable data are hard to acquire. However, a ought not to be much smaller than the corresponding bond distance. As an approximation, therefore, in view of the absence of the required data, one can use the equilibrium bond distance (measured in the gas by electron-diffraction or microwave techniques) for a.

The surface tensions computed with this choice should be slightly larger than the observed ones, since these values of radius are larger than the distance of closest approach. Table II collects the results of these calculations and compares them with experiment. Uni-univalent, di-univalent, and tri-univalent salts have been included. Column 5 lists the observed surface tension, while Column 6 shows the results calculated using Eq. (133) and employing the data on the corresponding line in Columns 2–4. We

TABLE II. Calculated and Observed Surface Tensions of Molten Salts

Molten salt	Temp., °C	Density,[a] g/cm³	Gas equilibrium distance, A	$\sigma_{obs.}$ d/cm	$\sigma_{calc.}$ d/cm
		Uni-univalent			
NaCl	1000	1.448	2.361[c]	98[d]	111
NaBr	900	2.217	2.502[c]	91[d]	95
NaI	700	2.704	2.712[c]	84[d]	85
KCl	900	1.452	2.667[c]	90[d]	96
KBr	800	2.073	2.821[c]	85[d]	87
KI	800	2.334	3.048[c]	69[d]	76
RbCl	828	2.024	2.787[c]	89[b]	81
RbBr	831	2.542	2.945[c]	81[b]	83
RbI	772	2.719	3.177[c]	72[b]	72
CsF	826	3.456	2.345[c]	96[b]	85
CsCl	830	2.592	2.906[c]	78[b]	79
CsBr	808	2.915	3.072[c]	72[b]	76
CsI	821	2.953	3.315[c]	63[b]	70
		Uni-divalent			
$BaCl_2$	1000	2.982[e]	2.82[i]	157[j]	152
$CdCl_2$	600	3.366[f]	2.21[i]	83[k]	79
$HgCl_2$	293	4.670[g]	2.29[i]	56[l]	36
$HgBr_2$	241	5.113[h]	2.41[i]	65[l]	37
		Uni-trivalent			
$BiCl_3$	271	3.554[b]	2.48[m]	66[b]	52
$BiBr_3$	281	4.099[b]	2.63[m]	64[b]	47

[a] Yaffe, L., and Van Artsdalen, E. R., *J. Phys. Chem.* **60,** 1125 (1956).

[b] Jaeger, F. M., *Z. Anorg. Allgem. Chem.* **101,** 1 (1917).

[c] Honig, A., Mandel, M., Stitch, M. L., and Townes, C. H., *Phys. Rev.* **96,** 629 (1954).

[d] Bloom, H., and Bockris, J. O'M., *Modern Aspects of Electrochemistry,* Academic Press, New York, 1959, Vol. 2, p. 198.

[e] Huber, P. W., Potter, E. V., and St. Clair, H. W., U.S. Bureau of Mines Report of Investigations, No. 4858 (1952).

[f] Boardman, N. K., Dorman, A., and Heymann, E., *J. Phys. Chem.* **53,** 375 (1949).

[g] Janz, G. J., Solomons, C., and Gardner, H. J., *Chem. Rev.* **58,** 481 (1958).

[h] Janz, G. J., and McIntyre, J. D., *Ann. N.Y. Acad. Sci.* **79,** 790 (1960).

[i] Akishin, P. A., and Spiridonov, V. P., *Kristallografiya* **2,** 475 (1957).

[j] Peake, J. S., and Bothwell, M. R., *J. Am. Chem. Soc.* **76,** 2656 (1954).

[k] Boardman, N. K., Palmer, A., and Heymann, E., *Trans. Faraday Soc.* **51,** 277 (1955).

[l] Prideaux, E. B. R., and Jarett, J. R., *J. Chem. Soc.*, 1203 (1938).

[m] Skinner, H. A., and Sutton, L. E., *Trans. Faraday Soc.* **36,** 681 (1940).

observe that agreement is good in all cases except in the case of the mercury salts.

Both $HgBr_2$ and $HgCl_2$ are known to have anomalously low conductivities in the molten state and are thought to be un-ionized. This means that the radius of the separate particle is that of the molecule rather than the ion, i.e., it is larger than would be expected for an ion. Thus the actual surface tension is larger and the value computed with Eq. (133) assuming ionic radii is relatively too small. It appears as though the anomalous state of ionization of the mercuric salts is reflected in the present data.

The agreement between theory and experiment in the case of fused salts is better than for the case of the simple fluids discussed in connection with Table I. The vapor densities of the salts are much lower than those of the simple fluids, so that the discontinuity at the interface is more abrupt. Hence it is not surprising in view of our earlier discussion that agreement should be better.

VIII. COMPRESSIBILITIES AND EXPANSIVITIES

The treatment of the hard sphere fluid contained in Sections II, III, and IV yields an expression for the equation of state for hard spheres of the following form:

$$\frac{\pi a^3 p}{6kT} = \frac{y(1 + y + y^2)}{(1 - y)^3} \tag{135}$$

where p is the pressure and y is the reduced concentration defined by Eq. (132). Equation (135) is Eq. (77), derived earlier. Now in Section VII the point was made that a liquid is usually sufficiently dense for the soft part of the intermolecular potential to act primarily to establish the volume V and that the liquid then behaves as though it were a hard sphere fluid confined to this volume. The success of scaled particle theory when applied to real fluids has been linked to this situation. One might be led to suspect therefore that Eq. (135) would find some use in the computation of the isothermal compressibilities of real fluids.

The isothermal compressibility is defined by

$$\beta = -\frac{1}{V}\left(\frac{\partial V}{\partial p}\right)_T \tag{136}$$

When the fluid is as dense as most liquids, any further isothermal compression may be regarded as a process of repacking hard spheres so that they occupy a smaller average volume. Thus the incremental pressure ought to be the same as that for repacking hard spheres. If this is so,

$$dp = dp_h \tag{137}$$

where we now denote the pressure of a hard sphere fluid by p_h. The quantity p_h denotes the pressure to be expected for a hard sphere fluid occupying the same volume as the real fluid whose pressure is p at the temperature T in question. The hard sphere fluid and the real fluid, of course, are assumed to occupy the same volume V. As a result of Eq. (137)

$$\left(\frac{\partial V}{\partial p}\right)_T = \left(\frac{\partial V}{\partial p_h}\right)_T \tag{138}$$

so that

$$\beta = -\frac{1}{V}\left(\frac{\partial V}{\partial p_h}\right)_T \tag{139}$$

and the compressibility of the real fluid should be the same as that of a hard sphere fluid at the same volume and temperature. Thus one should be able to evaluate β from Eq. (135) in which p is replaced by p_h, adopting the new notation for the pressure of the hard sphere system. Substituting (135) into (139) we obtain

$$\beta = \frac{\pi a^3}{6kT}\frac{(1-y)^4}{y(1+2y)^2} \tag{140}$$

Yosim and Owens[43] have applied this formula to a number of liquids. Unfortunately, one is somewhat limited by the experimental data which are available to facilitate the comparison. It is difficult to find data on liquids whose molecules are sufficiently spherical to suit the requirements of the present theory. Except for helium and argon, all the non-ionic fluids to which Yosim and Owens addressed themselves were either non-spherical or highly polar or both. On the average, experiment and theory differed by a factor of about 1.5, but in one or two cases disagreement by as large a factor as 2 was observed. The results are sensitive to the choice of a and so this is not surprising, especially for the non-spherical molecules.

Table III lists Yosim's and Owens' results for helium and argon. Although these molecules are spherical, helium is a quantum fluid,

TABLE III. Comparison of Calculated and Experimental Compressibilities of Helium and Argon

Substance	Temperature, °K	$\beta_{calc.}$, atm^{-1} $\times$ 10^4	$\beta_{exp.}$, atm^{-1} $\times$ 10^4
He	4.2	256	254
Ar	87.4	1.12	2.18

and in both cases compressibility data (which are difficult to obtain) are derived from the velocity of sound measurements. In view of the quantum nature of helium, the excellent agreement between theory and experiment must be somewhat fortuitous. We shall not pursue the issue of the compressibilities of non-ionic fluids any further because of the non-existence of truly spherical molecules and the paucity of data on compressibility.

Insofar as sphericity is concerned, we enter a much broader field with the consideration of fused salts. Here a wide variety of salts may be found possessing truly spherical ions. On the other hand, the treatment of fused salts involves the theorem leading to corresponding states[42] which has been used in Section VII to facilitate the treatment of fused salt surface tensions. This theorem states that the thermodynamic properties of a fused salt depend upon the sum of the anion and cation radii and not upon the individual diameters of the ions. An examination of ref. 42 reveals that this is an approximation. It is based upon the assumption that the average configuration of the fused salt derives very little contribution from configurations in which large numbers of ions of like sign are in contact with one another. This approximation is excellent when the anion and cation radii are not too different. On the other hand, when there is a large disparity in size, it may be expected that ions of like sign will be in contact. For example, if the anion is very large and the cation small a typical configuration might have the anions forming a quasi-lattice in which they contact one another with the small cations occupying the interstices.

Thus, in proceeding to fused salts we gain the advantage of sphericity, but must accept the approximation contained within the application of the law of corresponding states. Fortunately, ref. 42 is devoted chiefly to matters not involving scaled particle theory, and an independent experimental confirmation of the law appears there. It seems to fail for the salts of lithium where the situation just described occurs, i.e., the anion is large and the cation small.

The first application of Eq. (140) to real liquids was made by Stillinger,[44] who in fact applied it to fused salts. As in the case of surface tension the long-range Coulomb forces associated with the ions represent no added difficulty for the present theory. The soft part of the intermolecular potential appears implicitly by establishing the reduced concentration y. Since data on the hard core diameter a (now the sum of the anion and cation radii) are uncertain, Stillinger elected to work in reverse. He substituted measured values of β, y, and T into Eq. (140) and computed a for a number of salts at several different temperatures. If the theory is correct, the value of a computed in this manner should be relatively insensitive to temperature and always less (but not much less) than the sum of the crystal ionic radii (since a is a distance of closest approach). Such temperature dependence as there is should reveal a as a decreasing function of temperature, since in accordance with the discussion of Section VII the more energetic particles available at higher temperatures penetrate one another's repulsive fields more effectively. Stillinger's results are reproduced in Table IV.

The requirements we have just set down are well met in Table IV; thus it is clear that a is relatively insensitive to temperature. For example, in the case of lithium bromide a change in temperature amounting to 400° between 600° and 1000° only causes a to change from 2.55 to 2.31 angstroms. It is also observed in Table IV that a is a *decreasing* function of temperature, as required. Furthermore, all measured values of a are less than the corresponding sums of the Pauling radii, which appear in parentheses in the second column. This is as it should be, since a represents a distance of closest approach while the sum of the Pauling radii is the distance between equilibrium positions in the crystal.

Yosim and Owens[45] have adopted an approach opposite to that of Stillinger, and have used the gas equilibrium distance appearing in Table II to calculate the compressibility from Eq. (140). As in Stillinger's work the experimental compressibilities quoted are

TABLE IV. Values of the Anion–Cation Collision Diameter Computed from the Fused Salt Isothermal Compressibilities

Substance	Pauling sum	600°	700°	800°	900°	1000°
LiCl	(2.41)		2.31	2.26	2.20	2.14
NaCl	(2.76)			2.47	2.41	2.34
KCl	(3.14)			2.70	2.63	2.55
CsCl	(3.50)		3.01	2.93	2.84	2.74
LiBr	(2.55)	2.55	2.49	2.43	2.38	2.31
NaBr	(2.90)			2.63	2.58	2.51
KBr	(3.28)			2.85	2.78	2.70
CsBr	(3.64)		3.06	2.98	2.88	2.75
NaI	(3.11)		2.88	2.82	2.74	2.66
KI	(3.28)		3.09	3.01	2.93	2.82
$CsCl_2$	(2.78)	2.50	2.45	2.39		

The corresponding sums of Pauling ionic crystal radii appear in parentheses. All distances are in angstrom units.

those which have been obtained from velocity of sound measurements. An exhaustive list of these is not available, and only ten salts appear in Table V, which has been assembled from their work. All values which appear in Table V correspond to a temperature 10% higher than the melting point of the salts in question. It will be noticed that, except for the salts of lithium, agreement is remarkably good, the average disparity between theory and experiment being of the order of a few per cent. In the case of the salts of lithium, an error of more than a factor of 2 appears. This is not surprising because, as we mentioned earlier, the law of corresponding states fails in the case of these salts.

TABLE V. Comparison of Calculated and Experimental Compressibilities of Fused Salts. Temperature = $1.1T_m$

Substance	Density, g/cm³	$\beta_{\text{calc.}}$, cm²/dyne × 10¹²	$\beta_{\text{exp.}}$, cm²/dyne × 10¹²
LiCl	1.462	45	22
LiBr	2.475	56	23
NaCl	1.499	38	34
NaBr	2.176	45	35
NaI	2.654	58	45
KCl	1.464	42	44
KBr	2.037	47	47
KI	2.352	55	57
CsCl	2.699	53	45
CsBr	3.028	55	59

Equation (135) has also been used to compute the thermal expansivities of real liquids. The thermal expansivity, α, is defined by the following relation:

$$\alpha = \frac{1}{V}\left(\frac{\partial V}{\partial T}\right)_p \tag{141}$$

Since we are now using p_h in place of p in Eq. (135) we have for the thermal expansivity of a hard sphere fluid

$$\alpha_h = \frac{1}{V}\left(\frac{\partial V}{\partial T}\right)_{p_h} \tag{142}$$

At high enough densities we assume that α may be approximated by α_h. A minimum requirement for the validity of this approximation is that the constancy of p_h should imply the constancy of p. Some examination of this requirement is possible.

The pressure p is given by the following exact thermodynamic relation:

$$p = T\left(\frac{\partial p}{\partial T}\right)_V - \left(\frac{\partial E}{\partial V}\right)_T \tag{143}$$

in which E is the internal energy. Now the internal energy may be separated into the average kinetic energy and the average potential energy. Thus

$$E = \tfrac{3}{2}kT + \psi \tag{144}$$

in which ψ represents the average potential energy. Notice that ψ will be much more sensitive to volume than temperature, since the average potential energy is sensitive to the average distance between molecules. Thus, to a first approximation we may assume that ψ is independent of temperature.

The first derivative on the right of Eq. (143) may be approximated by the derivative of p_h in the following manner:

$$\left(\frac{\partial p}{\partial T}\right)_V \approx \left(\frac{\partial p_h}{\partial T}\right)_V \tag{145}$$

This follows from the fact closely associated with our statement concerning ψ that the average attraction between molecules depends more upon volume than temperature. If we consider the model discussed in Section VII in which a liquid behaves as a hard sphere fluid confined to a container whose volume is determined by the soft part of the intermolecular potential, then an increase of temperature during which the volume is maintained constant must lead to an increase of pressure due essentially to the pressure increase of the confined hard sphere fluid. Thus we have the result (145).

Now Eq. (135) has the following form:

$$p_h = Tf(V) \tag{146}$$

where $f(V)$ is a function of volume alone. From Eq. (146) we derive the relation:

$$\left(\frac{\partial p_h}{\partial T}\right)_V = f(V) = \frac{p_h}{T} \tag{147}$$

Substituting (147) into (145) then yields

$$\left(\frac{\partial p}{\partial T}\right)_V \approx \frac{p_h}{T} \tag{148}$$

and finally substitution of (148) and (144) into (143) yields

$$p = p_h - \left(\frac{\partial \psi}{\partial V}\right)_T \tag{149}$$

Since ψ is approximately independent of temperature, its derivative with respect to volume, appearing in Eq. (149), must also be approximately independent of temperature. Thus, Eq. (149) assures us that to a first approximation the difference between p and p_h is a function of volume alone. As a matter of fact, it is plausible to make an even stronger statement. The rate of change of ψ with volume ought to depend less strongly upon volume than ψ itself. Thus, for the small changes of volume involved in Eqs. (141) and (142) the derivative in Eq. (149) might be regarded as approximately independent of both temperature and volume, so that the constancy of p_h assures the approximate constancy of p. Under this circumstance α_h should be an excellent approximation for α.

Substituting Eq. (135) into Eq. (142) and equating α_h with α we obtain

$$\alpha = \frac{1 - y^3}{T(1 + 2y)^2} \tag{150}$$

More data are available on expansivities than on compressibilities. In fact, the expansivities for all the rare gas liquids are available. Employing values of a given in ref. 36, Yosim and Owens computed α for all the rare gases by means of Eq. (150). Calculations were made both at the melting and boiling points of these substances. The results appear in Table VI. Considering the simplicity of the theory, agreement between theory and experiment is satisfactory. The values of α examined in Table VI

TABLE VI. Comparison of Calculated and Experimental Expansivities of the Rare Gases

	At melting point		At boiling point	
Substance	$\alpha_{\text{calc.}}$, $°\text{K}^{-1} \times 10^3$	$\alpha_{\text{exp.}}$, $°\text{K}^{-1} \times 10^3$	$\alpha_{\text{calc.}}$, $°\text{K}^{-1} \times 10^3$	$\alpha_{\text{exp.}}$, $°\text{K}^{-1} \times 10^3$
He			134	119
Ne	11.6	12.4	10.9	13.4
Ar	3.09	4.08	2.99	4.14
Kr	2.31	3.02	2.26	3.07
Xe	1.53	2.24	1.49	2.26

vary by a factor of 100. The same may be said of our earlier considerations regarding surface tension and compressibility. The same simple theory is able to account for the data over a very large range of application.

Yosim and Owens have also applied Eq. (150) to fused salts. Again it was necessary to employ the law of corresponding states so as to define a single meaningful parameter a. The results assembled from their paper[45] appear in Table VII. All values of the expansivity refer to a temperature 10% in excess of the melting point.

TABLE VII. Comparison of Calculated and Experimental Expansivities of Fused Salts. Temperature = $1.1T_m$

Salt	Density, g/cm^3	$\alpha_{calc.}$, $°K^{-1} \times 10^3$	$\alpha_{exp.}$, $°K^{-1} \times 10^3$
LiF	1.747	0.48	0.27
LiCl	1.462	0.55	0.30
LiBr	2.475	0.59	0.26
LiI	3.069	0.65	0.30
NaF	1.870	0.39	0.30
NaCl	1.499	0.41	0.36
NaBr	2.176	0.43	0.36
NaI	2.654	0.46	0.36
KF	1.830	0.42	0.36
KCl	1.464	0.40	0.40
KBr	2.037	0.40	0.41
KI	2.352	0.41	0.41
RbF	2.818	0.45	0.36
RbCl	2.158	0.41	0.41
RbBr	2.618	0.41	0.41
RbI	2.804	0.42	0.41
CsF	3.552	0.50	0.36
CsCl	2.699	0.44	0.39
CsBr	3.028	0.43	0.40
CsI	3.078	0.43	0.38

Comparison of Table VII with Table V makes it clear again that more data are available concerning expansivity than compressibility. Agreement between theory and experiment is very good except for the salts of lithium. This is consistent with the behavior of the compressibilities in Table V. As in that case, the disparity should probably be traced to the failure of the law of corresponding states because of the small size of the lithium ion.

IX. HEATS OF VAPORIZATION AND FUSION

Reiss, Frisch, Helfand, and Lebowitz[37] made an early attempt to employ scaled particle theory in the calculation of the heats of vaporization of real liquids. Their method, which did not make use of Eq. (135), was reasonably successful, and the reader is referred to it for whatever interest it may possess as an alternative procedure. More recently, Yosim and Owens[46] have devised a method which uses Eq. (135) and is both simple and physically transparent. Furthermore, this method, which is based upon the use of an ingenious cycle, makes it possible to compute heats of fusion as well. The development of the present section will be based upon this method.

Before describing it, a simple approximate argument can be given which demonstrates that it is plausible for the heat of vaporization to be related to the properties of a hard sphere fluid confined to the same volume as the liquid. We recall that an adequate model for a liquid seems to be one in which a hard sphere fluid is confined to a container whose volume is established by the soft part of the intermolecular potential. Substituting Eq. (144) into Eq. (149) we obtain

$$p = p_h - \left(\frac{\partial E}{\partial V}\right)_T \tag{151}$$

[$(\partial E/\partial V)_T$ is obviously the pressure of the pseudo container (the internal pressure).] For ordinary liquids $p \ll p_h$, but for densities approaching that of the vapor in equilibrium with the liquid $p = p_h$. Thus Eq. (151) may be approximated by

$$\left(\frac{\partial E}{\partial V}\right)_T = p_h - \nu(V - V_g)p_h \tag{152}$$

where $\nu(V - V_g)$ is the unit step function (153) and V_g is the volume of one mole of vapor in equilibrium with the liquid at temperature T

$$\begin{aligned}\nu(V - V_g) &= 0, \quad V < V_g \\ &= 1, \quad V > V_g\end{aligned} \tag{153}$$

Integrating (152) yields

$$E_l = E(\infty) + \int_{V_g}^{V_l} p_h \, dV \tag{154}$$

where E_l is the internal energy of the liquid, $E(\infty)$ is the internal energy at infinite volume (translational energy alone), and V_l is the volume of one mole of liquid. Now

$$E(\infty) = \tfrac{3}{2}RT \tag{155}$$

where R is the gas constant. Thus

$$E_l = \tfrac{3}{2}RT + \int_{V_g}^{V_l} p_h dV \tag{156}$$

The heat of vaporization, ΔH_v, is given by

$$\Delta H_v = H_g - H_l = E_g + RT - E_l - pV_l \tag{157}$$

Since V_l is small, the term pV_l may be ignored. Furthermore, if the gas is ideal, $E_g = E(\infty) = \frac{3}{2}RT$. Introducing these approximations into Eq. (157) and substituting (156) we obtain

$$\Delta H_v = RT + \int_{V_l}^{V_g} p_h \, dV \tag{158}$$

This relation is approximate, but it suggests that ΔH_v may be evaluated by reference to the properties of a hard sphere fluid alone. The most serious part of the approximation involves the application of Eq. (151) to highly attenuated systems; i.e., an error is introduced when the integral in Eq. (158) is carried to values of V_g characteristic of a dilute vapor.

The cycle devised by Yosim and Owens[46] overcomes this difficulty and yields an expression for ΔH_v which is not too different from (158). We shall illustrate their technique by calculating the heat of vaporization of a liquid at its boiling point, i.e., when its vapor pressure is one atmosphere. The cycle involves four steps:

1. At a temperature T_b such that its vapor pressure is one atmosphere, one mole of liquid is vaporized by isothermal

reversible expansion under its own vapor pressure. The entropy of vaporization is

$$\Delta S_v = \Delta H_v / T_b \tag{159}$$

where ΔH_v is now the heat of vaporization at the boiling point.

2. The vapor so obtained is now held at constant volume V_g (the molar volume of the vapor at the boiling point) while the soft parts of the intermolecular potentials are discharged (formally, of course). The result is a hard sphere fluid possessing the molar volume of the real vapor. If the vapor is ideal, or almost so, the entropy change involved in this process should be zero. In any event, we represent it by

$$\Delta S_{\text{disch.}} \tag{160}$$

3. The hard sphere fluid is now compressed reversibly and isothermally to the molar volume V_l of the real liquid. If Eq. (143) is specialized to the case of hard spheres, and Eq. (146) is substituted into it, we obtain

$$p_h = T\left(\frac{\partial p_h}{\partial T}\right)_V - \left(\frac{\partial E}{\partial V}\right)_T = p_h - \left(\frac{\partial E}{\partial V}\right)_T$$

or

$$\left(\frac{\partial E}{\partial V}\right)_T = 0 \tag{161}$$

Thus for a hard sphere fluid E is independent of volume. It then becomes a requirement of the first law of thermodynamics that the entropy of compression in the present step be given by

$$\Delta S_{\text{compr.}} = \frac{\int_{V_g}^{V_l} p_h \, dV}{T_b} \tag{162}$$

since with constant internal energy the heat absorbed must be equal to the work performed.

4. The intermolecular potentials of the compressed hard sphere fluid are now recharged at constant volume V_l so that the original liquid is regained. During this step the change in entropy is small. It must be associated with the change of internal structure of the fluid, and at these high densities the structure is determined predominantly by the packing of hard

cores. Thus the structure in the real liquid is almost the same as that in the hard sphere fluid possessing the same density. Nevertheless, we indicate the entropy change of this step by

$$\Delta S_{\text{ch.}} \tag{163}$$

Summing the entropy changes around the cycle we have

$$0 = \frac{\Delta H_v}{T_b} + \Delta S_{\text{disch.}} + \frac{\int_{V_g}^{V_l} p_h \, dV}{T_b} + \Delta S_{\text{ch.}} \tag{164}$$

Now $\Delta S_{\text{disch.}}$ and $\Delta S_{\text{ch.}}$ are both small and tend to cancel one another. If these terms are ignored in Eq. (164) the result is

$$\Delta H_v = \int_{V_l}^{V_g} p_h \, dV \tag{165}$$

which is identical with Eq. (158) except for the small term RT. However, Eq. (165) has been derived from a sounder basis than Eq. (158).

Introducing Eq. (135) into Eq. (165) and evaluating the integral, we obtain

$$\Delta H_v = RT_b \left\{ \ln \left[\frac{\frac{V_g}{V_l} - y}{1 - y} \right] - \frac{3y}{2} \left[\frac{\frac{2V_g}{V_l} - y}{\left(\frac{V_g}{V_l} - y \right)^2} - \frac{(2 - y)}{(1 - y)^2} \right] \right\} \tag{166}$$

Since $(V_g/V) \gg 1 > y$, the term $[(2V_g/V_l) - y]/[(V_g/V) - y]^2$ may be considered negligible. Thus we have

$$\Delta H_v = RT_b \left\{ \ln \left[\frac{V_g}{V_l(1 - y)} \right] + \frac{3y(2 - y)}{2(1 - y)^2} \right\} \tag{167}$$

Equation (167) was applied by Yosim and Owens to many substances:

(1) the rare gases,
(2) diatomic molecules,
(3) triatomic molecules and hydrocarbons,
(4) substituted hydrocarbons.

The molecular "hard sphere" diameters a were taken from the tabulation of Hirschfelder, Bird, and Curtiss.[36] In cases where

TABLE VIII. Heats of Vaporization of Non-ionic Liquids

Liquid	Boiling point T_B, °K	Diameter a, A[f]	Density of liquid, g/cm³	Heat of vaporization, cals/mole calc.	Heat of vaporization, cals/mole obs.	$\frac{\Delta H_v(\text{calc.})}{\Delta H_v(\text{obs.})}$
He	4.22[a]	2.556	0.1249[a]	26.8	19.4[a]	1.38
Ne	27.07[a]	2.749	1.207[a]	427	414[a]	1.03
Ar	87.29[a]	3.400	1.400[a]	1611	1558[a]	1.03
Kr	119.8[a]	3.597	2.413[a]	2188	2158[a]	1.01
Xe	165.1[a]	3.963	3.057[a]	3264	3020[a]	1.08
H_2	20.39[b]	2.87	0.07085[g]	227.5	215.8[b]	1.05
N_2	77.2[c]	3.681	0.8091[g]	1434	1333[c]	1.08
O_2	90.13[c]	3.433	1.145[g]	1749	1628[c]	1.07
F_2	85.20[c]	3.65	1.108[g]	1581	1562[c]	1.01
Cl_2	239.0[b]	4.115	1.560[g]	5154	4878[b]	1.06
Br_2	331.4[b]	4.268	2.98[g]	6869	7170[b]	0.96
I_2	456[b]	4.982	3.744[h]	12673	9970[b]	1.27
NO	121.4[c]	3.470	1.269[g]	3139	3293[c]	0.95
CO	81.61[c]	3.59	0.799[g]	1409	1444[c]	0.98
HCl	188.1[c]	3.304	1.193[g]	3363	3860[c]	0.87
HI	237.8[c]	4.123	2.79[g]	5104	2724[c]	1.08
CS_2	319.4[d]	4.437	1.224[g]	6324	6400[d]	0.99
N_2O	184.6[c]	3.816	1.22[g]	3990	3958[c]	1.01
SO_2	263.1[c]	4.289	1.460[g]	7086	5960	1.19
CH_4	111.7[c]	3.796	0.415[i]	2098	1988[c]	1.06
C_2H_4	169.4[e]	4.232	0.566[i]	3499	3237[e]	1.08
C_2H_6	184.5[e]	3.954	0.546[g]	2987	3157[e]	0.95
C_4H_{10}	272.7[e]	4.971	0.601[g]	4676	5352[e]	0.87
C_6H_6	353.2[e]	5.270	0.8128[g]	7312	7353[e]	0.99
C_6H_{12}	353.8[e]	6.093	0.7199[g]	10430	7190[e]	1.45
CH_3Cl	248.9[d]	3.375	0.991[i]	3634	5150[d]	0.71
$CHCl_3$	334.4[d]	5.430	1.408[g]	9539	7020[d]	1.36
CCl_4	349.9[d]	5.881	1.482[g]	10685	7170[d]	1.49
CH_3OH	337.9[d]	3.585	0.7484[g]	5923	8430[d]	0.70
C_2H_5OH	351.7[d]	4.455	0.7455[g]	7148	9220[d]	0.78

[a] Cook, G. A., *Argon, Helium, and the Rare Gases*, Interscience Publishers, New York, 1961.

Footnotes continued on facing page.

more than one value for the diameter is given the smallest has been used.

The results appear in Table VIII, and in most cases the agreement between calculated and observed values is remarkably good. In the case of the rare gases, helium was the only system to yield poor results. This is not surprising in view of the fact that helium is a quantum fluid. In the case of the eleven diatomic systems, the agreement for nine of them was within 10% iodine being the only substance which departed from this trend. Even in the third group, which consists of more complex molecules, agreement on the whole was very good, the only serious discrepancy occurring in the case of cyclohexane. In the final group consisting of hydrocarbons containing highly polar bonds, agreement is not satisfactory. This is not surprising since the values for a listed by Hirschfelder *et al.* are calculated on the assumption that the intermolecular potential can be described by the Lennard-Jones potential.

The last column in Table VIII lists the ratio of calculated-to-observed heats of vaporization. The ratio is always close to unity, reaffirming the excellence of the agreement between theory and experiment. It is again worth noting that the data cover about two orders of magnitude so that agreement is obtained throughout a wide range. This was the case in Sections VII and VIII where surface tension, compressibility, and expansivity were discussed.

Yosim and Owens[46] have also used scaled particle theory to compute heats of fusion. Once again they employ a cycle which

---

[b] Stull, D. R., and Sinke, G. C., "Thermodynamic Properties of the Elements", Advances in Chemistry Series, No. 18 (1956).

[c] Reference 3.

[d] Gray, D. E., *American Institute of Physics Handbook*, McGraw-Hill, New York, 1957.

[e] Rossini, F. D., *et al.*, "Selected Values of Properties of Hydrocarbons", U.S. National Bureau of Standards Circular 461 (1947).

[f] Reference 5.

[g] *International Critical Tables*, McGraw-Hill, New York, 1926.

[h] Mellor, J. W., *Comprehensive Treatise on Inorganic and Theoretical Chemistry*, Vol. II, Suppl. I, "The Halogens", John Wiley and Sons, New York, 1962.

[i] *Handbook of Chemistry and Physics*, 44th Edition, Chemical Rubber Publishing Company, Cleveland, 1962.

involves charging and discharging the soft part of the intermolecular potential. This time the cycle is conducted at the melting point, T_m, of the substance in question. It consists of the following steps:

1. One mole of solid is melted. The entropy of fusion is:

$$\Delta S_f = \Delta H_f / T_m \tag{168}$$

where ΔH_f is the heat of fusion.

2. The liquid is discharged at constant volume so that we have a compressed fluid of hard spheres occupying the molar volume V_l of the liquid.

3. This hard sphere fluid is compressed until it occupies the volume V_s of the solid.

4. The compressed gas of hard spheres is recharged at constant volume.

If we again make the assumption that $\Delta S_{\text{ch.}} + \Delta S_{\text{disch.}} = 0$

$$\Delta H_f = \int_{V_s}^{V_l} p_h \, dV \tag{169}$$

Substitution of Eq. (135) into Eq. (169) then yields

$$\Delta H_f = RT_m \left\{ \ln \left(\frac{1-y}{\dfrac{V_s}{V_l} - y} \right) - \frac{3y}{2} \left[\frac{2-y}{(1-y)^2} - \frac{\dfrac{2V_s}{V_l} - y}{\left(\dfrac{V_s}{V_l} - y \right)^2} \right] \right\} \tag{170}$$

Yosim and Owens[46] applied Eq. (170) to all the rare gases except helium and the results appear in Table IX. Again striking agreement is obtained between the calculated and observed values. The agreement is especially dramatic in view of the fact that the term

$$\frac{\dfrac{2V_s}{V_l} - y}{\left(\dfrac{V_s}{V_l} - y \right)^2} - \frac{2-y}{(1-y)^2}$$

is the difference between two large quantities. For example, in the case of argon this value is $6.8 - 4.8 = 2.0$.

The theory should not work, in the case of fusion, for molecules more complex than the simple rare gases. This follows from the fact that rotational degrees of freedom not present in a solid appear in the liquid. The step which involves charging the soft part of the intermolecular potential, so that the compressed hard sphere fluid becomes the normal solid, then involves the disappearance of these rotational degrees of freedom, and the entropy of the process becomes more complex than we had imagined it to be. The same change in rotational degrees of freedom does not occur in the discharging process, and so the two formal processes cannot be made to compensate one another, to say nothing of the fact that the entropy of charging is no longer negligible. Indeed, application of the theory to complex molecules yields poor results for the heats of fusion.

TABLE IX. Heats of Fusion of the Rare Gases

Liquid	Boiling point[a] T_m, °K	Density of liquid, g/cm³	Density of solid,[a] g/cm³	Heat of fusion, cal/mole calc.	obs.[a]	$\frac{\Delta H_f\,(\text{calc.})}{\Delta H_f\,(\text{obs.})}$
Ne	24.55	1.247	1.444	65.5	80.1	0.82
Ar	83.78	1.409	1.623	269.9	280.8	0.96
Kr	116.0	2.441	2.826	363.4	390.7	0.93
Xe	161.3	3.084	3.540	602.2	548.5	1.10

[a] G. A. Cook, *Argon, Helium and the Rare Gases*, Interscience Publishers, New York, 1961.

It has been assumed that $\Delta S_{\text{ch.}} + \Delta S_{\text{disch.}} = 0$. Now, $\Delta S_{\text{ch.}}$ ought to involve the negative of the communal entropy[47] since we pass from a fluid to the more ordered solid, while $\Delta S_{\text{disch.}}$ should not contain this quantity. The fact that theory agrees with experiment indicates that the sum of these two entropy changes is zero, as we have assumed. This means that the communal entropy does not appear immediately in the liquid at the melting point, but only develops as the temperature is increased beyond T_m. Pople[48] has carried out an investigation based upon free volume theory which predicts this result; namely, that the communal entropy need not necessarily appear immediately at the melting point.

Yosim and Owens[45] have used Eq. (170) to calculate the heats of fusion of the alkali halide salts. Again, it is necessary to employ the law of corresponding states to define a single parameter a. Their results appear in Table X, where they employ the equilibrium

TABLE X. Heats of Fusion of Alkali Halides, $T = T_m$

Salt	Melting point, °K	Gas equil. distance, A	Liquid density, g/cm³	V_s/V_l	$\int_{V_s}^{V_l} P\,dV$, kcal/mole	Heat of fusion, kcal/mole calc.	Heat of fusion, kcal/mole expt.
LiF	1120	1.527	1.797	0.764	2.62	7.07	6.47
LiCl	887	2.022	1.462	0.801	1.93	5.46	4.72
LiBr	825	2.170	2.528	0.815	1.69	4.97	4.22
LiI	713	2.392	3.135	0.838	1.35	4.18	3.50
NaF	1268	1.840	1.945	0.791	2.95	7.98	8.03
NaCl	1073	2.361	1.557	0.812	2.72	6.98	6.69
NaBr	1020	2.502	2.342	0.821	2.45	6.50	6.24
NaI	935	2.712	2.740	0.830	2.19	5.91	5.64
KF	1130	2.129	1.906	0.839	2.10	6.59	6.75
KCl	1043	2.667	1.527	0.851	2.26	6.41	6.27
KBr	1015	2.821	2.121	0.859	2.14	6.17	6.10
KI	955	3.048	2.445	0.864	1.98	5.78	5.74
RbF	1048	2.246	2.924	0.846	1.85	6.02	5.82
RbCl	990	2.787	2.247	0.887	1.60	5.53	5.67
RbBr	950	2.945	2.721	0.885	1.64	4.47	5.57
RbI	911	3.177	2.908	0.900	1.46	5.08	5.27
CsF	955	2.345	3.675	0.876	1.26	5.07	5.19
CsCl	915	2.906	2.789	0.909	1.20	4.83	4.84
CsBr	905	3.072	3.137	0.789	3.41	6.84	5.64
CsI	894	3.315	3.183	0.778	4.07	7.62	5.64

distance observed in the gaseous diatomic molecule for a. Agreement between theory and experiment is again excellent, but it should be pointed out that such agreement is obtained only after a quantity $2RT_m$ is added to the value of the heat of vaporization specified by (170). This quantity is precisely the contribution to the heat of fusion which would be made by the communal entropy. Thus, in the case of salts, it seems as though the communal entropy appears *immediately* upon fusion.

Yosim and Owens[45] also treat the heats of vaporization of fused salts and obtain excellent results. However, additional complications are involved which are not present in the case of simple liquids. Thus whereas the fused salt liquid is considered to be an assembly of individual ions, it is well known that the vapor contains diatomic molecules and even more complex species. A correction must be introduced to take proper account of this circumstance, and the analysis becomes somewhat involved. We shall not discuss these results here, and the reader is referred to the original paper for details.

X. THE DEPENDENCE OF *a* UPON TEMPERATURE

It has been emphasized throughout this article that the effective hard core diameter a should exhibit a small temperature dependence. This dependence is evident in Stillinger's results, which are presented in Table IV. It originates in the abilities of molecules to penetrate one another's repulsive fields more effectively following an increase in average thermal energy. Since molecules do not really possess hard cores the parameter a is an average diameter, although for truly spherical molecules the average ought to be confined to a very narrow range at low temperatures.

On the other hand, the real problem devolves not so much upon the existence of a but rather on the selection of numerical values for it in special cases. (It has been emphasized that a has a well defined physical meaning and should not be treated as an adjustable parameter. In fact, one of the dramatic features of scaled particle theory is its ability to predict the results of experiments without involving adjustable parameters.) In this review several techniques have been used to establish the magnitude of a. For example, in the case of rare gases a has been chosen through an appeal to second virial coefficients for slightly imperfect gases. Thus it has been assumed that the intermolecular potential has the Lennard-Jones form. The parameters in this form are then fitted to the second virial coefficient and from these parameters it is possible to determine a value for a as the point at which the repulsive part of the potential passes through zero. In the same way, viscosity data can be used for the selection of a in conjunction with the Lennard-Jones potential. Alternatively, in connection with fused salts, a has been chosen as a sum of Pauling radii or as the equilibrium distance in the diatomic molecule present in the vapor state.

In other words, one phenomenon has been employed for the determination of an a to be used in connection with another phenomenon. The phenomena involved in selection and use have been interpreted by different theories, and so the best correspondence between the values of a may not have been achieved. On the other hand, the fact that such excellent results are achieved illustrates the fundamental validity of the theory.

Nevertheless, it is of interest to determine whether or not it is possible to select a through an application of scaled particle theory to one class of measurements, and to use it successfully to predict the results of another class of measurements. If scaled particle theory is correct, the effective diameter a should be the same in each class. As a matter of fact, the temperature dependence should be automatically accounted for. Furthermore, even for non-spherical molecules (if they are not too badly non-spherical) a correspondence may be possible. Thus rotational averaging, just like penetration averaging, should yield an effective value for a. If the averaging is the same in each class of phenomena a satisfactory correspondence should result.

This problem has been attacked recently by Mayer,[49] who correlated compressibility with surface tension. Thus it is possible to eliminate a between Eq. (133) (ignoring the term in the pressure) and Eq. (140). Then *surface tension* may be expressed as a function of temperature, concentration, and compressibility, or *compressibility* may be expressed as a function of temperature, concentration, and surface tension. Alternatively, the correspondence between the values of a obtained from the two formulas may be examined by a direct calculation of a from Eq. (133) or (140) when the surface tension or compressibility is known, respectively. Mayer has denoted the values of a derived in this manner by a_σ and a_β, respectively. His results are summarized in Tables XI through XVI.

None of the measured isothermal compressibilities in these

TABLE XI. Computed and Measured Surface Tensions and Compressibilities for Hydrocarbon Liquids

Liquid	Temp., °K	a_σ, 10^{-8} cm	a_β, 10^{-8} cm	Surface tension[a]		Compressibility	
				Measured, dynes/cm	Computed, dynes/cm	Measured, 10^{-11} cm²/dyne	Computed 10^{-11} cm²/dyne
Benzene	293.1	5.05	5.02	29	28	9[b]	8
Benzene	303.1	5.02	5.00	28	27	10	10
Benzene	323.1	4.94	4.96	25	25	12	12
Toluene	293.1	5.44	5.42	28	27	9[b]	8
Toluene	303.1	5.41	5.40	27	27	10	10
Toluene	323.1	5.33	5.36	25	26	11	12
Hexane	293.1	5.57	5.66	18	20	15[c]	18
Hexane	303.1	5.51	5.63	17	19	17	21
Hexane	313.1	5.45	5.61	16	19	18	25
Heptane	273.1	6.04	6.03	22	22	12[c]	12
Heptane	293.1	5.95	5.98	20	21	14	15
Heptane	303.1	5.90	5.96	19	20	15	17
Octane	273.1	6.36	6.36	24	24	10[c]	10
Octane	293.1	6.29	6.32	22	23	11	12
Octane	303.1	6.25	6.31	21	22	12	14
m-Xylene	293.1	5.79	5.77	29	28	8[d]	8
m-Xylene	303.1	5.76	5.75	27	27	9	9
m-Xylene	313.1	5.73	5.74	26	26	10	10
Cyclohexane	304.1	5.33	5.33	24	24	12[e]	12

For footnotes see the end of Table XVI.

TABLE XII. Halogen-Containing Compounds

Liquid	Temp., °K	a_σ, 10^{-8} cm	a_β, 10^{-8} cm	Surface tension		Compressibility[a]	
				Measured, dynes/cm	Computed, dynes/cm	Measured, 10^{-11} cm^2/dyne	Computed, 10^{-11} cm^2/dyne
Carbon tetrachloride	293.1	5.18	5.16	27	26	10[b]	10
Carbon tetrachloride	303.1	5.14	5.14	26	26	11	11
Carbon tetrachloride	313.1	5.10	5.11	24	24	12	12
Chloroform	283.1	4.82	4.80	29	28	9[b]	9
Chloroform	293.1	4.78	4.78	27	27	10	10
Chloroform	303.1	4.74	4.76	26	27	11	11
Ethyl bromide	293.1	4.55	4.55	24	24	13[d]	13
Ethyl iodide	293.1	4.84	4.81	29	28	10[d]	9
Ethylene dichloride	293.1	4.84	4.81	32	31	8[d]	7
Tetrachloroethane	298.1	5.56	5.54	35	34	6[f]	6
Chlorobenzene	293.1	5.46	5.40	33	31	7[b]	6
Chlorobenzene	303.1	5.43	5.38	32	30	8	7
Chlorobenzene	323.1	5.37	5.36	30	29	9	9
Brombenzene	298.1	5.58	5.51	36	33	7[g]	6
Sulfuryl chloride	336.1	4.75	4.75	26	26	13[h]	13
Germanium tetrachloride	303.1	5.43	5.45	22	22	13[h]	13
Phosphorus trichloride	303.1	4.96	4.93	28	27	10[h]	9
Phosphorus tribromide	303.1	5.46	5.30	45	35	6[h]	4
Phosphorus oxychloride	303.1	5.16	5.09	31	28	9[h]	8
Arsenic trichloride	303.1	5.12	5.04	38	34	7[h]	6
Antimony pentachloride	303.1	5.91	5.90	30	30	7[h]	7

For footnotes see the end of Table XVI.

TABLE XIII. Liquids Composed of Polar Organic Molecules

Liquid	Temp., °K	a_σ, 10^{-8} cm	a_β, 10^{-8} cm	Surface tension		Compressibility[a]	
				Measured, dynes/cm	Computed, dynes/cm	Measured, 10^{-11} cm²/dyne	Computed, 10^{-11} cm²/dyne
Methanol	283.1	3.36	3.50	23	27	11[b]	15
Methanol	303.1	3.28	3.46	22	27	13[b]	19
Ethanol	273.1	4.06	4.17	24	28	10[b]	13
Ethanol	293.1	4.00	4.13	23	27	11[b]	15
n-Propanol	301.1	4.50	4.63	23	27	11[e]	15
Isopropanol	303.1	4.43	4.64	21	26	11[e]	18
Butanol	302.4	4.96	5.13	24	29	9[e]	13
Cyclohexanol	304.9	5.49	5.51	32	33	7[e]	7
Glycol	293.1	4.44	4.45	48	48	4[i]	4
Glycol	353.1	4.29	4.35	42	46	5[i]	6
Ether	283.1	5.00	5.08	18	20	17[b]	20
Ether	293.1	4.93	5.04	17	19	18[b]	23
Ether	303.1	4.87	5.01	16	18	21[b]	27
Acetic acid	293.1	4.08	4.14	28	30	9[d]	11
Acetic acid	303.1	4.06	4.14	27	30	10[d]	12
Acetic anhydride	303.1	5.24	5.18	31	29	9[e]	8
Ethyl acetate	293.1	5.16	5.17	24	24	11[e]	11
Acetone	293.1	4.47	4.51	23	24	13[b]	14
Acetone	313.1	4.37	4.44	21	23	16[b]	18
Acetophenone	303.1	5.95	5.84	39	33	6[e]	4
Aniline	293.1	5.39	5.34	43	40	5[d]	4
Aniline	323.1	5.32	5.30	39	38	5[d]	5
Aniline	363.1	5.22	5.25	34	36	7[d]	8
Nitrobenzene	293.1	5.70	5.60	44	38	5[d]	4
Nitrobenzene	313.1	5.66	5.57	41	37	6[d]	5
Carbon disulfide	293.1	4.33	4.27	32	30	9[b]	8
Carbon disulfide	323.1	4.21	4.19	28	27	12[b]	11

For footnotes see the end of Table XVI.

TABLE XIV. Computed and Measured Surface Tensions and Compressibilities of Non-metallic Elements

Liquid	Temp., °K	a_σ, 10^{-8} cm	a_β, 10^{-8} cm	Surface tension		Compressibility[a]	
				Measured, dynes/cm	Computed, dynes/cm	Measured, 10^{-11} cm²/dyne	Computed, 10^{-11} cm²/dyne
Oxygen	89.5	3.24	3.23	14	14	18[j]	17
Oxygen	63.1	3.48	3.33	20	15	11[j]	8
Nitrogen	77.1	3.41	3.41	9	9	32[k]	32
Nitrogen	76.1	3.41	3.38	9	9	32[j]	30
Nitrogen	70.1	3.50	3.41	11	10	28[j]	22
Chlorine	293.1	3.50	3.61	18	20	20[j]	25
Bromine	293.1	4.20	4.13	42	38	6[j]	5
Argon	87.1	3.15	3.19	11	12	22[j]	25
Argon	84.1	3.18	3.20	12	12	21[j]	22

For footnotes see the end of Table XVI.

TABLE XV. Molten Salts

Liquid	Temp., °K	a_σ, 10^{-8} cm	a_β, 10^{-8} cm	Surface tension[a]		Compressibility[a]	
				Measured, dynes/cm	Computed, dynes/cm	Measured, 10^{-11} cm²/dyne	Computed, 10^{-11} cm²/dyne
NaCl	1073	2.46	2.47	116	119	29	30
NaCl	1143	2.40	2.43	110	114	32	34
NaCl	1173	2.37	2.41	107	112	34	37
NaCl	1273	2.27	2.33	98	105	40	45
NaBr	1048	2.56	2.64	98	108	32	39
NaBr	1073	2.55	2.63	96	107	34	41
NaBr	1103	2.52	2.60	94	103	36	43
NaBr	1173	2.47	2.58	90	102	39	48
NaI	933	2.87	2.91	88	93	37	41
NaI	973	2.83	2.88	84	90	40	45
NaI	1043	2.74	2.83	78	87	45	55
NaI	1073	2.70	2.81	75	86	47	59
KI	973	3.08	3.09	78	78	50	51
KI	1033	3.00	3.04	72	75	56	60
KI	1073	2.94	3.00	69	74	60	67
KI	1113	2.88	2.97	66	72	65	75

[a] J. O. Bockris, *Modern Aspects of Electrochemistry*, Butterworth Publications, London, 1959, Vol. 2, pp. 196, 198.

TABLE XVI. Liquid Metals and Water

Liquid	Temp., °K	a_σ, 10^{-8} cm	a_β, 10^{-8} cm	Surface tension[q]		Compressibility[a]	
				Measured, dynes/cm	Computed, dynes/cm	Measured, 10^{-11} cm²/dyne	Computed, 10^{-11} cm²/dyne
Zinc	693	2.72	2.52	795	386	2[m]	0.6
Cadmium	594	3.06	2.87	560	292	3[m]	1
Mercury	313	3.23	3.01	470	193	3.8[n]	0.6
Mercury	433	3.18	2.97	442	210	4.4[n]	1.0
Gallium	303	3.01	2.79	735	250	2[m]	0.3
Indium	429	3.20	3.10	350	243	3[m]	2
Thallium	575	3.34	3.17	410	242	4[m]	1
Sodium[p]	371	2.77	2.44	206	91	21[m]	4
Potassium[p]	337	3.32	3.04	88	56	40[m]	16
Gallium[p]	303	2.33	2.14	735	280	2[m]	0.4
Indium[p]	429	2.44	2.37	350	271	3[m]	2
Thallium[p]	575	2.53	2.40	410	271	4[m]	2
Lead[p]	600	2.59	2.46	455	298	4[m]	2
Bismuth[p]	544	2.65	2.51	395	253	4[m]	2
Tin[p]	505	2.54	2.38	526	297	3[m]	1
Water	283.1	2.93	2.71	74	49	48[d]	21
Water	303.1	2.88	2.72	71	52	45[d]	25
Water	323.1	2.83	2.71	68	55	44[d]	29
Water	353.1	2.76	2.69	63	56	46[d]	36
Water	373.1	2.71	2.67	59	56	48[d]	43

Footnotes on facing page.

tables was obtained by high-pressure volume-change methods since such measurements do not provide data at atmospheric pressure. Instead, the results of sound velocity methods[50] were used throughout. All the hydrocarbon liquids for which measurements of the isothermal compressibility could be found are listed in Table XI. Table XII covers all the halogen-containing organic and inorganic non-electrolytes for which data could be found. Table XIII treats alcohols and other organic liquids. The results for cryogenic and liquified gases are presented in Table XIV. In Table XV a comparison of measured and computed σ and β is made for representative molten salts. (Here the law of corresponding states must be involved.) Finally, in Table XVI liquid metals near their melting points as well as water are treated. The results are poor, as expected, since scaled particle theory does not properly account for the effect of the metallic electrons; and this table is presented in order to emphasize the fact that the theory fails where it should.

---

[a] Values for the measured surface tensions throughout the tables were obtained from *Landolt–Borstein Tabellen,* II Band, 3. Teil, Springer-Verlag, Berlin, 1956. a_σ was calculated from the corresponding value of the surface tension.

[b] Fryer, E. B., Hubbard, J. C., and Andrews, D. H., *J. Am. Chem. Soc.* **51,** 759 (1929).

[c] Eduljee, H. E., Newitt, D. M., and Weale, K. E., *J. Chem. Soc.* 3086 (1951).

[d] Tyrer, D., *J. Chem. Soc.* **105,** 2534 (1914).

[e] Philip, N. M., *Proc. Indian Acad. Sci.* **9A,** 109 (1939).

[f] Newitt, D. M., and Weale, K. E., *J. Chem. Soc.* 3092 (1951).

[g] Gibson, R. E., and Loeffler, O. H., *J. Am. Chem. Soc.* **61,** 2515 (1939).

[h] Weissler, A., *J. Am. Chem. Soc.* **71,** 1271 (1949).

[i] Gibson, R. E., and Loeffler, O. H., *J. Am. Chem. Soc.* **63,** 898 (1941).

[j] Bergmann, L., *Der Ultraschall,* Herzl Verlag, Zurich, 1954, 4th edition, p. 372 ff.

[k] Van Itterbeek, A., de Bock, A., and Verhaegen, L., *Physica* **15,** 624 (1949).

[m] Kleppa, O. J., *J. Chem. Phys.* **18,** 1331 (1950).

[n] Harrison, D., and Moelwyn-Hughes, E. A., *Proc. Roy. Soc. London* **A239,** 230 (1957).

[p] In these cases the computations were based on an ionic-salt model for liquid metals instead of a monatomic model.

[q] *Liquid Metals Handbook,* NAVEXOS P-733 (Rev.), 1952, R. N. Lyon, editor, pp. 40–56.

In these tables the "computed" surface tension means computed from compressibility and the "computed" compressibility means computed from the surface tension. Results for each substance are presented at several temperatures, and it may be seen from the excellence of the agreement that the temperature dependence of a seems to be automatically accounted for. In general, agreement between theory and experiment is exceptionally good. The results for compressibility are somewhat inferior to those for surface tension, but this is to be expected since compressibility, in view of the factor $(1 - y)^4$ which appears in Eq. (140), is more sensitive to small errors in a than is surface tension.

The substances treated in Table XIII do not exhibit as good agreement between theory and experiment as do the substances included in Tables XI and XII. This is to be expected since the polar nature of alcohol and oxi-organic molecules leads to an associated structure which tends to hamper rotational averaging. In this connection the effects which can be ascribed to the polar nature of a molecule are clearly evident in Table XVI, where results for water are tabulated throughout a temperature range extending from 283.1° to 373.1°K. At low temperatures the results are very poor. For example, at 283°K the measured and calculated compressibilities differ by more than a factor of 2. On the other hand, at higher temperatures, for example at 373°K, the disparity between the measured and calculated values of the compressibility is only a few per cent. Thus perhaps the associated structure of water breaks down with an increase in temperature, and sufficient rotational averaging sets in so that scaled particle theory may be applied.

The extent of agreement in these tables is sufficient to suggest that measurements of surface tension might be used to predict compressibilities when no compressibility data are available. Furthermore, because of the relationship of the surface tension to the parachor[51] and of the empirical additive dependence of the parachor on molecular composition, it may be possible to establish a crude relationship between the thermodynamic properties of liquids and molecular structure.

As we have mentioned, the results in Table XVI for liquid metals are poor. This is to be expected since the behavior of the electrons must be considered explicitly in metals, and scaled particle theory takes no account of this. In fact, scaled particle theory assumes that the total potential energy of the fluid arises from pair-wise interaction, an assumption which is invalid for metals.

As a result of the work described in this section, we must conclude that scaled particle theory is internally consistent, and that a fairly precise meaning may be given to the hard core diameter a. This point receives even more support when the results of the preceding sections are taken into account, for in those sections a was not determined with the aid of scaled particle theory, and yet agreement between theory and experiment was excellent. The results are extremely sensitive to small errors in a since a enters the formulas raised to various powers.

XI. CONCLUSION

The evidence presented in this article drives us to the following conclusions:

Scaled particle theory treats the problem of the packing of molecular hard cores in fluids very accurately. Insofar as the agreement between theory and experiment can be made to depend upon our ability to treat this packing problem, excellent agreement can be obtained. For real fluids this means that we are confined to densities sufficiently high so as to be characteristic of liquids. The theory which has been applied so far does not yield the equation of state but expresses thermodynamic quantities as functions of the measured density. The soft part of the intermolecular potential enters the theory implicitly through its role in establishing this density. The theory is distinguished by the fact that it provides numerical values for the thermodynamic properties of real fluids without invoking a single adjustable parameter.

In the case of the hypothetical system consisting of hard spheres, scaled particle theory makes possible the derivation of an excellent expression for the equation of state itself. Furthermore, this expression has also been obtained by the exact solution of the Percus–Yevick integral equation.[3]

In connection with the hard sphere problem, scaled particle theory provides a larger number of exact relations than has been used in deriving the equation of state. It is desirable to evolve a method of development which will permit one to utilize all of the exact relations and consequently all of the information which the theory is capable of supplying. A significant advance would result from the discovery of a method for expressing Q of Eq. (54) as a function of G in a form which would facilitate the determinate solution of Eq. (50). The most fruitful line of attack at this point may well involve a detailed consideration of the properties of Q.

The systematic extension of scaled particle theory to real fluids will have to proceed along the lines set forth in Section V. Further evolution of these methods may render it feasible to derive the equation of state for a real fluid. At this point, it appears as though an examination of the θ formalism in the high density region where it bears a close relation to the G formalism is in order.

The scaled particle theory suggests a useful model for a liquid. This model views the liquid as having its volume or density determined by the soft part of the intermolecular potential. Once this volume is established one may consider the liquid as a hard sphere fluid confined within it. Thus the soft part of the intermolecular potential provides a container of specified capacity, and for what remains the liquid may be treated as a hard sphere fluid.

References

1. Mayer, J. E., and Mayer, M. G., *Statistical Mechanics,* John Wiley and Sons, New York, 1940.
2. Kirkwood, J. G., *J. Chem. Phys.* **3,** 300 (1935); Yvon, J., *Actualités Scientifiques et Industrielles,* Hermann and Cie, Paris, 1935; Born, M., and Green, H. S., *A General Kinetic Theory of Liquids,* Cambridge University Press, London, 1949; Hill, T. L., *Statistical Mechanics,* McGraw-Hill, New York, 1956, Chapter 6.
3. Percus, J. K., and Yevick, G. J., *Phys. Rev.* **110,** 1 (1957).
4. Prigogine, I., *The Molecular Theory of Solutions,* North-Holland Publishing Company, Amsterdam, 1957.
5. Hill, T. L., *Statistical Mechanics,* Chapter 8.
6. Carlson, C. M., Eyring, H., and Ree, T., *Proc. Natl. Acad. Sci. U.S.* **46,** 333 (1960); Thomson, T. R., Eyring, H., and Ree, T., *Proc. Natl. Acad. Sci. U.S.* **46,** 336 (1960); Eyring, H., and Ree, T., *Proc. Natl. Acad. Sci. U.S.* **47,** 526 (1961).
7. Majumdar, R., *Bull. Calcutta Math. Soc.* **21,** 107 (1929).
8. Hirschfelder, J. O., Bird, R. E., and Curtiss, C. F., *Molecular Theory of Gases and Liquids,* John Wiley and Sons, New York, 1954, pp. 156–7.

9. Rosenbluth, M. N., and Rosenbluth, A. W., *J. Chem. Phys.* **22,** 881 (1954).
10. Wood, W. W., and Jacobsen, J. D., *J. Chem. Phys.* **27,** 1207 (1957).
11. Wainwright, T. E., and Alder, B. J. Report written under AEC contract W-7405-eng-48, Lawrence Radiation Laboratory.
12. Alder, B. J., and Wainwright, T., "Molecular Dynamics by Electronic Computers", *Proceedings of the International Symposium on Transport Process in Statistical Mechanics,* Interscience Publishers, New York, 1958, p. 97.
13. Reiss, H., Frisch, H. L., Helfand, E. and Lebowitz, J. L., *J. Chem. Phys.* **31,** 369 (1959).
14. Wertheim, M. S., *Bull. Am. Phys. Soc.* **8,** 329 (1963).
15. Kirkwood, J. G., *J. Chem. Phys.* **3,** 300 (1935).
16. Hill, T. L., *Statistical Mechanics,* p. 184.
17. Hill, T. L., *Statistical Mechanics,* p. 183, eq. (29.17).
18. Hill, T. L., *J. Chem. Phys.* **28,** 1179 (1958).
19. Tolman, R. C., *The Principles of Statistical Mechanics,* Oxford University Press, London, 1938, Chapter 14.
20. Fowler, R. H., and Guggenheim, E. A., *Statistical Thermodynamics,* Cambridge University Press, London, 1939, p. 271.
21. Hill, T. L., *Statistical Mechanics,* p. 190.
22. Helfand, E., Reiss, H., Frisch, H. L., and Lebowitz, J. L., *J. Chem. Phys.* **33,** 1379 (1960).
23. Macdougall, F. H., *Thermodynamics and Chemistry,* John Wiley and Sons, New York, 1939, p. 147.
24. Green, H. S., *The Molecular Theory of Fluids,* North-Holland Publishing Company, Amsterdam, 1952, p. 171.
25. Tolman, R. C., *J. Chem. Phys.* **16,** 758 (1948); **17,** 118 and 333 (1949).
26. Kirkwood, J. G., and Buff, F. P., *J. Chem. Phys.* **17,** 338 (1949); Buff, F. P., *J. Chem. Phys.* **19,** 1591 (1951); Buff, F. P., *J. Chem. Phys.* **23,** 419 (1955).
27. Hirschfelder, J. O., Curtiss, C. F., and Bird, R. B., *Molecular Theory of Gases and Liquids,* John Wiley and Sons, New York, 1954, Chapter 5.
28. Kirkwood, J. G., *J. Chem. Phys.* **7,** 919 (1939).
29. Langmuir, I., *Chem. Rev.* **13,** 147 (1933).
30. La Mer, V. K., and Pound, G. M., *J. Chem. Phys.* **17,** 1337 (1949).
31. Percus, J. K., *Phys. Rev. Letters* **8,** 462 (1962).
32. Stell, G. Private communication to M. S. Wertheim.
33. Lebowitz, J. L., *Bull. Am. Phys. Soc.* **8,** 329 (1963).
34. Helfand, E., Frisch, H. L., and Lebowitz, J. L., *J. Chem. Phys.* **34,** 1037 (1961).
35. Helfand, E., and Stillinger, F. H., *J. Chem. Phys.* **37,** 2646 (1962).
36. Hirschfelder, J. O., Bird, R. B., and Curtiss, C. F., *Molecular Theory of Gases and Liquids,* Appendix Table I-A.
37. Reiss, H., Frisch, H. L., Helfand, E., and Lebowitz, J. L., *J. Chem. Phys.* **32,** 119 (1960).

38. Fowler, R. H., and Guggenheim, E. A., *Statistical Thermodynamics*, p. 340.
39. *International Critical Tables*, McGraw-Hill, New York, 1926.
40. Lange, N. A., *Handbook of Chemistry*, Handbook Publishers, Sandusky, 1941, 4th edition.
41. Reiss, H., and Mayer, S. W., *J. Chem. Phys.* **34,** 2001 (1961).
42. Reiss, H., Mayer, S. W., and Katz, J. L., *J. Chem. Phys.* **35,** 820 (1961).
43. Yosim, S. J., and Owens, B. B., to be published.
44. Stillinger, F. H., *J. Chem. Phys.* **35,** 1581 (1961).
45. Yosim, S. J., and Owens, B. B., to be published.
46. Yosim, S. J., and Owens, B. B., to be published.
47. Hirschfelder, J. O., Stevenson, D., and Eyring, H., *J. Chem. Phys.* **5,** 896 (1937).
48. Pople, J. A., *Phil. Mag.* **42,** 459 (1951).
49. Mayer, S. W., *J. Chem. Phys.*, to be published.
50. Fryer, E. B., Hubbard, J. C., and Andrews, D. H., *J. Am. Chem. Soc.* **51,** 759 (1929).
51. Glasstone, S., *Textbook of Physical Chemistry*, Van Nostrand, Princeton, 1940, p. 516.

* *Note Added in Proof* This manuscript was completed more than a year ago and just missed being published in Volume VI of this series. It still represents a worthwhile contribution, especially because it remains the only review devoted specifically to scaled particle theory. However, in the interim, several relevant papers have been published. These should be noted, especially in connection with the work of Yosim and Owens discussed in Sections VIII and IX. The work of these authors reported there with the promise that it would be published was never so published in precisely that form, but rather in connection with a more ambitious theory described in the new papers. The articles in question are:

(a) Yosim, S. J., and Owens, B. B., *J. Chem. Phys.* **39,** 2222 (1963).
(b) Yosim, S. J., *J. Chem. Phys.* **40,** 3069 (1964).
(c) Yosim, S. J., and Owens, B. B., *J. Chem. Phys.* **41,** 2032 (1964).

In addition, an interesting paper dealing with the application of scaled particle theory to mixtures of hard spheres has been published since the writing of this manuscript. It is:

Lebowitz, J. L., and Rowlinson, J. S., *J. Chem. Phys.* **41** (1), 133 (1964).

QUANTUM CHEMISTRY OF CRYSTAL SURFACES

JAROSLAV KOUTECKÝ, *Institute of Physical Chemistry, Czechoslovak Academy of Sciences, Prague, Czechoslovakia*

CONTENTS

I. INTRODUCTION

Phenomena on the surface of solids are so important to so many branches of physics and chemistry that it is not necessary to emphasize how desirable it would be to obtain any information which in principle could be obtained by the solution of Schrödinger's equation for the system consisting of electrons in a finite crystal or in a finite crystal which is in interaction with molecules of the gaseous phase.

In the study of surface properties, however, the difficulties which are encountered by the program for the derivation of physical and chemical properties from the treatment of the quantum-mechanical problem in the physics of solid substances combine with the difficulties which are encountered by this program in quantum chemistry. The systems to be studied do not show a high symmetry and do not seem to be suitable for semi-empirical methods. Therefore, more promising than the attempts at quantitative or semi-quantitative calculations are the efforts to get qualitative results which could supply basic terms which could be used for the discussion and understanding of surface phenomena. It is well known that in the chemistry of organic as well as of complex compounds quantum chemistry succeeded in underlying with theory some important terms, indispensable for the understanding of these regions. Therefore the first task of the quantum chemistry of surfaces is to obtain qualitative information about the possibility of bonding atoms or ions in the surface layers of the crystal, about the appearance of free valencies due to the formation of the surface, and about which bonds can appear during the interaction of molecules of the surface of a crystal with molecules from the gaseous phase, etc. We can further ask to what qualitative changes is the spectrum of energies of the electron states in the crystal subject when passing from an infinite crystal either to a finite one or to the crystal whose surface is in interaction with the molecules.

In studying the problems of the quantum chemistry of surfaces it is advantageous to start from the solution of simple models using the simplest methods. If the quantitative calculations do not seem to be promising, it is reasonable to choose a procedure which permits results which are intelligible and easily interpreted to be obtained. It can therefore be considered admissible to use the one-electron approximation, in spite of its known imperfections. It is well known that in some regions of the theory of solids as well as in quantum chemistry this approximation supplies useful qualitative information. Experience in the interpretation of the results obtained by the MO LCAO method in the theory of chemical bonding in organic as well as in inorganic compounds can in particular be used for the interpretation of results in the quantum chemistry of surfaces. Verification that the qualitative conclusions derived from the results of this method do not depend upon the procedure used can be obtained by the application of other methods, e.g. of the method starting from Kronig–Penney's model.[11,16,24,68,69]

In spite of all the simplifications the solution of the model of a finite crystal is rather difficult. It is therefore necessary to use as much as possible the translation symmetry of a semi-infinite crystal (i.e. a crystal filling up a half-space). Such a crystal is most useful for the study of surface properties as it contains a single surface plane.

Subjecting one-electron wave functions to the Born–Kármán conditions, a molecular orbital in an infinite crystal (which is then in reality a sort of cyclic crystal) has the following form:

$$\psi^{\alpha}_{\xi_1,\xi_2,\xi_3} = \sum_{m_1,m_2,m_3=-\infty}^{+\infty} [\sum_{\beta} \sum_{\nu=1}^{\nu_\beta} c^{\alpha}_{\beta,\nu}(\xi_1, \xi_2, \xi_3)\chi_{\beta,\nu}(\mathbf{r} - \sum_{k=1}^{3} m_k \mathbf{a}_k)]\, e^{i \sum_{k=1}^{3} m_k \xi_k} \quad (1)$$

where the m_i are integers and ξ_k the components of the wave vectors attributed in the usual way to the vectors of elementary translations $\mathbf{a}_k$ of the lattice $(0 \leqslant \xi_i \leqslant 2\pi)$. The summation is performed over all cells of the crystal. $\chi_{\beta,\nu}(\mathbf{r} - \sum_{k=1}^{3} m_k \mathbf{a}_k)$ is a localized function on the β-th atom in the elementary cell, characterized by the vector $\mathbf{m} \equiv (m_1, m_2, m_3)$. Wanier's functions,[21,91,104] atomic orbitals or delta atomic orbitals (DAO)[24] can be used for these localized functions according to the method used. Several functions $(\nu = 1, \ldots, \nu_\beta)$ can be localized on one atom. $c^{\alpha}_{\beta,\nu}(\xi_1,\xi_2,\xi_3)$ does not depend upon $\mathbf{m}$. Equation (1) can be rewritten in the form

$$\psi^{\alpha}_{\xi_1,\xi_2,\xi_3} = \sum_{m_3=-\infty}^{+\infty} \sum_{\beta} \sum_{\nu=1}^{\beta_\nu} e^{i m_3 \xi_3} c^{\alpha}_{\beta,\nu} (\xi_1, \xi_2, \xi_3)\, a_{m_3;\beta,\nu} (\xi_1, \xi_2) \quad (2)$$

where

$$a_{m_3;\beta,\nu}(\xi_1, \xi_2) = K \sum_{m_2,m_3=-\infty}^{\infty} e^{i(m_1\xi_1 + m_2\xi_2)} \chi_{\beta,\nu}(\mathbf{r} - m_1\mathbf{a}_1 - m_2\mathbf{a}_2 - m_3\mathbf{a}_3) \quad (3)$$

K is a normalizing factor. The functions $a_{m_3;\beta,\nu}(\xi_1, \xi_2)$ will be called layer molecular orbitals.

Also, when the surface parallel to the elementary translations $\mathbf{a}_1$, $\mathbf{a}_2$ is formed, translation symmetry in the direction of these vectors is maintained. For that reason the wave functions of a semi-infinite crystal must belong to the irreducible representations of a translation group of the surface lattice. These representations are characterized by wave vectors ξ_1, ξ_2. If the surface of the crystal is occupied by chemisorbed* atoms or molecules in such a

* In this article chemisorption is understood to mean the exchange interaction of the surface crystal with an atom or a molecule. For a strong interaction, the formation of surface compounds occurs rather than chemisorption.

way that the translation symmetry of the surface is not changed (which occurs with a fully occupied surface), the above statement is still valid. Because the components of the wave vectors remain equally good quantum numbers for a semi-infinite crystal, a one-electron wave function can be written in the form:

$$\psi^{\alpha}_{s,\xi_1,\xi_2} = \sum_{m_3=0}^{\infty} \sum_{\beta} \sum_{\nu=1}^{\beta_\nu} d^{\alpha}_{s;\beta,\nu,m_3}(\xi_1, \xi_2) a_{m_3;\beta,\nu}(\xi_1, \xi_2) \tag{4}$$

Expressing the wave function for a semi-infinite crystal (or a similar expression for a semi-infinite crystal with chemisorption when fully covered) in this way, the three-dimensional problem is in principle reduced to a one-dimensional one for constant values of ξ_1 and ξ_2.

The energy spectrum breaks down into the partial spectra for constant values of the wave vectors lying in the surface of the crystal. Superposition of these partial spectra gives the overall spectrum of allowed electron energies in a finite crystal.

The wave functions of an electron in a finite crystal are of two kinds:

(a) delocalized states with an infinitely low probability of finding an electron in the space limited by the surface of the crystal and the plane parallel to the surface in the finite distance $S|a_3|$,

(b) localized states where this probability is finite:[9,48]

$$\Big(\sum_{\beta,\nu} \sum_{m_3=0}^{S} |d^{\alpha}_{s;\beta,\ ,m_3}(\xi_1, \xi_2)|^2\Big) \Big/ \Big(\sum_{\beta,\nu} \sum_{m_3=0}^{\infty} |d^{\alpha}_{s;\beta,\nu,m_3}(\xi_1, \xi_2)|^2\Big) > M > 0 \tag{5}$$

where M is a finite number.

These states are called surface states.*[90,95] For chemisorption with a fully covered surface, chemisorption localized states can be defined in an analogous way. With one atom or molecule chemisorbed on the surface of a semi-infinite crystal such states

* According to the possibility mentioned of reducing the more-dimensional problem of a finite crystal to a one-dimensional problem, we are also using the name surface states for the states localized in the neighbourhood of the end of a semi-infinite chain. These states could more precisely be called "end-states".

can exist where the probability of finding an electron in the finite surrounding of the chemisorbed particle is finite.[12,14,100]

States localized near the surface of the crystal are obviously the expression of specific properties of the surface,[74] and that is why the main interest of quantum mechanics in its application to the theory of surface phenomena and especially to the theory of chemisorption and catalysis was up until this time directed towards the discussion of these localized states. The existence of the localized states is associated with the occurrence of the levels or bands of energy in one-electron spectra (in the band scheme). It is therefore possible to utilize the analogy between these energy levels and the energy levels exhibited by the disturbances in an ideal crystal.[3,38,39,102] It must nevertheless be emphasized that the reasoning starting from the given term diagram and from the corresponding statistics and exploiting the theories worked out widely for semi-conductors may be dangerous, since the results of this procedure depend essentially on the plausibility of the term diagram mentioned. It is therefore important to first obtain a survey of the different possibilities of the occurrence of surface and chemisorption states. First, it is necessary to judge under what conditions and in what type of crystal the surface states appear, of what kind they are and how they are changed during the changes of the surface. Only such considerations will supply a qualitative prescription for the construction of a band scheme for a finite crystal of a given substance.

In this article the interpretation of the individual surface states is therefore especially emphasized. For the reasons mentioned the simplest model is used to start with. An explanation of how the given one-electron energy spectrum can be connected with physical and chemical properties using corresponding statistics was published recently in several reviews.[39,102,97,123] Energies of localized states alone cannot supply sufficient information concerning the changes of the total electron energies resulting from the interaction of the surface of the crystal with the particle from the gaseous phase.[30,81]

As the changes of the chemisorption heat with the degree of covering of the surface are very characteristic, special attention is given to the methods of calculating the electronic contribution to the chemisorption heat.

II. LOCALIZED SURFACE AND CHEMISORPTION STATES

A. Theory of the Interaction of Molecules Where the Disturbance Reaches only Part of Some of the Molecules. Koster–Slater's Method[9,46,48,51,61,62,73]

In different models of systems of interacting molecules it is often supposed that the shape of the effective potential is not changed in some parts of the interacting molecules. The use of this supposition is especially important if one of these interacting "molecules" is a crystal or, more generally, at least a very big molecule. In these cases it is advantageous to begin the solution of the problem from the results obtained in the solution of the system of non-interacting molecules (or crystals). It is convenient for the calculations to have the wave functions of the starting system as simple as possible. For crystals, it is therefore useful to start from the solution for an infinite crystal. Necessary planes limiting the crystals considered should be created only during the transition from the system of non-interacting molecules and crystals to the actual system. Thus our theory also includes the theory of finite crystals.[62] It is convenient to use the projection operator formalism (cf. ref. 77).

Let us consider a system Σ_0 of non-interacting molecules (or crystals) in the MO LCAO approximation. We designate the non-interacting parts of the system Σ_0 by the numbers $i = 1, \ldots, N$. The corresponding one-electron Hamiltonian $\mathscr{H}_0$ has the following properties: the eigenfunctions of the operator $\mathscr{H}_0$ form a basis of a function space P_0.*

* Let us introduce the following definitions:

(a) The projection operator projecting into the space A is designated by $\mathscr{A}$.

(b) When we have an operator $\mathscr{A}_i$, with the subscript (or superscript) i corresponding to the i-th sub-system (molecule or crystal), we write

$$\cup \mathscr{A}_i = \mathscr{A}.$$

(c) Among the functions belonging to P_0 a set of functions can be chosen which is complete in all following considerations. The space P_0 is therefore complete. If it were not complete, it could be completed by additional subspaces having the property (7).

In the space P_0 subspaces P_0^i exist belonging to the corresponding molecules:

$$\mathscr{P}_0 = \mathscr{I} = \cup \mathscr{P}_0^i; \quad \mathscr{P}_0^i \mathscr{P}_0^j = 0 \quad (i \neq j) \tag{6}$$

$$\mathscr{P}_0^i \mathscr{H}_0 = \mathscr{H}_0 \mathscr{P}_0^i = \mathscr{H}_0^i \tag{7}$$

Consequently it holds that*

$$\mathscr{H}_0^i |i,k\rangle = \varepsilon_{ik} |i, k\rangle \tag{8}$$

Let us further consider a system Σ of interacting molecules or crystals whose Hamiltonian is given by

$$\mathscr{H} = \mathscr{H}_0 + \mathscr{V} \tag{9}$$

These molecules need not necessarily be the same as in Σ_0.

The space P of the eigenfunctions of the operator $\mathscr{H}$ has the property

$$\mathscr{P} = \mathscr{P}_0 - \mathscr{S} \tag{10}$$

Let us designate

$$\mathscr{S}^i = \mathscr{P}_0^i \mathscr{S}, \quad \mathscr{P}^i = \mathscr{P}_0^i \mathscr{P} \tag{11}$$

From the definition of S it follows that

$$\mathscr{S}^i |E\rangle = 0, \quad \mathscr{S}^i \mathscr{H} = \mathscr{H} \mathscr{S}^i \tag{12}$$

if

$$\mathscr{H} |E\rangle = E |E\rangle \tag{13}$$

An important special case occurs if the following relationship holds

$$\mathscr{H} \mathscr{S} = 0 \tag{14}$$

Generally there exists a subspace Q^i of P^i (cf. Fig. 1)

$$\mathscr{P}^i = \mathscr{Q}^i + \bar{\mathscr{Q}}^i \tag{15}$$

with the property

$$\bar{\mathscr{Q}}^i \mathscr{V} = \mathscr{V} \bar{\mathscr{Q}}^i = 0 \tag{16}$$

If Eq. (14) is valid then it necessarily follows that

$$\mathscr{H}_0 \mathscr{S} = -\mathscr{V} \mathscr{S} \tag{17}$$

$$\bar{\mathscr{Q}} \mathscr{H}_0 \mathscr{S} = -\bar{\mathscr{Q}} \mathscr{V} \mathscr{S} = 0$$

* In this section and in Section III-A Dirac's notation is used as it is convenient for general reasoning.

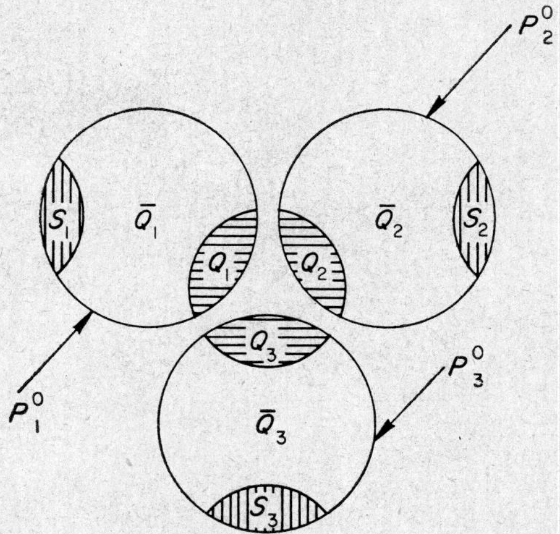

Fig. 1. Diagrammatic representation of the subspaces Q_i, $\bar{Q}_i$ and S_i for systems Σ_0 and Σ.

Let us define the operator

$$\mathscr{L}(E) = (E\mathscr{I} - \mathscr{H}_0)^{-1} = \sum_j \sum_k \frac{|j,k\rangle \langle j,k|}{(E - \varepsilon_{jk})}$$
$$= \sum_j \mathscr{L}_j(E) \tag{18}$$

According to the relations (9), (13), (16), and (17) this operator has the following property:

$$|E\rangle = \mathscr{L}\mathscr{V}|E\rangle = -\mathscr{L}\mathscr{S}\mathscr{H}_0\mathscr{Q}|E\rangle + \mathscr{L}\mathscr{Q}\mathscr{V}\mathscr{Q}|E\rangle \tag{19}$$

Let us denote atomic orbitals, forming the basis in the space P_0^j, by $|j, \xi_k\rangle$.

If $|j, \xi_k\rangle \in P^j$ then

$$\langle j, \xi_k|E\rangle = \langle j, \xi_k|\mathscr{L}_j(\mathscr{Q}\mathscr{V} - \mathscr{S}\mathscr{H}_0)\mathscr{Q}|E\rangle \tag{20}$$

If on the contrary $|j, \xi_k\rangle \in S^j$, Eq. (19) takes the form

$$0 = \langle j, \xi_k|\mathscr{L}_j(\mathscr{Q}\mathscr{V} - \mathscr{S}\mathscr{H}_0)\mathscr{Q}|E\rangle \tag{21}$$

The condition necessary for solving the system of Eq. (20) is the disappearance of the following determinant:*

$$|\langle j, \xi_k|\mathcal{Q}^j\mathcal{L}_j(\mathcal{Q}\mathcal{V} - \mathcal{S}\mathcal{H}_0 - \mathcal{I})\mathcal{Q}^l|l, \xi_m\rangle| = 0 \qquad (22)$$

During the rearrangement of the determinant on the left-hand side of Eq. (22) we used the argument that the following relations hold:

$$\begin{vmatrix} I & O \\ J & K \end{vmatrix} = |K| \qquad (23)$$

Here J and K designate arbitrary matrices of suitable order, I a unit matrix and O a zero matrix.

B. One-Dimensional Chain of Atoms in the Simple MO LCAO Method

(1) *Derivation of General Relations*

Fundamental types of surface states, arising from the formation of the surface of a crystal, as well as chemisorption states can already be found in very simple models. Let us start with a model so simple that on the one hand it can be easily treated but, on the other hand, is so general that it contains in a simple form all the fundamental types of localized states.

Let us consider a one-dimensional chain of atoms of the species A and B (see Fig. 2).[59] The atoms of both species are arranged

—— A —— B —— A —— B ——

Fig. 2. Model of an atomic chain of different atoms with different bond lengths.

in such a way that A atoms have B atoms for neighbours and vice versa. Let us assume that each atom has only one atomic orbital with a Coulomb integral α_M, where M = A or B according to the kind of atom. Further, it is assumed to be sufficient to

* Here and in the following text we designate by $|\langle j, \xi_k|\mathcal{B}\mathcal{A}\mathcal{B}|l, \xi_m\rangle|$ the determinant of the matrix elements of an operator $\mathcal{A}$ on the basis of the atomic orbitals belonging to the space B only. The order of this determinant is equal to the number of atomic orbitals in B.

consider the resonance integrals between the nearest neighbours only and to use the approximation of zero overlap. Generally, in the model mentioned, the bonds appearing in the system are not equivalent: let us therefore denote the resonance integral characterizing the bond to the left of an A atom by β and the resonance integral of the bond connecting an A atom with a B atom, which is on the right, by β'. The calculations should be performed using Hückel's approximation.

As the generalized method of Koster and Slater[46,48,62] will be used let us mention first the equations for the energy of the molecular orbitals of the volume states of an infinite chain and for the coefficients in these molecular orbitals:

$$x_{\pm} = \pm\kappa$$
$$\kappa = [1 + \gamma^2 + \delta^2 + 2\delta \cos \xi]^{\frac{1}{2}}$$
$$c_{j\pm}^{\mathrm{A}}(\xi) = \sqrt{\frac{1}{2N}}\sqrt{\frac{\pm\kappa - \gamma}{\pm\kappa}}\, e^{ij\xi} c_{j\pm}^{\mathrm{B}} = \sqrt{\frac{1}{2N}}\sqrt{\frac{\pm\kappa - \gamma}{\pm\kappa}}\, \frac{(1 + \delta)e^{ij\xi}}{\pm\kappa - \gamma} \tag{24}$$

In these equations the following symbols were introduced:

$$x = E/\beta - \bar{\alpha}, \quad \bar{\alpha} = (\alpha_{\mathrm{A}} + \alpha_{\mathrm{B}})/2\beta, \quad \gamma = (\alpha_{\mathrm{B}} - \alpha_{\mathrm{A}})/2_{\beta} \tag{25}$$
$$\beta'/\beta = \delta$$

Let us now consider the case when one B atom is removed so that from a cyclic chain a finite chain with A atoms at its ends is created. Further, let us assume that the Coulomb integral of one of the final atoms is changed by the quantity $\alpha\beta$ and that a bond is formed between the atomic orbital of this final atom and the atomic orbital of the atom C, characterized by the Coulomb integral $\beta(\bar{\alpha} + \rho)$. The resonance integral between the atomic orbital of the atom C and the final atom of the chain should be $\beta\sigma$.

As we shall first discuss the existence of the states localized in the neighbourhood of one atom at one end of the chain, we need not mention what is happening at the other end of the chain. We shall therefore assume that the chain is semi-infinite. The coefficients in the molecular orbital of a localized state

$$\phi = \sum_{\mathrm{M=A,B}} \sum_{j=0}^{\infty} d_j^{\mathrm{M}}\, \chi_j^{\mathrm{M}} + d^c \chi^c \tag{26}$$

are given by the following equations (cf. Eqs. (20) and (21)):

$$d_j^{\mathrm{M}} = (-L_{j,-1}^{\mathrm{MB}} + \alpha L_{j,0}^{\mathrm{MA}})d_0^{\mathrm{A}} + \sigma L_{j,0}^{\mathrm{MA}} d^c$$

$$d^c = \frac{\sigma}{W - \rho} d_0^{\mathrm{A}}, \quad d_{-1}^{\mathrm{B}} = 0 \tag{27}$$

$$L_{jk}^{\mathrm{MN}} = \langle \chi_j^{\mathrm{M}} | L | \chi_k^{\mathrm{N}} \rangle$$

W being the energy of the localized state in the scale with origin $\bar{\alpha}$ and unit β. From the above equations we can derive two conditions which determine the energy of the localized molecular orbitals. It is obvious that these two conditions must not be in mutual contradiction:

$$-L_{0,-1}^{\mathrm{AB}} + \left(\alpha + \frac{\sigma^2}{W - \rho}\right) L_{0,0}^{\mathrm{AA}} = 1$$

$$-L_{-1,-1}^{\mathrm{BB}} + \left(\alpha + \frac{\sigma^2}{W - \rho}\right) L_{-1,0}^{\mathrm{BA}} = 0 \tag{28}$$

In Eq. (27) the supposition was made that the extension coefficients of the atomic orbitals of the atoms at the second end of the chain are very small and in the second Eq. (28) it was assumed that dB_1 vanishes. By a simple calculation (using the integration in the complex plane), we find that

$$L_{j,0}^{\mathrm{AA}} = \frac{W - \gamma}{\delta} \frac{z_1^j}{z_2 - z_1}, \quad L_{j,-1}^{\mathrm{AB}} = \frac{1 + z_1 \delta}{\delta(z_2 - z_1)}$$

$$L_{-1,0}^{\mathrm{BA}} = L_{0,-1}^{\mathrm{AB}}$$

$$L_{j,0}^{\mathrm{BA}} = \frac{\delta + z_1}{\delta} \frac{z_1^j}{z_2 - z_1} \ (j \geq 0), \quad L_{j,-1}^{\mathrm{BB}} = \frac{W + \gamma}{\delta} \frac{z_1^{j+1}}{z_2 - z_1} \tag{29}$$

where

$$z_{1,2} = \frac{\mu}{2}\left[1 \mp \left(1 - \frac{4}{\mu^2}\right)^{\frac{1}{2}}\right] \tag{30}$$

and

$$\mu = \delta^{-1}[W^2 - 1 - \delta^2 - \gamma^2] \tag{31}$$

It is obvious that the inequality

$$\mu^2 > 4 \tag{32}$$

must be fulfilled, which is equivalent to the requirement that the energy of the localized state at the surface is outside the bands of the allowed energies of the volume states. The condition for the level of the energy of the localized state being inside the gap between the bands of energies of the volume states is

$$\mu > 2 > 0 \tag{33}$$

If, on the other hand

$$\mu < -2 < 0 \tag{34}$$

the localized state is outside the system of the two bands of energies of the volume states.

Further, let us note that the following relations are valid

$$|z_1| < 1, \quad z_1 z_2 = 1 \tag{35}$$

The equations for the calculation of the energies of the localized states therefore obtain the forms:

$$z_2(\mu) = \delta^{-1}\left[-1 + \left(\alpha + \frac{\sigma^2}{W - \rho}\right)(W - \gamma)\right]$$

$$z_1(\mu) = \delta^{-1}\left[-1 + (W + \gamma)\Big/\left(\alpha + \frac{\sigma^2}{W - \rho}\right)\right] \tag{36}$$

Multiplying them together as in Eq. (35) we finally obtain:

$$W^2 + 1 - \gamma^2 - \delta^2 = \left(\alpha + \frac{\sigma^2}{W - \rho}\right)(W - \gamma) + \frac{W + \gamma}{\left(\alpha + \dfrac{\sigma^2}{W - \rho}\right)} \tag{37}$$

For the ratios of the coefficients the following equations result:

$$j \geq 0: \quad d_j^{\mathrm{M}}/d_0^{\mathrm{M}} = z_1^j$$

$$d_j^{\mathrm{B}}/d_j^{\mathrm{A}} = \frac{(W + \gamma)z_1}{1 + z_1\delta} = \left(\alpha + \frac{\sigma^2}{W - \rho}\right) z_1 \tag{38}$$

(2) *Surface States*

(a) *Specialization of the general relations for surface states*

Inserting in the final relations the condition $\sigma = 0$, we obtain equations describing the surface states. The energy is then given by the equation

$$W = (\alpha + \alpha^{-1})/2 \pm \text{Sign}\,(\alpha)\,[x^2 + \delta^2]^{\frac{1}{2}} \tag{39}$$

where

$$x = \gamma + (\alpha^{-1} - \alpha)/2 \tag{40}$$

Combining equations (38) we realize that the ratio of the coefficient of the atomic orbital of a B atom in the j-th elementary cell and the coefficient of the atomic orbital of an A atom in the $(j + 1)$-th elementary cell is a constant

$$d_0^{\mathrm{B}}/d_1^{\mathrm{A}} = \alpha \tag{41}$$

The second equation (38) then assumes the form:

$$d_0^{\mathrm{B}}/d_0^{\mathrm{A}} = f_{\pm}(x) \tag{42}$$

where

$$\text{Sign}\,(\alpha)\,f_{\pm}(x) = x \pm \delta\,[x^2 + \delta^2]^{\frac{1}{2}} \tag{43}$$

From the condition for the existence of a surface state (cf. Eq. (35))

$$|z_1| = |\alpha^{-1}f_{\pm}(x)| < 1 \tag{44}$$

finally follows the condition for the fulfilment of the inequality

$$(\alpha^2 - 1)(\delta \pm 1) > \pm 2\alpha\gamma \tag{45}$$

where the upper sign must be used if in Eq. (39) the plus sign is considered and the lower sign if the minus sign is considered. From the inequalities (33) and (34) we can determine whether the energy of the corresponding localized state is inside the gap of forbidden energies or outside the system of both bands of energies of the volume states. This can be determined from the sign of the quantity μ, which can be expressed in the form:

$$\mu = (\alpha^{-1} - \alpha) \pm (|\alpha| + |\alpha|^{-1})[\delta^2 + x^2]^{\frac{1}{2}} \tag{46}$$

This equation gives for the upper sign $\mu > 2$, and for the lower sign $\mu < -2$. Therefore those surface states whose energy is given by Eq. (39) with the plus sign have an energy value outside the system of the two allowed bands of energy of the volume states. On the other hand, Eq. (39) with the minus sign determines the energy gap.

(b) *Zero change of the Coulomb integral of a border atom: the simplest model of Shockley's surface state*[90]

Zero change of the Coulomb integral ($\alpha = 0$) means that the formation of the surface does not change the Coulomb potential inside the semi-infinite crystal. The condition for the existence of a surface state is as follows: the bond broken by the formation of the surface, in the neighbourhood of which we wish to study the surface states, must be that of the stronger type:

$$\delta < 1 \tag{47}$$

According to expression (39) the coefficients of the atomic orbitals of the B atoms are equal to zero. The above condition means that as long as the difference between the two kinds of atoms is negligible the probability of finding an electron on the atomic orbital of an odd numbered atom, if we gradually number atoms from the surface, the surface atom having the number zero, disappears. The surface state has an energy which is within the energy gap (cf. Eqs. (31) and (34)) given by:

$$W = -\gamma \tag{48}$$

We can see that these surface states are directly connected with the existence of bonds of different strengths in a one-dimensional chain, originally infinite. The energy of these states is precisely equal to the energy of an electron on the atomic orbital of atom A. This is due to the fact that the probability of finding an electron on the atomic orbitals of the B atoms is equal to zero. The surface state thus obtained is therefore non-bonding and because of its properties it is analogous to the non-bonding states known from the theory of odd alternant hydrocarbons with conjugated bonds.[75] As long as the number of electrons supplied by the atoms of the atomic chain is equal to number of these atoms, the surface state of this system is occupied by a single electron. It can therefore be seen that this surface state expresses unsaturated valency which originated as a result of breaking a strong localized bond.[53,54] It is also the expression of some radical character of the end of the one-dimensional atomic chain considered.

(c) *A very great change in the Coulomb integral of a border atom. The simplest model of Shockley's subsurface state.*

Another interesting case appears if the absolute value of the Coulomb integral at the border atom becomes infinite: $|\alpha| \to \infty$. One energy level of the localized state can always be obtained. Its value is given by the relation

$$W = \alpha \tag{49}$$

In this state the electron is fully localized on the border atom of the chain. Apart from this a surface state can occur, the energy of which is given by the equation

$$W = \gamma \tag{50}$$

The condition for the existence of this state is given by the inequality

$$\delta > 1 \tag{51}$$

It means, therefore, that in this case a surface state exists in the energy gap if by the formation of the surface a weaker bond is broken. In this state the probability of finding an electron on an A atom is zero (cf. ref. 64).

These results are easily understood because the occurrence of a potential greatly differing from that inside the crystal makes the surface atom act as if it were eliminated from the crystal. This gives rise to a state whose energy is essentially different from that of volume states.[55] All the other electrons are, owing to this potential, restricted to the region of the crystal from which the surface atom appears to be excluded. We have, therefore, a surface state in which an electron is localized exclusively on the B atoms. In fact the bond whose breaking creates the crystal is the bond situated to the left of the first B atom from the surface of the crystal.

(d) *Chain with atoms of the same kind and with alternating strength of bonds*

Let us now consider a chain in which stronger and weaker bonds alternate but in which only one kind of atom is present ($\gamma = 0$).[4,85,99] The existence condition (45) in this case is

$$(\alpha^2 - 1)(\delta \pm 1) > 0 \tag{52}$$

We see that as long as the change in the Coulomb integral on the surface atom is so large that (the upper sign in existence condition (52))[8,27]

$$|\alpha| > 1 \tag{53}$$

then a surface state exists apart from the system of bands of the volume states. If the equation $\delta = 1$ holds, the energy is given by the relation[27]

$$W = \alpha + \alpha^{-1} \tag{54}$$

Inside the energy gap surface states appear if either the conditions

$$|\alpha| > 1, \quad \delta > 1 \tag{55}$$

or, on the contrary, the conditions

$$|\alpha| < 1, \quad \delta < 1 \tag{56}$$

are fulfilled.

Let us now note the behaviour of the extension coefficients of the surface states in the three cases mentioned above. Surface states for which the only condition for their existence is a sufficiently large change in the Coulomb integral of the surface atom (condition (53)) have, according to Eqs. (41) and (42), an absolute value of the ratio d_1^A/d_0^B, as well as of the ratio d_0^B/d_0^A, smaller than unity. For the second ratio, this is concluded from the fact that Sign x = —Sign α. For this surface state the probability of finding an electron decreases monotonously from atom to atom as one proceeds further from the surface. The surface states whose existence requires the simultaneous fulfilment of conditions (56) have an absolute value of the ratio d_0^B/d_0^A smaller than unity and of the ratio d_1^A/d_0^B larger than unity. The probability of finding an electron therefore oscillates as one proceeds further from the surface into the bulk. At the same time the amplitude of the wave function decreases exponentially when equivalent places in individual elementary cells are compared. On the contrary, the surface states whose existence is given by the fulfilment of conditions (55) have an absolute value of the ratio $d_0^{\mathrm{B}}/d_0^{\mathrm{A}}$ larger than unity and of the ratio $d_1^{\mathrm{A}}/d_0^{\mathrm{B}}$ smaller than unity. The oscillations again appear but with maxima and minima which are interchanged when compared with surface states with the existence condition (56).

There is a clear connection between the three kinds of surface states which we have already obtained and the surface states found in previous cases, where the change in the Coulomb integral of the surface reached its limiting values. The surface state given by the existence condition (53) is similar in its properties to the surface state whose energy is outside the system of the levels of energies of volume states when $|\alpha| \to \infty$. In these states the electron is localized near the surface as the change in the local potential inside the crystal (i.e. in the neighbourhood of the border atom) is considerable. These surface states can be called Tamm's surface states.[8,14,22,27,87,93,94,95]

Surface states given by the existence condition (56) are obviously related to the surface states whose existence condition is expressed by the relation (47) when $\alpha = 0$. For their existence it is obviously favourable for the change in the Coulomb integral of a surface atom to be as small as possible. They are characterized by oscillations in the amplitude, the maximum amplitude occurring at the surface atom. These states, which are qualitatively different from those mentioned above, can be called Shockley's surface states.[26,78,79,82,90,94] These states are caused by the rupture of a localized bond during the formation of the surface and are the expression of bonding properties (i.e. radical properties) of the surface atoms.[53,54]

Finally, the third type of surface state, whose existence requires the fulfilment of relations (55), is also characterized by oscillations in the amplitude and by the occurrence of an energy level in the energy gap. From the properties mentioned, these states are analogous to Shockley's surface states, but as the maximum amplitude occurs at the second atom from the surface, it is possible to call these states Shockley's subsurface states. These states resemble the states with the existence condition (51).

(e) *Chain with atoms of two kinds and with the same bond strengths*

Let us suppose that all the bonds in the chain are equivalent ($\delta = 1$) but that the atoms concerned are of two kinds ($\gamma \neq 0$). For the surface states of Tamm's type, situated outside the system of bands of allowed energies of the volume states, we obtain the existence condition[4]

$$\alpha^2 - \alpha\gamma > 1 \tag{57}$$

The occurrence of surface states of this type is therefore made more difficult if the sign of α is equal to the sign of γ and is facilitated if the signs of these parameters are opposite. According to the definition (cf. Eq. (25)) this means that the occurrence of Tamm's states is facilitated if the difference between the Coulomb integral of the surface atom and the Coulomb integral of the B atom, given by the expression $(\alpha - 2\gamma)\beta$, is intensified. Inside the energy gap the energy level of a surface state occurs if the condition[59]

$$\alpha\gamma > 0 \tag{58}$$

is fulfilled.

These surface states can only arise therefore if the difference between the Coulomb integral of the surface atom and the Coulomb integral of a B atom is, in its absolute value, smaller than the difference between the Coulomb integral of the A atom in the middle of the chain and the Coulomb integral of the B atom in the middle of the chain, or if the difference mentioned changes its sign.

These surface states with energies inside the gap are obviously states of a new type. Their existence is conditioned by the occurrence of atoms of two different kinds in the chain. If the difference between the Coulomb integrals of A and B atoms decreases to zero, the width of the energy gap also decreases to zero. If the quantity α approaches zero, the energy level of a surface state of this type approaches the edge of the band of energies of the volume states which originate because of the broadening of the energy level for the isolated A atom. If α is very large, the energy level of the surface state which fulfils naturally the existence condition (58) approaches the edge of the volume states energy band which was created by the broadening of the degenerate level for an electron isolated on the B atom.

It is nevertheless evident that these surface states with energy levels inside the energy gap show some resemblance to Shockley's surface states (as long as $|\alpha|$ is small), or to Shockley's subsurface states (as long as $|\alpha|$ is large). This close relationship between different surface states with energies in the energy gap will become clear from the discussion of the general type of chain.

(f) *General type of chain*

From the existence condition (45) it follows that for the surface states with energies outside the system of bands of the volume states, the existence condition can be written in the form

$$(\alpha^2 - 1)(1 + \delta) > 2\alpha\gamma \tag{59}$$

The surface state in question is obviously of Tamm's type. The same statements are valid as were made in Section II-B-(2e) about the help or hindrance of its formation by the presence of two kinds of atoms in the chain.

Choosing the lower signs in the existence condition (45) the following relation can be derived

$$\alpha^2 - 1 > -\frac{2\alpha\gamma}{\delta - 1} \tag{60}$$

if, by the formation of the surface, a weaker bond was broken $(1 < \delta)$. For the occurrence of the surface state, a high value of the change in the Coulomb integral of the surface atom is advantageous. This makes the states in question similar to Shockley's subsurface states. This resemblance is also helped by the form of the existence condition, relating to the strength of the broken bond. The appearance of the surface state is facilitated if the signs of α and γ are the same.

If by the formation of the end of the chain the stronger bond is broken $(\delta < 1)$ then the existence condition for the surface state with its energy level inside the energy gap is

$$1 - \alpha^2 > -\frac{2\alpha\gamma}{1 - \delta} \tag{61}$$

A low value of the change in the Coulomb integral of the surface atom is now advantageous for the formation of the surface state.

This fact, and the circumstance that for the occurrence of the surface state the breaking of a stronger bond is advantageous, shows the relationship of these states to Shockley's surface states. In this case too, the formation of the surface state is made easier if the quantities α and γ are of the same sign.

For $\delta \to 1$, the existence conditions (59) as well as (61) change into (57) and (58) respectively. This fact illustrates the connection between the surface states discussed in the last two paragraphs.

(3) *Chemisorption States, Resulting from Adsorption at the End of a Semi-infinite Chain*

(a) *Chemisorption of an atom C with a Coulomb integral equal to the Coulomb integral of atoms of the kind B*

Using the special case of formulae (37) and (38) in which $\alpha = 0$, $\rho = \gamma$, $\sigma \neq 0$, we obtain the required case.[43] Atom C is generally bound by a bond which is different from those appearing in the middle of the chain ($\sigma \neq 1$). Thus a semi-infinite chain is obtained in which the strength of the bond between the border atom and its neighbour is different from that in an ideal chain. It can be imagined that the change in the strength of the bond is caused by the difference in the length of the bond between the surface atom and its neighbour. This case can also be treated as a special case of chemisorption. The energy of the localized state is given by the equation

$$W = \pm[\gamma^2 + \sigma^2 + \sigma^2\delta^2/(\sigma^2 - 1)]^{\frac{1}{2}} \tag{61}$$

The occurrence of the surface state is controlled by the existence condition:

$$|\sigma^2 - 1| > \delta \tag{62}$$

If the bond between the surface atom and its neighbour is strengthened ($\sigma > 1$) the existence condition (62) assumes the form

$$\sigma^2 > 1 + \delta \tag{63}$$

In order to obtain the pair of surface states with energy levels outside the system of energy bands of the volume states, the bond of the surface atom must be sufficiently strengthened. The quantity μ, defined by Eq. (31), acquires, after inserting expression (61) for the energy, the form

$$\mu = (\sigma^2 - 1)\delta^{-1} + \delta(\sigma^2 - 1)^{-1} \tag{64}$$

from which the correctness of the statement concerning the position of the energy of the surface state is immediately evident. A considerable change in the resonance integral makes the stay of an electron between the surface atom and its neighbour from the viewpoint of the energy very advantageous in the bonding state and very disadvantageous in the antibonding state (plus or minus signs on the right-hand side of Eq. (61)). The fact that such a considerable change is essential in the existence condition shows the similarity between these surface states and Tamm's surface states, also caused by the appreciable change in the Coulomb integral of the atomic orbital of the border atom. This similarity is even more obvious from the position of the energy levels in question outside the system of allowed energies of the volume states as well as from the fact that the amplitude of the wave function in question decreases monotonously. If the bond between a surface atom and its neighbour is weakened ($\sigma < 1$), the existence condition (62) requires

$$\sigma^2 < 1 - \delta \tag{65}$$

The bond must be sufficiently weak for the energy level of the surface state to be lying in the energy gap $\mu < 0$. The existence condition (65) can nevertheless be fulfilled only when $\delta < 1$. The bond which was broken for the formation of a semi-infinite chain must be a weaker one than those appearing in the chain with alternating bonds.

By the weakening of the bond between the surface atom and its neighbour compared with the bonds inside the chain, the border atom is partially isolated from the rest of the linear chain. In this way we are passing to a semi-infinite chain in which the present border atom is the atom which was originally a neighbour of a border atom of the original chain. The border atom of the primary chain is isolated from the rest of our model of the crystal. From this discussion it is clear that sufficient weakening of the bond which in an ideal infinite chain is a strong one leads to the localized states which gradually change into Shockley's surface states. From the existence conditions given it follows that for the appearance of surface states a relatively great change in the resonance integral of the bond between the border atom and its neighbour is necessary. The physical interpretation of the model studied which explains the changes in the resonance integral only by the formation of the surface of the crystal and by the related changes in the lengths of the bonds is certainly not realistic. Nevertheless, treating the surface atom as a chemisorbed atom, with a Coulomb integral not too different from the Coulomb integral of an atom of the B type, this model shows us some of the most important properties of localized chemisorption states, even for more general cases. It is remarkable that two kinds of chemisorption states are obtained which due to their properties remind us of Tamm's and Shockley's surface states and which can therefore be called Tamm's or Shockley's chemisorption states respectively. Shockley's chemisorption states occur if the chemisorption bond is weak and if in the absence of chemisorption Shockley's surface states exist. Tamm's chemisorption states are, on the other hand, caused by a very strong bond between the chemisorbed atom and the end of the chain.

(b) *General case of chemisorption localized states*

Let us rewrite the first Eq. (36) in the form:

$$D(W) = G(W) \tag{66}$$

where

$$G(W) = [\delta z_2(\mu) + 1]/(W - \gamma) \tag{67}$$

and

$$D(W) = \alpha + \sigma^2/(W - \rho) \tag{68}$$

μ is given by Eq. (31). The function G (cf. Figs. 3 and 4) is an increasing one: in the interval from $-\infty$ to W_1 ($W_{1,2} = [\gamma^2 + (1 \pm \delta)^2]^{\frac{1}{2}}$) it increases from $-\infty$ to $-(1 + \delta)/(W_1 + \gamma)$, and in the interval from W_1 to $+\infty$ it increases from the value $(1 + \delta)/(W_1 - \gamma)$ to $+\infty$. The behaviour in the interval $(-W_2, W_2)$, which corresponds to the energy gap, is different according to whether δ is larger or smaller than unity. If $\delta \leq 1$, the function G increases in this interval from $(\delta - 1)/(x + W_2)$ to $(1 - \delta)/(W_2 - \gamma)$. If $\delta = 1$, $\gamma < 0$, the value of the function G for

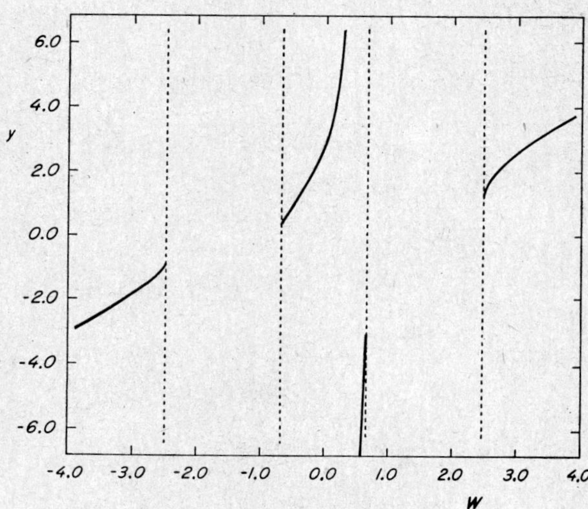

Fig. 3. Plot of the function G for $\gamma = 0.5$, $\delta = \sqrt{2}$ (cf. Eq. (66)). The limits of the energy bands of non-localized states are marked by dashed lines.

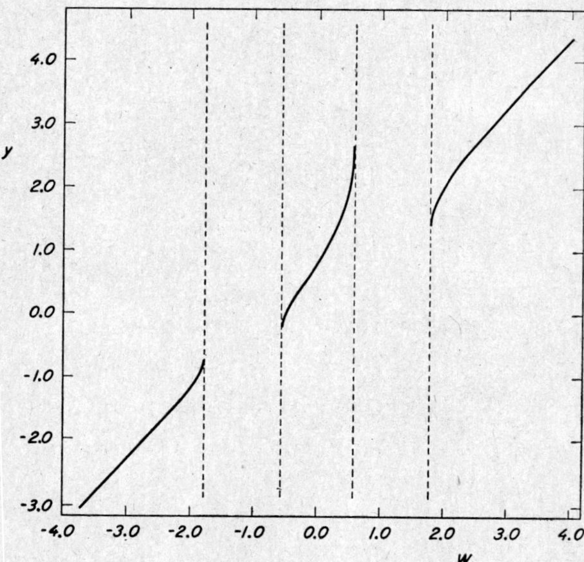

Fig. 4. Plot of the function G for $\gamma = 0.5$, $\delta = 2^{-\frac{1}{2}}$ (cf. Eq. (66)). The limits of the energy bands of non-localized states are marked by dashed lines.

$W = -W_2$ decreases to $-\infty$. For $\delta = 1$, $\gamma > 0$, the value of G increases for $W = W_2$ beyond all limits. If $\delta > 1$, the function G increases in the interval $\langle -W, \gamma)$ from the value $(\delta - 1)/(\gamma + W_2)$ to $+\infty$ and in the interval $(\gamma, W_2\rangle$ from $-\infty$ to $(1 - \delta)/(W_2 - \gamma)$.

The energies of the surface states are obtained as crossing points of the function G with the line $y = \alpha$. From the plot of the function G we can easily understand the existence conditions for the occurrence of surface states which were previously mentioned in detail.

The energies of the localized chemisorption states are obtained as crossing points of the function G with the hyperbola whose asymptotes are given by the equations

$$x = \rho, \quad y = \alpha \tag{69}$$

From the plot of the function G we can follow the conditions for the existence of the chemisorption states. If the exchange integral which characterizes the chemisorption band is very large we always obtain chemisorption states of Tamm's type.[31,100,101] They appear as crossing points of the hyperbola with those branches of the function G from which the Tamm's surface states also originate. Shockley's chemisorption states for $\delta < 1$ can be created only if, on the contrary, σ^2 is very small. Apart from this chemisorption states can appear from the crossing points with the branches which supply Shockley's subsurface states. In this case it is especially advantageous if σ^2 is very large or if α has an absolute value larger than unity. If the chain contains atoms of two kinds, chemisorption states can appear inside the "energy gap". These states can be created because a branch is present in this energy region which results from the presence of atoms of different kinds. The presence of heteroatoms thus facilitates the formation of the surface states.

Shockley's surface states, which are produced by the breaking of the stronger bond during the formation of the surface, therefore disappear during the formation of the chemisorption bond, which is comparable with bonds inside the crystal, composed of one kind of atom only. Strong chemisorption always causes the presence of Tamm's chemisorption states. Strong chemisorption does not prevent the existence of Shockley's subsurface states on a semi-infinite chain if they existed beforehand without chemisorption.

The existence of a Tamm's bonding chemisorption state can be interpreted as a localized bond between the chemisorbed atom and a surface atom to which the chemisorbed atom is bonded.[100,101] This bond is more or less homopolar, as long as the probabilities of finding an electron on both these atoms are comparable.[30,51] If the localization is appreciable, a molecular orbital is formed surrounding only these two atoms. This orbital is then completely uninfluenced by the rest of the crystal, with the exception of purely Coulomb interaction. Because of this, the right-hand sides of the equations for the calculation of the energy of chemisorbed localized states in different models (cf. Eqs. (66), (73), (176) and (180)), where $|W|$ is always large, are directly proportional to W or to $-W$ respectively, which transforms these equations into equations for the energy of the molecular orbital of a bi-atomic molecule. If, however, such a molecular orbital is not formed, it does not mean that a chemisorption bond does not exist but only that this bond has a metallic character.[30]

Analogous reasoning is also valid, *mutatis mutandis*, for the chemisorption of a molecule.

A review of the existence conditions for $\delta = 1$, $\gamma = 0$ is given in Table I (cf. refs. 30, 63). According to the values of the

TABLE I. The Existence Conditions for Different Numbers of Chemisorption States. Chemisorption at the End of a Semi-infinite Chain ($\delta = 1$, $\gamma = 0$). BB two bonding, AA two anti-bonding states, BA one bonding and one anti-bonding state, B one bonding, A one anti-bonding state, 0 no localized states. ($\alpha' = \alpha - 1$, $\alpha'' = \alpha + 1$, $\rho' = \rho - 2$, $\rho'' = \rho + 2$)

BB	$\rho' > 0$, $\sigma^2 < \alpha'\rho'$
AA	$\rho'' < 0$, $\sigma^2 < \alpha''\rho''$
AB	$\rho' < 0 < \rho''$, $\sigma^2 > \alpha'\rho'$, $\sigma^2 > \alpha''\rho''$
AB	$\rho' < 0$, $\rho'' \leq 0$, $\sigma^2 > \alpha'\rho'$
AB	$\rho' \geq 0$, $\rho'' > 0$, $\sigma^2 > \alpha''\rho''$
B	$\rho' > 0$, $\alpha'\rho' > \sigma^2 > \alpha''\rho''$
A	$\rho'' > 0$, $\alpha''\rho'' > \sigma^2 > \alpha'\rho'$
0	$\rho' < 0 < \rho''$, $\sigma^2 < \alpha'\rho'$, $\sigma^2 < \alpha''\rho''$

Coulomb integrals α and ρ and the resonance integral σ the number of chemisorption states (whether bonding or antibonding) may be zero, one or two. Taking into account overlap integrals may introduce some complications, but even here three chemisorption states may be obtained.[96]

(4) *Chemisorption on a One-Dimensional Infinite Crystal*

Let us consider a model of an infinite chain of atoms as mentioned in Section II-B-(1). We shall use the same notation. Let an atom C be chemisorbed on an atom of type A in the elementary cell denoted by the number zero. This means that a bond is created between the A atom considered and the atom C, the bonds between the A atom and its neighbours of the type B not being disturbed. Let us denote the parameters characterizing the atomic orbitals of the atom C, the bond between the atom C and the A atom and finally the change of potential at the A atom which is bound to the atom C in the same way as in Section II-B-(1).

The equations for estimating the extension coefficients of the molecular orbital, which is assumed to be of the form

$$\phi = \sum_{\mathrm{M=A,B}} \sum_{j=-\infty}^{+\infty} d_j^{\mathrm{M}} \chi_j^{\mathrm{M}} + d^c \chi_c \tag{70}$$

are (cf. Eq. (67))

$$d_j^{\mathrm{M}} = D(W) L_{j0}^{\mathrm{MA}} d_0^{\mathrm{A}} \tag{71}$$

and from these it follows that

$$D(W) = (L_{00}^{\mathrm{AA}})^{-1} \tag{72}$$

After the indicated operations have been carried out (cf. Eqs. (29)), we obtain the equation for the calculation of the energy in a form suitable for discussion and graphical solution:

$$D(W) = F(W) = \frac{\mathrm{Sign}\,(\mu)}{(W-\gamma)} [(W^2-\gamma^2)^2 - 2(1+\delta^2)(W^2-\gamma^2) + (1-\delta^2)^2]^{\frac{1}{2}} \tag{73}$$

Obviously relation (32) must be fulfilled, guaranteeing that the energy of the localized state will not be inside one of the bands of allowed energies of the delocalized states.

The function F has zero values at the edges of the bands of the volume states, i.e. for $W = \pm W_1$, $\pm W_2$ (cf. Section II-B-(3)). For $W < -W_1$ the function F grows from zero to $+\infty$, and for $W < -W_2$ it grows from $-\infty$ to zero. For W between $-W_2$ and γ, F increases from zero to $+\infty$. In the interval of W from γ to W_2, F increases from $-\infty$ to zero. For $\delta = 1$ and $\gamma < 0$ the discontinuity of the function F coincides with the lower edge of the conduction band, and for $\delta = 1$ and $\gamma < 0$ with the upper edge of the valency band. In the first case, the function F increases in the interval of the values of W characteristic of the energy gap from $-\infty$ to zero, in the second case it increases in this interval from zero to the value $+\infty$. From this analysis of the function F the following conclusions can be drawn about the localized states in the neighbourhood of the elementary cell denoted by the number zero (cf. Fig. 5).

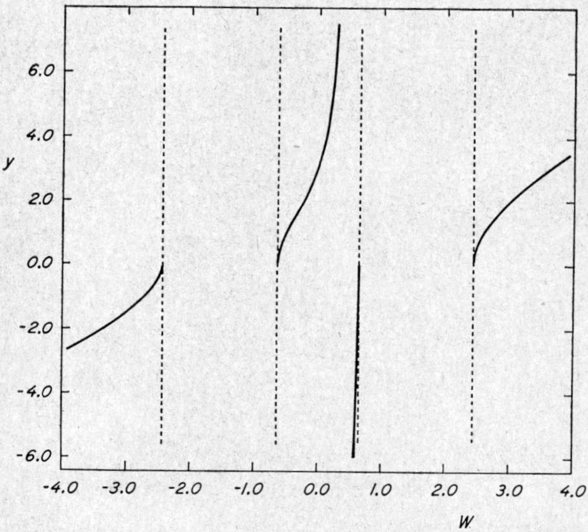

Fig. 5. Plot of the function F for $\gamma = 0.5$, $\delta = \sqrt{2}$ (cf. Eq. (73)). The limits of the energy bands of non-localized states are marked by dashed lines.

Chemisorption localized states are obtained as intersections of the function F with a hyperbola having asymptotes given by Eq. (69). At least one chemisorption state always appears, either below the valency band or above the conduction band, if α is larger or smaller than zero respectively. If $\delta \neq 1$ at least one chemisorption state appears with an energy lying inside the energy gap. If $\delta = 1$ sufficient (but not necessary) conditions for the appearance of one chemisorption state in the energy gap are given by the relations

$$\alpha > 0, \quad \gamma > 0, \quad \rho < \gamma \tag{74}$$

or the relations

$$\alpha < 0, \quad \gamma < 0, \quad \rho > \gamma \tag{75}$$

In general, three chemisorption localized states may appear.

A special case of the model considered arises if $\sigma = 0$. If the Coulomb integral of one of the A atoms is changed at least one localized state is always obtained outside the system of both bands of allowed energies of the delocalized states. Further, if the conditions Sign α = Sign γ, $\gamma \neq 0$ hold, a localized state whose energy lies inside the energy gap is obtained.

Let us emphasize the difference between the results obtained here and the results obtained with the model of chemisorption where the atom C is bound at the end of a semi-infinite chain: for chemisorption on a cyclic infinite chain, in contrast to the chemisorption at the end of a semi-infinite chain, localized states always exist.[12,63] The interpretation of the chemisorption states with energies in the energy gap is, in the present model, different from the interpretation of Shockley's chemisorption states at the end of a semi-infinite chain. Due to the change of potential in the neighbourhood of the A atom the probability of finding an electron on the A atom in the elementary cell "zero" for the orbital considered decreases. Because of this, the electron on such a molecular orbital behaves in the same way as in a semi-infinite chain, resulting in the removal of the A atom from the cell "zero". In a semi-infinite chain with a weaker bond at the end, surface states of the Shockley's type occur. The change in the Coulomb integral α and the bond formed with the atom C therefore create Shockley's surface states whose appearance is connected with the difference of the bonds in the surface. An analogous consideration is valid for localized states in a chain with different atoms. Chemisorption states of this kind appear more readily, the larger the disturbance on the A atom in the cell "zero". The probability of finding an electron in these states on the A atom in the cell "zero" and on the atom C is a small one. This probability is large for B atoms which are neighbours of the A atom considered.

C. Simple Models of Two-Dimensional and Three-Dimensional Crystals in the Tight-Binding Approximation

(1) *Transition from a One-Dimensional to Two- and Three-Dimensional Crystals*

In as far as we are studying either surface states on the clean surface of a crystal or chemisorption with a fully covered surface such that the chemisorbed layer has the same plane symmetry as a surface layer of a crystal, the whole problem can be formulated using the layer functions (3). In the calculations Coulomb and resonance integrals between these layer molecular orbitals must be used which are defined by the following relations

$$\int a_{m_3;\beta,\nu}(\xi_1, \xi_2)\mathscr{H} a_{m_3';\beta',\nu'}(\xi_1, \xi_2)d\tau = \mu_{m_3;\beta,\nu.m_3';\beta',\nu'} \tag{76}$$

Equations for the energy of the localized states and for the extension coefficients $d^{\alpha}_{s;\beta\nu,m_3}$ (cf. Eq. (4)) are obtained from the corresponding one-dimensional case by substituting for the Coulomb and resonance integrals of the atomic orbitals analogous quantities for the layer molecular orbitals. The energy spectrum is broken down into partial spectra for constant values of the wave vectors lying in the surface of the crystal. It was mentioned that the overall spectrum of allowed electron energies in a crystal originates from the superposition of these partial spectra. In this way a case can arise in which the energy of a localized state which can exist for the given values of the wave vectors lying in the surface can be in the region allowed for volume states characterized by different wave vectors ξ_1, ξ_2. By a linear combination of the wave functions for localized and non-localized states wave functions are formed which according to the definition (5) are also localized.

(2) *Two Examples of Simple Crystals*

(a) *Three-dimensional tetragonal crystal limited by basic face*

Let us consider a crystal built of atoms of two kinds, A and B. The A atoms are deposited in planes alternating with those occupied by the B atoms. The atomic orbital of an A atom has a non-zero exchange integral β with the atomic orbital of the B atom which is situated above it in the plane, nearer to the surface, and a resonance integral β' with the atomic orbital of the B atom situated below it in the plane, further from the surface. The atomic orbital of an A atom also has non-zero resonance integrals β'' with the four orbitals of its nearest neighbouring A atoms, which are at distances equal to the elementary translations $\pm\mathbf{a}_1$ and $\pm\mathbf{a}_2$ from the atomic orbital of the A atom considered. Similarly, the atomic orbital of a B atom has non-zero resonance integrals β''' with the atomic orbitals of the four nearest neighbouring B atoms, which are situated at distances equal to the elementary translations $\pm\mathbf{a}_1$ and $\pm\mathbf{a}_2$ from the B atom considered.

Keeping the symbols of Section II-B-(1) unchanged, and introducing further

$$a_{m_3}^{\mathrm{M}}(\xi_1, \xi_2) = K \sum_{m_1, m_2 = -\infty}^{+\infty} e^{i(\xi_1 m_1 + \xi_2 m_2)} \chi^{\mathrm{M}}(\mathbf{r} - \sum_{k=1}^{3} m_k \mathbf{a}_k) \tag{77}$$

$$\mathrm{M} = \mathrm{A}, \mathrm{B}$$

we obtain the following expressions for the Coulomb and resonance integrals of the layer molecular orbitals:

$$\begin{aligned} \mu_{m_3, m_3}^{\mathrm{A\ B}}(\xi_1, \xi_2) &= \beta', \quad \mu_{m_3, m_3 \pm 1}^{\mathrm{A\ \ \ B}} = \beta \\ \mu_{m_3, m_3}^{\mathrm{A\ A}} &= \alpha_{\mathrm{A}} + 2\beta''(\cos \xi_1 + \cos \xi_2) \\ \mu_{m_3, m_3}^{\mathrm{B\ B}} &= \alpha_{\mathrm{B}} + 2\beta'''(\cos \xi_1 + \cos \xi_2) \end{aligned} \tag{78}$$

The energies of volume states given by the equation

$$E/\beta - \bar{A} = \pm[1 + \Gamma^2(\xi_1, \xi_2) + 2\delta \cos \xi_3 + \delta^2]^{\frac{1}{2}} \tag{79}$$

The energies of localized surface states are given by Eq. (39) in which we insert the quantity Γ instead of γ and the quantity $\bar{A}$ instead of $\bar{\alpha}$. The quantities Γ and $\bar{A}$ are given by the equations

$$\begin{aligned} \Gamma(\xi_1, \xi_2) &= \gamma + \frac{\beta''' - \beta''}{\beta}(\cos \xi_1 + \cos \xi_2) \\ \bar{A}(\xi_1, \xi_2) &= \frac{\alpha_{\mathrm{A}} + \alpha_{\mathrm{B}}}{2\beta} + \frac{\beta'' + \beta'''}{2\beta}(\cos \xi_1 + \cos \xi_2) \end{aligned} \tag{80}$$

The quantity $\bar{A}$ is taken as the zero of the scale of energies. The ratios of the extension coefficients are given by Eqs. (41)–(43) and the existence conditions for the surface states are determined by the inequalities (45) in which the interchanges mentioned are carried out.

As can be seen from this analogy, the results of the discussion of the surface states of a one-dimensional chain can be applied without change to the system being studied. The values of the resonance integrals β'' and β''' determine whether the energies of the surface states calculated for ξ_1 and ξ_2 coincide with the energies of the volume states for other values of the wave vectors ξ_1' and ξ_2'.

(b) *Two-dimensional graphite lattice in the tight-bonding approximation*

Let us consider a two-dimensional graphite lattice (a system of regular hexagons) built from two kinds of atoms (A and B). The neighbours of A atoms are B atoms only, and vice versa. Let us assume each atom has one atomic orbital with non-zero resonance integrals with the atomic orbitals of its nearest neighbours. Let us again denote the Coulomb integrals of A and B atoms by α_A and α_B respectively. An elementary cell is built from one A atom and one B atom. A graphite lattice originates by shifting this elementary cell by whole multiples of the elementary translations $\mathbf{a}_1$ and $\mathbf{a}_2$ (cf. Fig. 10).

Let us now create the surface of a two-dimensional crystal by a line parallel with the elementary translation $\mathbf{a}_1$ in such a way that this line always cuts one bond of each surface atom of the kind A. For the resonance integrals between the layer molecular orbitals of the layer consisting of A atoms and the molecular orbitals of the layer consisting of the B atoms which are nearer to the surface we obtain

$$\mu^{\mathrm{A}\ \ \mathrm{B}}_{m_2,m_2-1} = \beta \tag{81}$$

as each A atom is bound to one B atom of the neighbouring plane.

The analogous resonance integral for two layers in which an A atom is bound to two B atoms and vice versa is given by:

$$\mu^{\mathrm{A}\ \mathrm{B}}_{m_2,m_2} = \beta(1 + e^{-i\xi_1}) = \beta|1 + e^{i\xi_1}|e^{-i\omega} \tag{82}$$

We thus obtain a complete analogy with the model of a linear chain if we put

$$\delta = a = |1 + e^{i\xi_1}|, \qquad \gamma = (\alpha_B - \alpha_A)/2 \tag{83}$$

and introduce new layer molecular orbitals

$$a'^{A}_{m_3} = a^{A}_{m_3} e^{-im_3\omega}, \qquad a'^{B}_{m_3} = a^{B}_{m_3} e^{-i(m_3+1)\omega} \tag{84}$$

Let us recall that the energies of the volume states are given by the equation:

$$x_{\pm} = \pm[\gamma^2 + 3 + 2\{\cos \xi_1 + \cos \xi_2 + \cos (\xi_1 - \xi_2\}]^{\frac{1}{2}} \tag{85}$$

If the graphite lattice contains two kinds of atoms, a gap of width 2γ is formed in the whole spectrum of allowed energies of the volume states. With a lattice formed from one kind of atom only no such gap occurs. This spectrum is the result of the superposition of the separate spectra for given values of the wave vector ξ_1.

As long as the change in the Coulomb integral of the surface atoms is negligible, the Shockley's surface states can exist only inside the gaps between the volume states (for a given ξ_1) whose energy equals[65]

$$W = -\gamma \tag{86}$$

and the existence condition is

$$|2 \cos (\xi_1/2)| < 1 \tag{87}$$

This level is $N'/3$ times degenerate (N' being the number of surface atoms) and is situated at the edge of the resulting energy band of the volume states which originated from the broadening of the energy level of the electron isolated on the atomic orbital of the A atom.

The question could arise as to what is the condition for the existence of the surface states inside the gap of the overall energy spectrum and outside the system of both bands of this overall spectrum, respectively, when $\alpha \neq 0$. For the first case, one obtains the existence condition (inserting $\delta = a = 0$ in Eq. (61))

$$1 - \alpha^2 > -2\alpha\gamma \tag{88}$$

For the second case, inserting $\delta = a = 2$ in Eq. (60), we find the existence condition

$$\alpha^2 - 1 > -\tfrac{2}{3}\alpha\gamma \qquad (89)$$

It is worth remarking that the existence conditions for Shockley's surface states can again be interpreted as due to the breaking of a stronger bond; however in this case the bonding between the layers of atoms, parallel with the surface of the two-dimensional graphite lattice, is meant (cf. Eqs. (81) and (82)).

Finally let us note for graphite the striking fact that inside the volume energy bands of the overall energy spectrum energy levels of the surface states can exist. This, however, is made possible by the fact that the overall spectrum results from the superposition of partial energy spectra. Wave functions of wave vectors of different values cannot interact as they belong to different irreducible representations of a translation group.

(3) *General Condition of Non-existence of Surface States of Shockley's Type in the Tight-Binding Approximation*[65]

As mentioned above, an infinite crystal with Born–Kármán's conditions is essentially a crystal cyclic in the direction of every elementary translation.

Let us consider therefore a closed chain of atomic orbitals of $2(N + 1)$ members, situated in the plane. Using the LCAO method and the tight-bonding approximation, it is possible to obtain from this closed chain two equal, finite chains by removing the atomic orbitals numbered 1 and $(N + 1)$ from the interaction. In certain cases, these chains can pass one into the other by reflection in the plane σ passing through the atomic orbitals 1 and $(N + 1)$ and perpendicular to the plane of the cyclic chain. Then N wave functions of the cyclic chain exist which are antisymmetric against the reflexion plane σ. The amplitudes of these functions at the atoms 1 and $(N + 1)$ are equal to zero, and therefore in the tight-bonding approximation these functions are also wave functions of the finite chains generated.[40] As the number of these wave functions is equal to the number of atomic orbitals in one chain, no further wave functions can exist. Because these functions are suitable for an infinite cyclic chain, they cannot be localized at the ends of the finite chains. This consideration does not lose its validity even if N is increasing beyond all limits.

This reasoning makes it clear that a linear, semi-infinite chain containing only one kind of atoms in which the atoms are connected by similar bonds, as well as a linear chain containing two kinds of atoms which alternate and are connected by bonds of one kind, cannot exhibit Shockley's surface states. Similarly, a two-dimensional graphite lattice, limited in the way indicated in Fig. 11, and a three-dimensional cubic lattice, where the A atoms have only B atoms for neighbours and are limited by the plane (1, 0, 0), cannot exhibit Shockley's surface states.

The condition derived is obviously related to Baldock's sufficient conditions for the non-existence of surface states of Shockley's type, associated with the plane $m_3 = 0$:[9,17]

(a) Every atom is surrounded by neighbours in the same relative positions.

(b) If we choose the coordinate system so that two axes lie in the face of the crystal, every neighbour of any atom lies in one or other of the planes m_3, $m_3 \pm 1$.

(4) *General Theory of Tamm's Surface States for Tetragonal Lattices*[48]

Let us consider a semi-infinite tetragonal crystal for which the energy of a volume state has the form

$$E = \varepsilon_0(\xi_1, \xi_2) + 2 \cos \xi_3 \tag{90}$$

where the value ε_0 may depend upon further wave vectors. We define

$$\lambda_{m_3,m_3'}(\xi_1, \xi_2) = \int a_{m_3}(\xi_1, \xi_2)[\mathscr{H} - \mathscr{H}_0]a_{m_3'}(\xi_1, \xi_2)d\tau \tag{91}$$

where $\mathscr{H}_0$ is the Hamiltonian of the infinite crystal and $\mathscr{H}$ is the Hamiltonian of the semi-infinite crystal considered. We suppose that these integrals are non-zero if m_3, $m_3' \leqslant (Z - 1)$. The subscripts β and ν (cf. Eq. (76)) have only one value and are therefore omitted. The equation for the coefficients $d_{m_3}(\xi_1, \xi_2)$ in the wave function (Eq. (4)) has the form (cf. Eqs. (29) and (30) for $\delta = 1$, $\gamma = 0$):

$$d_p = (z_2 - z_1)^{-1}[\sum_{m=0}^{Z} \sum_{m'=0}^{Z} z_1^{|p-m'|}\lambda_{m',m}d_m - z_1^{p+1}d_0] \tag{92}$$

where the letters p, m, etc. denote various values of m_3. $z_{1,2}$ is defined in Eq. (30).

The energy is given by the equation

$$\left|\sum_{m'=0}^{Z} z_1^{|p-m'|}\lambda_{m',m} - \delta_{m,0}z_1^{p+1} - \delta_{p,m}\right| = 0 \tag{93}$$

Let us multiply this equation by the determinant $|S|$, where the matrix elements are defined in the following way: $S_{00} = S_{ZZ} = z_2$; $j \neq 0, Z$: $S_{jj} = z_1 + z_2$; $S_{j,j+1} = S_{j,j-1} = 1$; $j \neq k$, $k+1$: $S_{jk} = 0$. Hence the equation for the calculation of the energy assumes the form

$$|U - A| = 0 \tag{94}$$

where the matrices U and A are defined in the following way:

$$\begin{aligned} U_{pm} &= \lambda_{pm} + \delta_{p-1,m} + \delta_{p+1,m} \\ A &= (z_1 + z_2)\delta_{pm} + z_1\delta_{pZ}\delta_{mZ} \end{aligned} \tag{95}$$

As U is a Hermitian matrix, a unitary matrix Q exists such that

$$QUQ^+ = K \tag{96}$$

where

$$K_{ij} = K_i\delta_{ij} \tag{97}$$

The diagonal elements of the matrix K in Eq. (96) have the meaning of the energies of the molecular orbitals for a finite chain consisting of $(Z+1)$ atomic orbitals whose Coulomb and resonance integrals are U_{mp}.

Equation (92) can be rearranged in the following way:

$$\begin{aligned} &|Q|\cdot|U - A||Q^+| \\ &= z_1 \prod_{j=0}^{Z} (K_j - W)\left[z_2 + \sum_{k=0}^{Z} \frac{\rho_k}{(K_k - W)}\right] = 0 \end{aligned} \tag{98}$$

so that

$$z_2 = -\sum_{j=0}^{Z} \frac{\rho_j}{(K_j - W)} \tag{99}$$

where

$$\rho_j = |Q_{jz}|^2 \tag{100}$$

From the form of Eq. (99) we can easily derive the following theorems:

(1) Not more than Z surface states exist.
(2) If s values of K_j exist, such that $|K_j| > 2$, at least s surface states exist.

From these theorems we can see that the number of surface states is directly connected with the depth to which the defect due to the formation of the surface reaches, or in other words, the number of layers in which the perturbation potential is sufficiently large.

If only the λ_{00}, λ_{01} and λ_{11} are different from zero this case can be transformed into the case discussed in Section II-B-(3b) by introducing

$$\lambda_{00} = U_{00} = \rho, \quad \lambda_{01} + 1 = U_{01} = \sigma, \quad \lambda_{11} = U_{11} = \alpha \tag{101}$$

This shows the connection between Tamm's surface and chemisorption states.

For given values of the wave vectors ξ_1, ξ_2 only two surface states can exist. One, with

$$|\lambda_{00}| > |\lambda_{11} + 1| \tag{102}$$

has a maximum amplitude in the second layer beneath the surface. Even the Tamm's states can therefore be "subsurface" ones (cf. ref. 84).

D. Goodwin–Artmann's Model and Related Models

(1) *Shockley's Surface States in Goodwin–Artmann's Model*

A deeper and wider understanding of the properties of Shockley's surface states can be obtained by using other models apart from the model already mentioned. In this paragraph a simple model of a crystal with several atomic orbitals on each atom is considered.[28] With this model we can follow the relation between the hybridization and the existence of localized bonds with the existence conditions of Shockley's surface states.[53,66]

For this treatment a method will be chosen which is usual for the solution of difference equations, although Koster–Slater's method could also be applied here. However, the method used gives faster results, and this case is used to illustrate the method of calculation. The calculation will be described in detail as it is extremely simple when using a suitable procedure.

Let us consider an infinite chain of atoms. On each atom we consider two orbitals, one of them, $\chi_s(\mathbf{r} - m\mathbf{a})$, being symmetric towards reflection in the plane which passes through the atom in question and is perpendicular to the linear chain. The other atomic orbital, $\chi_p(\mathbf{r} - m\mathbf{a})$, is antisymmetric towards this reflection. An *s*-orbital can therefore be chosen as the first orbital and

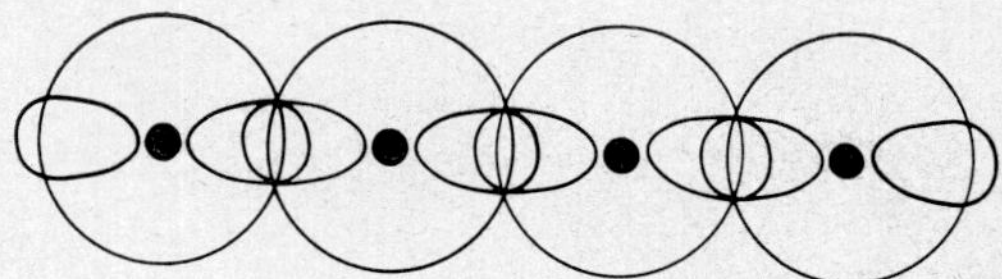

Fig. 6. Diagrammatic representation of the atomic orbitals in Goodwin–Artmann's model of a linear atomic chain.

a *p*-orbital as the second orbital (Fig. 6). As far as the resonance and Coulomb integrals are concerned the following will be required:

$$\int \bar{\chi}_s(\mathbf{r} - m\mathbf{a})\mathscr{H}\chi_s(\mathbf{r} - m'\mathbf{a})d\tau = \alpha_s\delta_{m,m'} + \gamma_s(\delta_{m,m'+1} + \delta_{m,m'-1})$$

$$\int \bar{\chi}_p(\mathbf{r} - m\mathbf{a})\mathscr{H}\chi_p(\mathbf{r} - m'\mathbf{a})d\tau = \alpha_p\delta_{m,m'} - \gamma_p(\delta_{m,m'+1} + \delta_{m,m'-1})$$

$$\int \bar{\chi}_s(\mathbf{r} - m\mathbf{a})\mathscr{H}\chi_p(\mathbf{r} - m'\mathbf{a})d\tau = \gamma_{sp}(\delta_{m,m'-1} - \delta_{m,m'+1}) \tag{103}$$

In Hückel's approximation we obtain for the extension coefficients the following system of equations:

$$\begin{aligned}(\alpha_s - E)c_s(m) + \gamma_s[c_s(m+1) + c_s(m-1)]& \\ + \gamma_{sp}[c_p(m+1) - c_p(m-1)] &= 0 \\ (\alpha_p - E)c_p(m) - \gamma_p[c_p(m+1) + c_p(m-1)]& \\ - \gamma_{sp}[c_p(m+1) - c_p(m-1)] &= 0\end{aligned} \tag{104}$$

From these follow the equations for the energies of the volume states:

$$F = (\gamma_s - \gamma_p)\cos\xi \pm 2[(q + q_0\cos\xi)^2 + \gamma_{sp}^2\sin^2\xi]^{\frac{1}{2}} \tag{105}$$

and for the extension coefficients:

$$\lambda = \frac{c_p(m)}{c_s(m)} = \frac{i}{\gamma_{sp}\sin\xi}\{q + q_0\cos\xi \pm [(q + q_0\cos\xi)^2 + \gamma_{sp}^2\sin^2\xi]^{\frac{1}{2}}\}$$

In the above equations we have used the following symbols:

$$q = (\alpha_s - \alpha_p)/4, \quad q_0 = (\gamma_s + \gamma_p)/2, \quad F = E - (\alpha_s + \alpha_p)/2 \qquad (107)$$

The energies of the volume states form two bands touching each other for $|q| = |q_0|$.

For the purpose of further discussion let us note that the resonance integrals between hybrid atomic orbitals

$$\phi_\pm = 2^{-\frac{1}{2}}[\chi_s(\mathbf{r} - m\mathbf{a}) \pm \chi_p(\mathbf{r} - m\mathbf{a})] \qquad (108)$$

are

$$\int\phi_+(m)\mathscr{H}\phi_-(m)d\tau = 2q, \quad \int\phi_+(m)\mathscr{H}\phi_\pm(m \pm 1)d\tau = \tfrac{1}{2}(\gamma_s - \gamma_p),$$

$$\int\phi_+(m)\mathscr{H}\phi_-(m-1)d\tau + \int\phi_+(m)\mathscr{H}\phi_-(m+1)d\tau = 2q_0 \qquad (109)$$

The interaction between the hybrid orbitals of neighbouring atoms becomes more important than the interaction between the hybrid orbitals of the same atom if $|q| < |q_0|$. This means that localized bonds between neighbouring atoms are formed.

If the chain considered is limited from one side, as well as the difference equations (104) (in which E is replaced by the energy of the localized state, W), valid for $m \geqslant 1$, the following boundary conditions must be satisfied:

$$(\alpha_s - W)c_s(0) + \gamma_s c_s(1) + \gamma_{sp}c_p(1) = 0$$
$$(\alpha_p - W)c_p(0) - \gamma_p c_p(1) - \gamma_{sp}c_s(1) = 0 \qquad (110)$$

For solving this system of difference equations (104) with the boundary conditions (110) let us assume that the form of the extension coefficients of the corresponding molecular orbitals will be

$$c_s(m) = c_s(0)(\tau_1^m + \kappa\tau_2^m)$$
$$c_p(m)/c_s(m) = \lambda \qquad (111)$$

where the quantities τ_1 and τ_2 are real and smaller than unity.

Let us assume that

$$j = 1,2\colon\ c'_{s,j}(m) = \tau_j^m, \quad c'_{pj}(m) = \lambda_j^m \tag{112}$$

is the particular solution of difference equations (104), so that we obtain the following relations:

$$\begin{aligned} (\alpha_s - W)(1 + \kappa) + (\gamma_s + \gamma_{sp}\lambda)(\tau_1 + \kappa\tau_2) &= 0 \\ (\alpha_p - W)(1 + \kappa)\lambda - (\gamma_p\lambda + \gamma_{sp})(\tau_1 + \kappa\tau_2) &= 0 \end{aligned} \tag{113}$$

and

$$\begin{aligned} j = 1,2\colon\ (\alpha_s - W) + \gamma_s(\tau_j + \tau_j^{-1}) + \gamma_{sp}\lambda(\tau_j - \tau_j^{-1}) &= 0 \\ (\alpha_p - W)\lambda - \gamma_p\lambda(\tau_j + \tau_j^{-1}) - \gamma_{sp}(\tau_j - \tau_j^{-1}) &= 0 \end{aligned} \tag{114}$$

Subtracting from the first equation of (113) the first equation of (114) for $j = 1$ and the first equation of (114) multiplied by κ for $j = 2$ we immediately obtain

$$\kappa = -\tau_2/\tau_1 \tag{115}$$

We now insert relation (115) into Eqs. (113). We subtract the first (the second) Eq. (113), divided by the expression $(\tau_1 - \tau_2)/\tau_1$, from the first (the second) Eq. (114) and we obtain

$$\tau_1\tau_2 = \frac{\gamma_s - \gamma_{sp}\lambda}{\gamma_s + \gamma_{sp}\lambda} = \frac{\gamma_p\lambda - \gamma_{sp}}{\gamma_p\lambda + \gamma_{sp}} \tag{116}$$

or

$$\lambda = (\gamma_s/\gamma_p)^{\frac{1}{2}} \tag{117}$$

where the assumption Sign (γ_s) = Sign (γ_p) = Sign (γ_{sp}) was used. Dividing the first equation (113) by the second equation of (113) we easily obtain the equation

$$W - (\alpha_s + \alpha_p)/2 = (\gamma_p - \gamma_s)q/q_0 \tag{118}$$

Finally, inserting these results into one of the equations of (113) we obtain

$$\tau_{1,2} = [1 + (\gamma_{sp}^2/\gamma_s\gamma_p)^{\frac{1}{2}}]^{-1} \cdot \{q/q_0 \pm [(q/q_0)^2 + \gamma_{sp}^2/\gamma_s\gamma_p - 1]^{\frac{1}{2}}\} \tag{119}$$

τ_1 and τ_2 are smaller than unity if

$$|q| < |q_0| \tag{120}$$

which is the existence condition for Shockley's surface states.

The energy of a Shockley's surface state therefore lies inside the energy gap and the condition for the existence of this state is the breaking of the strong bond between hybrid atomic orbitals belonging to neighbouring atoms and directed in the opposite sense.

As long as the hybridization is not sufficient, favourable conditions cannot occur for the formation of Shockley's surface states, because at insufficient interaction between *s*- and *p*-orbitals we actually have two almost independent chains of *s*- and *p*-orbitals respectively, with delocalized bonds. The probability of finding an electron on hybrid *sp*-orbitals (Eq. (108)) directed from the surface of the crystal to the interior of the crystal is substantially smaller than the probability of finding an electron on atomic orbitals directed in the opposite sense. The ratio of these probabilities is $|(\sqrt{\gamma_s} - \sqrt{\gamma_p})/(\sqrt{\gamma_s} + \sqrt{\gamma_p})|^2$. As the greatest probability is, naturally, to find an electron on an *sp*-hybrid orbital of a surface atom, directed from the surface of the crystal to the vacuum, we

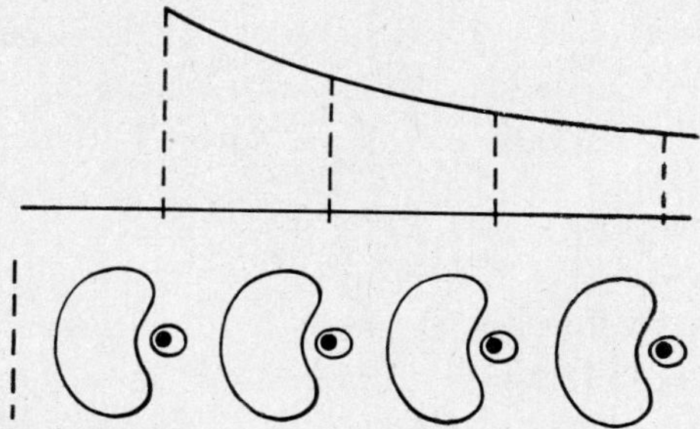

Fig. 7. Diagrammatic representation of the form of hybridized atomic orbitals by surface states for Goodwin–Artmann's model. The values of the electron densities on lines in different cells are different. The dependence of the wave amplitude upon the distance from the surface is shown schematically in the upper part of the figure.

can see that a Shockley's surface state expresses a directed localized unsaturated valency in the surface. Also in this model we can find all the essential properties which a Shockley's surface state must possess for it to be interpreted as an unsaturated bond in the surface (Figs. 7 and 8).

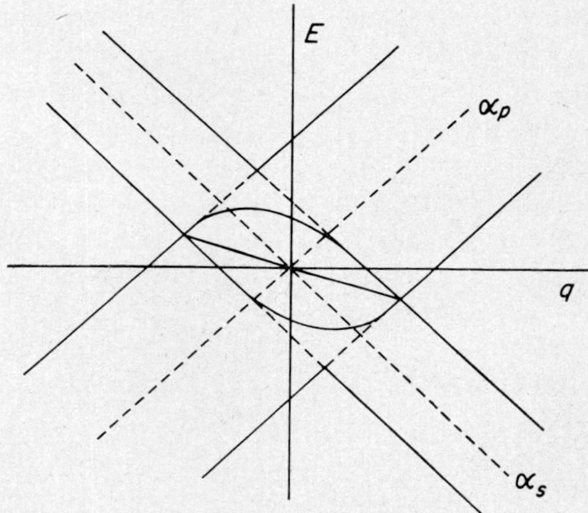

Fig. 8. The dependence of the energy bands on q for Goodwin–Artmann's model. Surface states occur in the gap between the bands of volume states. $\gamma_p = 2$, $\gamma_s = \gamma_{sp} = 1$.

(2) *Shockley's Surface States in Models for Interactions between Next-Nearest Neighbours*

(a) *Linear chain of atoms connected alternately by bonds of different strengths*[54]

The question arises as to how far the interpretation of Shockley's surface states is bound to the tight-binding approximation used up until now. It is therefore interesting to consider the following model of a chain of equal atoms: on each atom only one atomic orbital is considered. We assume that the resonance integrals between neighbouring atoms are different from zero and that between atoms bonds of different strengths alternate, with the corresponding resonance integrals β and $\delta\beta$. Let us denote the resonance integral between next-nearest neighbours by $\delta'\beta$. The m-th elementary cell contains two atomic orbitals, χ_m and χ_m^* (Fig. 9). Let us define:

$$\int \chi_m \mathscr{H} \chi_m^* d\tau = \delta\beta, \quad \int \chi_{m-1} \mathscr{H} \chi_m^* d\tau = \beta \tag{121}$$

The infinite chain can be changed into a semi-infinite one by breaking the bond corresponding to the resonance integral β. As we are considering the interaction between next-nearest neighbours, this breaking of the bond corresponds to the removing of two orbitals, χ_1, χ_1^*, from the interaction. For the origin of the energy scale we again use the value of the Coulomb integral of the atomic orbitals, and as energy unit the value β. Writing the system of difference equations with the corresponding boundary

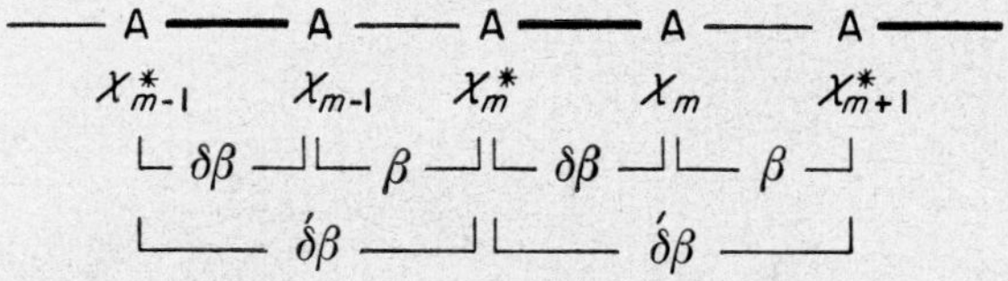

— A — A — A — A — A —

χ^*_{m-1} χ_{m-1} χ^*_m χ_m χ^*_{m+1}

$\delta\beta$ β $\delta\beta$ β

$\delta'\beta$ $\delta'\beta$

Fig. 9. Diagram of a chain of equal atoms with alternately strong bonds. The interaction between next-nearest neighbours is included.

conditions for the extension coefficients of the atomic orbitals in the molecular orbital

$$\psi = \sum_{m=0}^{\infty} (d_m\chi_m + d_m^*\chi_m^*) \tag{122}$$

we can see immediately that this system can be changed into the system of equations (104) and (110), in other words into the system of equations for Goodwin–Artman's model, by putting (cf. Eq. (108))

$$\varphi_s = 2^{-\frac{1}{2}}(\chi_m^* + \chi_m), \quad \varphi_p = 2^{-\frac{1}{2}}(\chi_m^* - \chi_m) \tag{123}$$

so that

$$\begin{aligned}
\int \varphi_s(m)\mathscr{H}\varphi_s(m')d\tau &= \delta\delta_{m,m'} + (\tfrac{1}{2} + \delta')(\delta_{m',m+1} + \delta_{m',m-1}) \\
\int \varphi_p(m)\mathscr{H}\varphi_p(m')d\tau &= -\delta\delta_{m,m'} + (-\tfrac{1}{2} + \delta')(\delta_{m',m+1} + \delta_{m',m-1}) \\
\int \varphi_s(m)\mathscr{H}\varphi_p(m')d\tau &= \tfrac{1}{2}\delta(\delta_{m',m+1} - \delta_{m',m-1})
\end{aligned} \tag{124}$$

As long as the resonance integrals δ' are not too large, or in other words as long as the following equation holds

$$2\delta' < 1 \tag{125}$$

it is possible to accept the result for Goodwin–Artmann's model putting

$$\alpha_s = -\alpha_p = \delta, \quad \gamma_s = \tfrac{1}{2} + \delta', \quad \gamma_p = \tfrac{1}{2} - \delta'$$
$$\gamma_{sp} = \delta/2, \quad q = \delta/2, \quad q_0 = \tfrac{1}{2} \tag{126}$$

The energy of the surface state is then given by the equation:

$$W = -2\delta\delta' \tag{127}$$

and the existence condition is:

$$\delta < 1 \tag{128}$$

The ratio of the coefficients of two atomic orbitals, both situated in the same elementary cell, is (cf. Eq. (117))

$$\frac{d_m}{d_m^*} = \frac{(1 - 2\delta')^{\frac{1}{2}} - (1 + 2\delta')^{\frac{1}{2}}}{(1 - 2\delta')^{\frac{1}{2}} + (1 + 2\delta')^{\frac{1}{2}}} \tag{129}$$

It is evident that the condition for the existence of Shockley's surface states is again the breaking of the stronger bond and that the probability of finding an electron is greater on the atomic orbitals of those atoms which are nearer to the surface in each elementary cell. Consideration of the further interaction represented by the integral δ' makes the probability of finding an electron on those atomic orbitals in each elementary cell which are further from the surface different from zero. Nevertheless, the whole character of Shockley's surface states as well as their interpretation is obviously independent of the generalized model mentioned.

(b) *Wallace's model of two-dimensional graphite lattice*[65]

If we take into consideration the interaction between next-nearest p_z-orbitals in the graphite lattice (i.e. if we consider the resonance integrals $\beta\delta'$ between atomic orbitals both with asterisks or both without asterisks (cf. Fig. 10)) we obtain Wallace's model of the graphite lattice.[103] Let our consideration be limited to the case where all the atoms in the graphite lattice are of the same species. Let us limit the graphite lattice by a line parallel to the elementary translation $\mathbf{a}_1$ (cf. Section II-C-(2b)).

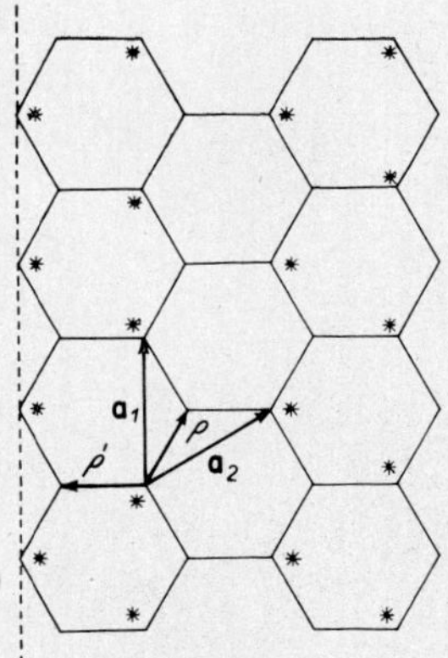

Fig. 10. Two-dimensional graphite crystal limited by a straight line parallel to the elementary translation $\mathbf{a}_1$. If there are two kinds of atoms in the lattice, the A atoms are labelled by asterisks.

Let us consider the system of difference equations with the corresponding boundary conditions for the coefficients in the molecular orbital

$$\psi = \sum_{m_2=0}^{\infty} [d_{m_2,s}\varphi_s(\mathbf{r} - m_2\mathbf{a}_2) + d_{m_2,p}\varphi_p(\mathbf{r} - m_2\mathbf{a}_2)] \tag{130}$$

where φ_s, φ_p indicate the following linear combinations of the layer molecular orbitals:

$$\varphi_s(\mathbf{r} - m_2\mathbf{a}_2) = 2^{-\frac{1}{2}}e^{-im_2\omega}[a^*_{m_2}(\xi_1) + e^{-i\omega}a_{m_2}(\xi_1)]$$

$$\varphi_p(\mathbf{r} - m_2\mathbf{a}_2) = 2^{-\frac{1}{2}}e^{-im_2\omega}[a^*_{m_2}(\xi_1) - e^{-i\omega}a_{m_2}(\xi_1)]$$

$$a_{m_2}(\xi_1) = N^{-\frac{1}{2}} \sum_{m_1=-\infty}^{+\infty} e^{im_1\xi_1}\chi(\mathbf{r} - m_1\mathbf{a}_1 - m_2\mathbf{a}_2) \tag{131}$$

$$a^*_{m_2}(\xi_1) = K \sum_{m_1=-\infty}^{+\infty} e^{im_1\xi_1}\chi^*(\mathbf{r} - m_1\mathbf{a}_1 - m_2\mathbf{a}_2)$$

For the definition of ω see Eq. (82). $\chi(\mathbf{r} - m_1\mathbf{a}_1 - m_2\mathbf{a}_2)$ and $\chi^*(\mathbf{r} - m_1\mathbf{a}_1 - m_2\mathbf{a}_2)$ denote two p_z orbitals in an elementary cell.

Putting

$$\alpha_s = -\alpha_p = |a|, \quad \gamma_s = \tfrac{1}{2} + |a|\delta', \quad \gamma_p = \tfrac{1}{2} - |a|\delta'$$
$$\gamma_{sp} = \tfrac{1}{2}, \quad q = |a|/2, \quad q_0 = \tfrac{1}{2} \tag{132}$$

the equations for the coefficients $d_{m_2,s}$, $d_{m_2,p}$ change into the Eqs. (104) and (110) obtained by a variation procedure for the Goodwin–Artmann's model. For a definition of a see Eq. (83).

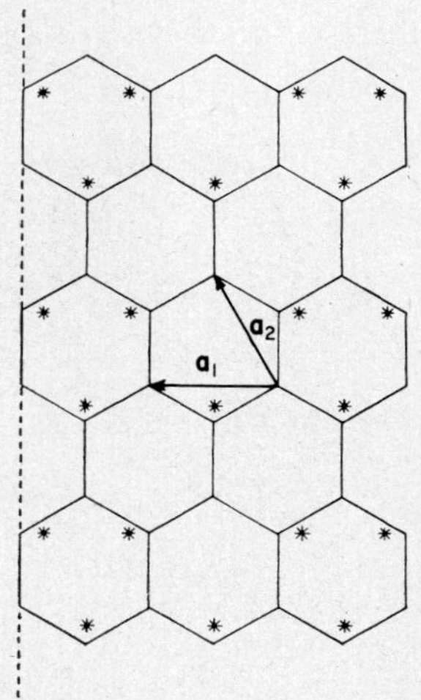

Fig. 11. Semi-infinite two-dimensional graphite crystal limited by a straight line perpendicular to the elementary translation $\mathbf{a}_1$.

From the facts mentioned, it follows that the condition for the existence of Shockley's states in Wallace's model of the graphite lattice is the simultaneous fulfilment of the inequalities

$$|a| < 1, \quad |a|\delta' < \tfrac{1}{2} \tag{133}$$

The energy of a Shockley's surface state is then given by the expression (cf. Fig. 12):

$$W = -2|a|^2\delta' = -4\delta'(\cos \xi_1 + 1) \tag{134}$$

and for the ratio of the coefficients of linear combinations of layer molecular orbitals with and without asterisks respectively, we obtain

$$\lambda = \frac{(1 - 2\delta'|a|)^{\frac{1}{2}} - (1 + 2\delta'|a|)^{\frac{1}{2}}}{(1 - 2\delta'|a|)^{\frac{1}{2}} + (1 + 2\delta'|a|)^{\frac{1}{2}}} \tag{135}$$

We can see therefore that even if in the graphite lattice limited by the line parallel to the translation $\mathbf{a}_1$ further resonance integrals

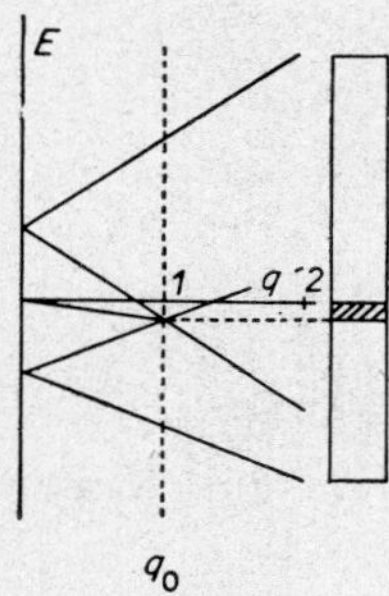

Fig. 12. The energy spectrum of π-electrons in Wallace's model of a semi-infinite two-dimensional graphite crystal limited by a straight parallel to the elementary translation $\mathbf{a}_1$. The straight line in the gap between the bands of energies of the volume states to the left of their intersection corresponds to the surface states. For the definition of q and q_0 see Eq. (132).

are considered, no qualitative changes in the behaviour of Shockley's surface states occur.

(3) *Tamm's States with Consideration of the Resonance Integrals between Next-Nearest Neighbours*[98]

Denoting the resonance integrals between nearest neighbouring layer molecular orbitals built up of the same kind of atoms by β and the resonance integrals between next-nearest layer molecular orbitals by $\delta'\beta$, we have for the energy of a Tamm's surface state in the energy scale used:

$$W = \alpha + |\delta'|^2/\alpha + \alpha/(\alpha - \delta')^2 \tag{136}$$

A condition for the existence of a surface state is the simultaneous fulfilment of the inequalities:

$$|\alpha| > |\delta'|, \quad |\alpha - \delta'|^2 > |\alpha| \tag{137}$$

The Tamm's surface state obviously retains the same character if further interactions are considered as long as the resonance integrals expressing these interactions are not too large.

E. Surface and Chemisorption States on Diamond-Like Crystal Faces

(1) *Surface and Chemisorption States on the Face* (1, 1, 1)[58,65]

A simple model of a diamond-like crystal is obtained by considering at each lattice point of a face-centered cubic lattice four atomic orbitals of the sp^3-hybrid type, their axes of maximum density being directed towards the nearest-neighbouring lattice points. The elementary cell contains two atoms and therefore eight different sp^3-hybrid atomic orbitals. The resonance integrals γ between sp^3-hybrids of neighbouring atoms which have their axes of maximum density in the line connecting these neighbouring atoms are considered to be different from zero. In addition, we consider as non-zero the resonance integrals γ' between sp^3-hybrid atomic orbitals belonging to the same atom. This model (cf. refs. 33, 34 and 92) requires a considerable localization of the bonds between neighbouring atoms, as it neglects a number of interactions. This is justified only when the sp^3-hybrids have a preferential position.

For the calculation it is possible to use as the scale zero the Coulomb integral of an sp^3-orbital and as the scale unit the resonance integral γ.

Let us limit the crystal by a plane parallel to the (1, 1, 1) plane so that the surface of the crystal always interrupts one bond of each surface atom. Because the surface states can be interpreted as special cases of chemisorption states with fully covered surfaces, we shall consider the localized states originating during chemisorption. An sp^3-hybrid belonging to a surface atom and perpendicular to the surface thus creates a bond with the atomic orbital of a chemisorbed atom (Fig. 13). On this atom we consider a single atomic orbital whose Coulomb integral, on the accepted scale, is α. The resonance integral between the orbital of a chemisorbed atom and an sp^3-hybrid perpendicular to the surface is β. The Coulomb integral of an sp^3-hybrid which is perpendicular to the surface and which belongs to a surface atom is denoted by α'.

If $\beta = 0$ we obtain the model of a clean-surface atomic sp^3-orbital, generally directed into a vacuum, whose Coulomb integral is changed, because the properties of this orbital are obviously different from the properties of sp^3-orbitals inside the crystal.

According to the results obtained with simple models we can also expect the occurrence of Tamm's surface states here as long as α' is sufficiently large. In the special case $\alpha' = 0$, it is possible on the contrary to obtain Shockley's surface states as the surface is interrupting strong localized bonds between neighbouring atoms.

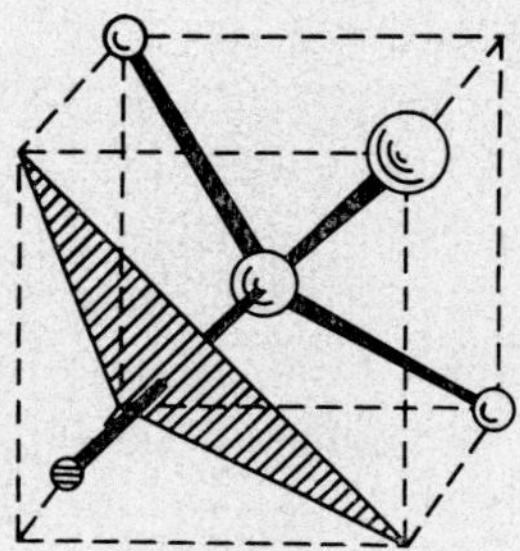

Fig. 13. Diagrammatic representation of chemisorption on the crystal face (1, 1, 1) of diamond.

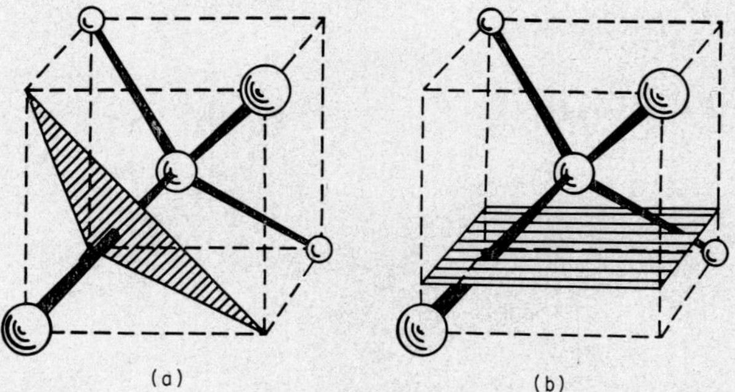

Fig. 14. Positions of the sp^3-wave functions of border atoms in diamond, limited by the planes (1, 1, 1) (a) and (1, 0, 0) (b).

A diamond lattice is created by shifting the elementary cell by whole multiples of the elementary translations $\mathbf{a}_1, \mathbf{a}_2, \mathbf{a}_3$ (Fig. 14). The vectors of the elementary translations $\mathbf{a}_1$, $\mathbf{a}_2$ lie in the plane (1, 1, 1). Let us number the sp^3-hybrids in such a way that the sp^3-orbital which is perpendicular to the plane (1, 1, 1) and directed out of the crystal into the vacuum has the number eight and the sp^3-hybrid which is perpendicular to the surface and directed in the opposite sense has the number one.

Let us consider the wave function in the form of a linear combination of atomic orbitals:

$$\psi_j(\xi_1, \xi_2) = \sum_{p=1}^{8} \sum_{m_3=1}^{\infty} d_{j;p,m_3}(\xi_1, \xi_2) a_{p;\xi_1,\xi_2}(\mathbf{r} - m_3\mathbf{a}_3) + d_j(\xi_1, \xi_2) a_{\xi_1,\xi_2} \tag{138}$$

where

$$a_{p;\xi_1,\xi_2} = K \sum_{m_1,m_2=-\infty}^{+\infty} e^{i(m_1\xi_1+m_2\xi_2)} \chi_p(\mathbf{r} - \sum_{k=1}^{3} m_k\mathbf{a}_k)$$

$$a_{\xi_1,\xi_2} = K \sum_{m_1,m_2=-\infty}^{+\infty} e^{i(m_1\xi_1+m_2\xi_2)} \chi(\mathbf{r} - m_1\mathbf{a}_1 - m_2\mathbf{a}_2 - \mathbf{c}) \tag{139}$$

The χ_p ($p = 1$–8) are sp^3-orbitals and χ is the atomic orbital of the chemisorbed atom being considered. The vector $\mathbf{c}$ measures the distance of the chemisorbed atom from the surface.

On passing from an infinite crystal to a semi-infinite one, when chemisorption is considered, the bond between the layer molecular orbitals $a_8(\mathbf{r} - \mathbf{a}_3)$ and $a_1(\mathbf{r})$ must be broken, the Coulomb integral of the layer molecular orbital $a_8(\mathbf{r} - \mathbf{a}_3)$ changed by the quantity α' and a bond between the layer molecular orbitals $a_8(\mathbf{r} - \mathbf{a}_3)$ and a created. Therefore the equation for the calculation of the coefficients $d_{j;p,m_3}$ and d_j has the following form (cf. Eqs. (20) and (27)):

$$d_{j;p,m_3} = -L_{p,m_3;1,0} d_{j;8,1} + \beta L_{p,m_3;8,1} d_j + \alpha' L_{p,m_3;8,1} d_{j;8,1}$$
$$d_j = \beta L d_{j;8,1} \tag{140}$$

where

$$L_{p,m_3;p',m_3'} = \frac{(-1)^{p+p'}}{2\pi} \int (\Delta_{p'p}/\Delta) e^{i(m_3-m_3')\xi_3} d\xi_3$$
$$L = 1/(W - \alpha) \tag{141*}$$

The determinant Δ in Eq. (141) is defined in the following way:

$$\Delta = \begin{vmatrix} MN^+ \\ NM \end{vmatrix} \tag{142}$$

where

$$j,k = 1,4\colon\ M_{jj} = -W,\ M_{j,k} = \gamma'\ (j \neq k)$$
$$N_{j,5-j} = e^{i\xi_{4-j}},\quad N_{j,k} = 0\ (j \neq 5 - k) \tag{143}$$

* The first equation (141) can easily be proved by a procedure analogous to that used in Section III-A. See also ref. 25.

$\Delta_{pp'}$ is a subdeterminant which can be obtained from Δ by omitting the p-th column and p'-th line. W is the energy of the state localized at the surface.

The equation determining the energy of the volume states is given by

$$\Delta = (1 - x^2)^2(\varphi^2 - \gamma'^2|A|^2) = 0 \tag{144}$$

where

$$\begin{aligned} \varphi &= x^2 - 4\gamma' x - 1 \\ x &= W + \gamma' \\ A &= 1 + \sum_{j=1}^{3} e^{i\xi_j} \end{aligned} \tag{145}$$

The first equation of Eq. (140) for $m_3 = 0$ and the second equation of (140) combined with the second equation of Eq. (141) give the fundamental relation for the calculation of the energy of the states localized at the surface:

$$\beta^2/(W - \alpha) + \alpha' = L_{1,0;1,0}/L_{1,0;8,1} \tag{146}$$

The right-hand side of the preceding relation, which reminds us in form of Eqs. (66) and (72), depends exclusively upon the properties of the undisturbed infinite crystal, but the left-hand side contains parameters which characterize the changes on the surface.

Performing the integration indicated in the definition equations (141) we can change relation (146) into the final form:

$$H = A/(B - C) = \alpha' + \beta^2/(x - \gamma' - \alpha) \tag{147}$$

where

$$A = x(1 - z_1^0 z_2^0), \quad B = 1 - (xz_2^0)^2, \quad C = z_1(x^2 - 1)/|a| \tag{148}$$

We have introduced the following definitions:

$$\begin{aligned} z_1^0 &= (x\varphi + \gamma')/|a|\gamma', \quad z_2^0 = (\varphi + x\gamma')/x|a|\gamma' \\ a &= 1 + e^{i\xi_1} + e^{i\xi_2} \\ \nu &= [\varphi^2 - \gamma'^2(1 + |a|^2)]/|a|\gamma' \end{aligned} \tag{149}$$

The function $z_1(\nu)$ is defined by Eq. (30). The solution of Eq. (147) can be performed graphically. The limiting values of the quantity $|a|$, expressing the dependence upon the wave vectors lying in the surface of the semi-infinite crystal, establish the boundaries of the volume states.

The existence condition for the states localized at the surface follows from the definition of the quantity z_1 and is

$$|\nu| \geqslant 2 \tag{150}$$

Comparing this condition with the expression for the calculation of the energy of a volume state, Eq. (144), we can see that it expresses the necessary requirement for the energies of the localized states to be situated, for a given $|a|$, outside the bands of energies of the volume states at the same $|a|$.

For a clean surface, i.e. for $\beta = 0$, $\alpha' = 0$, the condition necessary for the existence of Shockley's surface states is

$$z_1^0 z_2^0 = 1 \tag{151}$$

This condition is also sufficient as long as $B \neq C$. In Fig. 15 is shown a plot of the function H for $\gamma' = 0.25$ and $|a| = 2$. In the

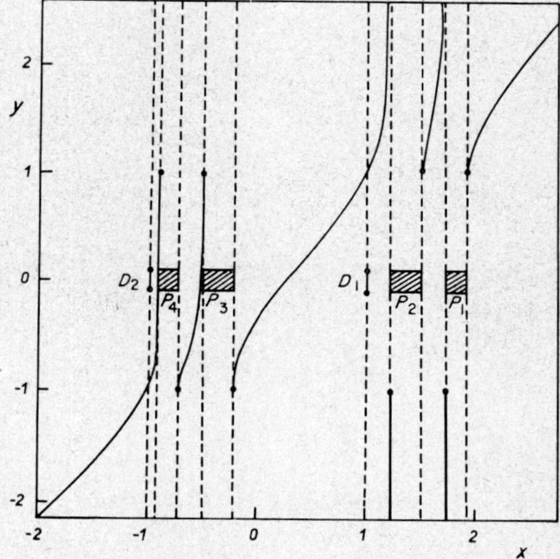

Fig. 15. H (cf. Eq. (147)) as a function of x (cf. Eq. (145)) for $\gamma' = 0.25$ and $|a| = 2$ (solid lines). Limits of the one-dimensional bands of allowed volume states for $\gamma' = 0.25$ and $|a| = 2$ (dashed lines).

same figure the bands of allowed energies of the volume states are plotted with wave vectors ξ_1, ξ_2 corresponding to $|a| = 2$. In the figure we can see the two bands of the bonding states (P_1 and P_2), two bands of the non-bonding states (P_3 and P_4) and the degenerate levels D_1 and D_2. States localized at the surface appear in the gaps between these bands of the partial spectrum for the given value of $|a|$.

In Fig. 16, the limiting values of the function H for $\gamma' = 0.25$ are plotted. As long as in a certain region the boundary is formed

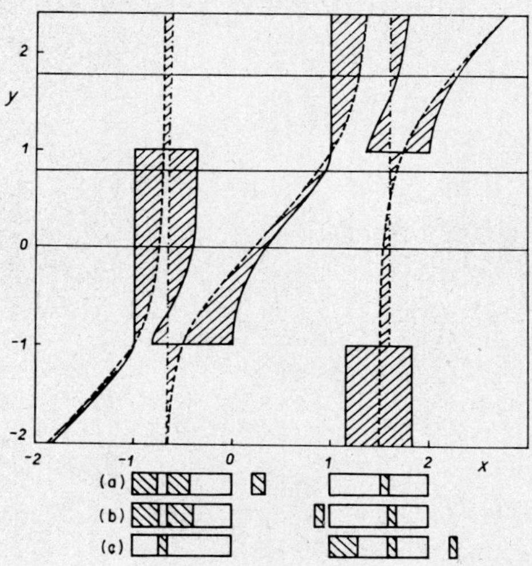

Fig. 16. Surface states on diamond face (1, 1, 1). Function $y = H(x)$ ($\gamma' = 0.25$) for $|a| = 0$ (– – –), $|a| = 3$ (——) and, as far as it forms the borderline delineating the region of allowed surface states, also for $|a| = 1$ (– · – · –). The lower part of the figure shows spectra of surface-state energies for $\beta = 0$: (a) $\alpha' = 0$; (b) $\alpha' = 0.8$; (c) $\alpha' = 1.75$.

by the function H for $|a| = 1$ (and therefore not for $|a| = 0$ or $|a| = 3$). The function H in this region has no physical meaning for $|a| < 1$ or $|a| > 1$ and the second border line is formed by the function H for $|a| = 3$ or $|a| = 0$, respectively. The qualitative change in the regions of possible occurrence of localized states at the surface obtained in this way for $H = 1$ is remarkable. Let us note that with the simple model we are using, the total energy spectrum of the volume states forms two bands: one covering the interval $1 \leqslant x \leqslant 2$ and belonging to the bonding states (the valency band) and the other covering the interval $-1 \leqslant x \leqslant 0$ (the conduction band).

In our diagram the intersections of the abscissa axis with the system of regions of possible occurrence of localized states create the set of energy bands of Shockley's surface states.

From the definition (141) and from the form of the determinant Δ, or in other words from the independence of Δ_{11} on ξ_3, it follows that the condition necessary for the existence of the Shockley's surface states is

$$\Delta_{11} = 0 \tag{152}$$

From the expression for the extension coefficients, Eq. (140), we further obtain the following relation

$$d_{j;1,m_3} = 0 \tag{153}$$

The molecular orbital of a Shockley's surface state is therefore a linear combination of "layer" molecular orbitals which do not interact with each other. These layer molecular orbitals are linear combinations of sp^3-hybrids χ_j, where $j = 2$–8. Here we encounter again a characteristic property of wave functions of Shockley's surface states, namely the alternation of the probability of finding an electron with increasing distance from the surface of a semi-infinite crystal.

Of great interest are the Shockley's states whose energies are in the neighbourhood of $x = 0.25$, or in other words near $W = 0$, i.e. lying inside the energy gap. These states are obviously non-bonding ones, as long as the "bonding character" is ascribed to the strong interaction creating strong, relatively well localized bonds between neighbouring atoms. The properties of the wave functions of surface states can easily be estimated by direct calculation according to Eq. (140) or by a discussion of the relations which are obtained from the usual variation procedure for the extension coefficients of the layer molecular orbitals of the corresponding two-dimensional crystals. As a result, we can see that in the wave function corresponding to the surface state whose energy is in the region of the "energy gap" the electron is localized primarily on the hybrid χ_8. Electrons in surface states whose energies are inside the bands of allowed energies of the volume states are, on the other hand, localized on the sp^3-hybrids χ_j, where $j = 2$–7. The Shockley's surface states whose energies are in the "energy gap" have properties fully analogous to the properties of the Shockley's surface states of simpler models. They are the expression of unsaturated valencies in the surface. The Shockley's surface states whose energies are inside the energy bands of the volume states are the expression of the changes in the bonding properties of the layers (or in other words of the two-dimensional crystals) oriented parallel to the surface and formed by the sp^3-hybrids χ_j, $j = 2$–7.

As long as the change in the Coulomb potential in the region of the sp^3-hybrids of surface atoms is not too large, Shockley's surface states are only modified. According to Eq. (147), we obtain the corresponding surface states for $\beta = 0$ as intersections of the line parallel to the abscissa axis with that set of regions shown in Fig. 16 where surface states can appear. The changes in the position of the energy band of the surface states as well as in their character are only quantitative as long as $|\alpha'| < 1$. On the other hand, if $\alpha' > 1$ or $\alpha' < -1$, respectively, the energy band of the surface states appears below the valency band or above the conduction band respectively. The corresponding surface states are obviously of Tamm's type. In the energy gap between the valency and conduction bands for $|\alpha'| > 1$ surface state energies cannot appear. Similarly, as with simpler models, a large change in the Coulomb integral of the atomic orbital which is nearest to the surface favours the appearance of Tamm's surface states and does not favour the appearance of Shockley's surface states.

Inside the band of the overall energy spectrum of the volume states we can find, even at large absolute values of α', bands of states localized at the surface. These states are related to Tamm's states and Shockley's subsurface states by the condition necessary for their existence, which is a high value of the Coulomb integral of the atomic orbital nearest the surface. For the interpretation of these states, which for the sake of brevity will be called L-states, the results of the theory of heteroanalogues of aromatic hydrocarbons will be of help. If the Coulomb integral of an atomic orbital of a heteroatom differs appreciably from the Coulomb integral of an atomic orbital of carbon, molecular orbitals exist in which an electron has zero probability of being found on the heteroatom. We can expect analogously that for L-states the following condition will hold

$$d_{j;8,m_3} \doteq 0 \tag{154}$$

If we create a semi-infinite crystal from an infinite one by removing the atomic orbitals $\chi_8(r - m_1\mathbf{a}_1 - m_2\mathbf{a}_2)$, the condition necessary for the existence of the surface states is again given by relation (152), and in fact the function F reaches an infinite value if Eq. (151) is fulfilled and at the same time the equation

$$B = C \tag{155}$$

holds (cf. the definition of the function H in Eq. (147)). From the symmetry of the problem towards orbitals χ_1 and χ_8 it follows that in L-states the electron is localized mainly on the sp^3-hybrids χ_1. L-states therefore resemble Shockley's subsurface states but with the difference that their energies are in the region of the valency or conduction bands respectively. This difference is due to the fact that L-states describe the changes of bonds in two-dimensional crystals parallel to the surface, created by the sp^3-hybrids χ_j, where $j = 1$–7.

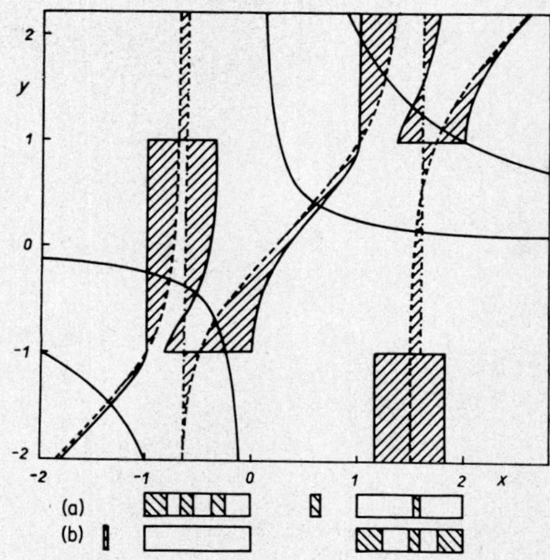

Fig. 17. Chemisorption states on diamond crystal face (1, 1, 1). For caption see Fig. 16. The lower part of the figure shows spectra of chemisorption localized states for $\alpha = \alpha' = 0$: (a) $\beta = 0.5$; (b) $\beta = 2$.

The energies of localized chemisorption states when the surface is completely covered are obtained as intersections of the function H with the hyperbola whose asymptotes are given by the equations

$$x = \gamma' + \alpha, \quad y = \alpha' \tag{156}$$

(cf. Fig. 17). According to the values of the parameters, we obtain chemisorption states of different character. It is clear from Fig. 17 that Shockley's and Tamm's chemisorption states cannot exist simultaneously. However, the simultaneous occurrence of Shockley's states and L-states is not excluded.

The condition for the appearance of Shockley's chemisorption states with energies inside the energy gap is the fulfilment of the inequality

$$(1 - \alpha')(1 - \gamma' - \alpha) > \beta^2 \tag{157}$$

assuming that

$$1 - \gamma' - \alpha > 0 > x_0 - \gamma' - \alpha \tag{158}$$

or the fulfilment of the inequality

$$(1 - \alpha')(1 - \gamma' - \alpha) > \beta^2 > (\alpha' - 1)(x_0 - \gamma' - \alpha) \tag{159}$$

assuming that

$$x_0 - \gamma' - \alpha > 0 \tag{160}$$

In these relations x_0 is the lower boundary of the conduction band (e.g. for $\gamma' = 0.25$, $x_0 = 0$).

As long as the electronegativity of the chemisorbed atoms, expressed by the Coulomb integral, does not differ too much from the electronegativity of atoms of the chemisorbent, the condition necessary for the existence of Shockley's chemisorption states with energies inside the energy gap is the weakness of the bonds between the chemisorbed atoms and the surface atoms of the chemisorbent. If the strength of the chemisorption bond is comparable to the strength of the bonds between the atoms of the chemisorbent, the appearance of localized states with energies lying inside the energy gap is improbable. Finally, if the chemisorption bond is very strong, Tamm's chemisorption states begin to appear with energies outside the system of the valency and conduction bands. We thus obtain full agreement with the qualitative results obtained with simple models. This fact is readily understood as the basic qualitative properties of different kinds of localized states are closely connected with their chemical interpretation.

Using for the diamond crystal $\gamma' = 0.25$,[32,45,88] and taking into account the fact that the C—H bond is usually considered to be stronger than the C—C bond and neglecting the difference in the electronegativities of carbon and hydrogen, our model does not supply Shockley's chemisorption states with energies inside the energy gap (cf. ref. 72).

It is obvious that the surface states created by the breaking of the bonds on the clean surface of the crystal begin to disappear as soon as the unsaturated valencies of the surface are being saturated by the formation of bonds with atoms of the substance to be chemisorbed. As long as these bonds are essentially weaker than the bonds inside the chemisorbent they are, due to their lability, different from the bonds inside the crystal. This difference enables Shockley's chemisorbed states to appear.

If the bonds between the surface atoms of the crystal and the chemisorbed atoms are comparable to or even stronger than the bonds inside the crystal, chemisorption begins to be irreversible: we can often speak about the formation of surface compounds rather than about chemisorption. From these considerations it is clear that Tamm's chemisorption states are important from the viewpoint of the formation of surface compounds and not in the theory of catalysis.

(2) *Shockley's Surface States on the Face* (1, 0, 0)[55,56]

The wave function of an infinite crystal can be written in the form

$$\psi_j(\eta_1, \eta_2, \eta_3) = \sum_{n_3=-\infty}^{\infty} [\sum_{p=1}^{8} c_{j;p}(\eta_1, \eta_2, \eta_3) u_{p;\eta_1,\eta_2}(\mathbf{r} - n_3\mathbf{b})]\, e^{in_3\eta_3} \tag{161}$$

where

$$u_{p;\eta_1,\eta_2}(\mathbf{r} - n_3\mathbf{b}) = \sum_{n_1,n_2=-\infty}^{+\infty} e^{i(n_1\eta_1+n_2\eta_2)} \chi_p(\mathbf{r} - \sum_{k=1}^{3} n_k\mathbf{b}_k) \tag{162}$$

The χ_p are again the sp^3-hybrid orbitals and the numbering of the orbitals is chosen in the same way as in the previous section. The elementary translations $\mathbf{b}_j$ are given by the relations

$$\mathbf{b}_1 = \mathbf{a}_1, \quad \mathbf{b}_2 = \mathbf{a}_2 - \mathbf{a}_3, \quad \mathbf{b}_3 = \mathbf{a}_3 \tag{163}$$

Their significance is shown in Fig. 18, where they are drawn inside the face-centered cubic lattice. The elementary translations $\mathbf{b}_1$, $\mathbf{b}_2$ are in the plane (1, 0, 0). An arbitrary translation by whole

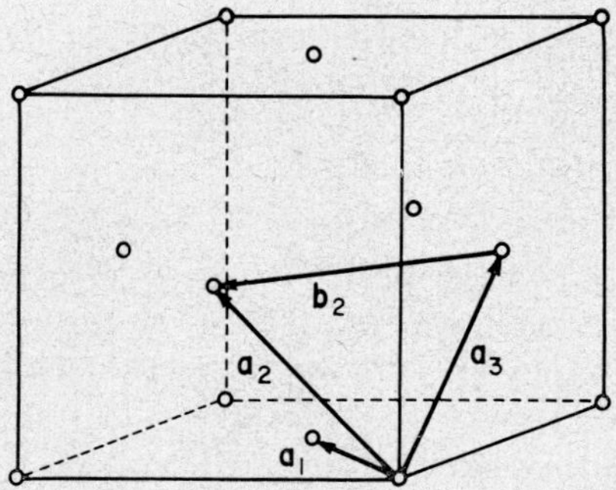

Fig. 18. Meaning of the vectors $\mathbf{a}_1$, $\mathbf{a}_2$, $\mathbf{a}_3$, $\mathbf{b}_2$. The vectors are drawn inside a face-centered cubic lattice.

multiples of the elementary translations can be written in the following way:

$$\sum_{j=1}^{3} m_j \mathbf{a}_j = \sum_{j=1}^{3} n_j \mathbf{b}_j \tag{164}$$

where

$$n_1 = m_1, \quad n_2 = m_2, \quad n_3 = m_2 + m_3 \tag{165}$$

Components of the wave vectors corresponding to the translations b_i are denoted by η_i. These components are related to the components of the wave vectors ξ_j which correspond to the elementary translations a_j. The following equations hold

$$\eta_1 = \xi_1, \quad \eta_2 = \xi_2 - \xi_3, \quad \eta_3 = \xi_3 \tag{166}$$

Applying the general method of Koster and Slater we obtain for the extension coefficients $d_{j;p,n_3}$ of the wave function

$$\psi_j(\eta_2, \eta_3) = \sum_{n_3=1}^{\infty} \sum_{p=1}^{8} d_{j;p,n_3}(\eta_1, \eta_2) u_{p;\eta_1,\eta_2}(\mathbf{r} - n_3 \mathbf{b}_3) \tag{167}$$

for an electron in a semi-infinite crystal limited by the plain (1, 0, 0) the following equations:

$$n_3 \geqslant 0: \; d_{j;p,n_3} = -L_{p,n_3;1,0} d_{j;8,1} - L_{p,n_3;2,0} e^{i\eta_2} d_{j;7,1} \tag{168}$$

In these equations

$$L_{p,n_3;r,n_3+k} = \frac{1}{2\pi}\int_0^{2\pi}\left[\sum_{j=1}^{8}\frac{\bar{c}_{jr}c_{jp}}{(x - x_j)}\right]e^{ik\eta_3}d\eta_3$$

$$= \frac{(-1)^{p+r}}{2\pi}\int_0^{2\pi}\frac{\Delta_{rp}}{\Delta}e^{ik\eta_3}d\eta_3 \tag{169}$$

In this definition the symbols have the same meaning as in Section II-E-(1) except that the determinant Δ and its subdeterminants are functions of η_j. x is again defined by the second relation of Eq. (145), where W is the energy of the surface state. The selection of the energy scale is the same as in Section II-E-(1). η_1 and η_2 are good quantum numbers for surface states and therefore they are not explicitly mentioned.

For the calculation of the energy W we use the condition for solving Eqs. (168) for $p = 1, 2$ and $n_3 = 0$:

$$L_{1,0;1,0}L_{2,0;2,0} = L_{1,0;2,0}L_{2,0;1,0} \tag{170}$$

After carrying out the operations indicated, we obtain two equations suitable for the numerical calculation. Either

$$x = 0 \tag{171}$$

or

$$\frac{\nu_2^2 - 4}{2\nu_2^2}A_1\left\{\left[1 - \left(\frac{2\nu_1\nu_2}{A_1}\right)^2\right]^{\frac{1}{2}} - 1\right\} = \phi^2 + 4 - \nu_1^2 + 2\phi(x + x^{-1}) \tag{172}$$

where

$$A_1 = \phi^2 - \nu_1^2 - \nu_2^2$$

$$1 + e^{i\eta_k} = \nu_k e^{i\kappa}, \quad \nu_k = \bar{\nu}_k, \quad \phi = \frac{\varphi}{\gamma'} \tag{173}$$

φ is defined by the relation (145).

The surface states whose energy is given by Eq. (171) are the states in which the electron is localized in the sp^3-hybrid orbitals χ_7 and χ_8, directed from the surface into the vacuum. The coefficients of the other hybrids are equal to zero. In this state no interaction between the orbitals of neighbouring atoms exists and therefore the corresponding energy level is fully degenerate and independent of the components of the wave vector, η_1 and η_2. As $W = -\gamma'$, the corresponding state is an antibonding one according to the interaction sp^3-hybrids χ_7 and χ_8. In every elementary cell a hybrid orbital is therefore formed having the form (Fig. 20)

$$\phi_{II} = 2^{-\frac{1}{2}}(\chi_7 - \chi_8) \tag{174}$$

The energy level of this state is near the edge of the conduction band of our model; for $\gamma' < 0.25$ or > 0.25 it is below or above this edge respectively.

For the surface states whose energy is given by Eq. (172) the following existence condition holds:

$$|\phi^2 - \nu_1^2 - \nu_2^2| \geqslant 2\nu_1\nu_2 \tag{175}$$

In Fig. 19 is shown an example of the spectrum of a semi-infinite

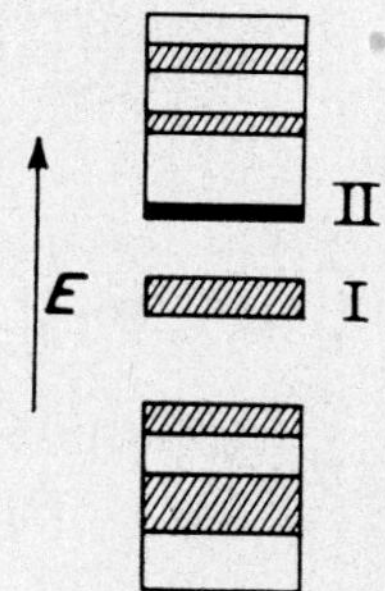

Fig. 19. Energy spectrum of a semi-infinite diamond crystal limited by the plane (1, 0, 0) ($\gamma' = 0.25$).

crystal of diamond-like type limited by the face (1, 0, 0) for $\gamma' = 0.25$. Inside the energy gap an energy band of surface states occurs situated in the vicinity of $x = 2\gamma'$ or $W = \gamma'$.

It is apparent that this energy is just the energy of an isolated hybrid orbital of the form (cf. Fig. 20)

$$\phi_I = 2^{-\frac{1}{2}}(\chi_7 + \chi_8) \tag{176}$$

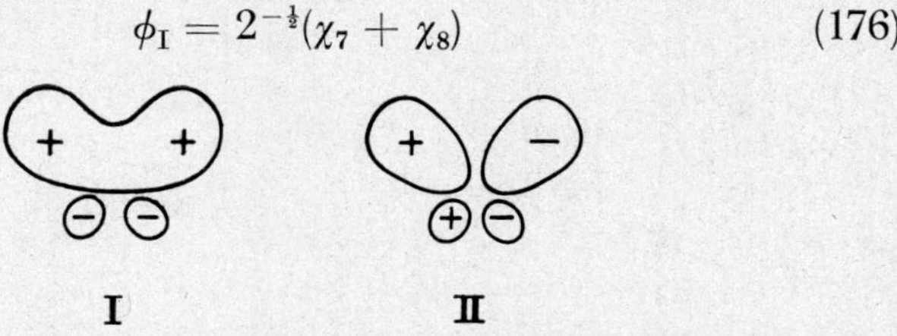

Fig. 20. Diagrammatic representation of hybridized wave functions ϕ_I and ϕ_{II} corresponding to the surface states (cf. Eqs. (174) and (176)).

In these surface states the electron is thus preferentially localized on hybrid orbitals of the form (176) in each elementary cell. The probability of its appearance on other orbitals is lower.

Apart from the surface states which we have already mentioned and discussed in detail, further surface states exist whose energy is inside the allowed energy bands of the volume states.

As long as the number of electrons supplied by each atom of the crystal is equal to four, the energy band of the surface state ϕ_{I} is fully occupied. From the character of the wave function as well as from the fact that the band is fully occupied it follows that the surface state in question cannot be interpreted as the expression of the radical character of the surface. On the hybrids ϕ_{I} are situated electron pairs, analogous to the electron pairs on nitrogen or oxygen. On the surface $(1, 0, 0)$ electron pairs exist, which makes this surface suitable for acceptor–donor reactions. When reacting with atoms of the gaseous phase during chemisorption, valence states may obviously appear in which the radical character connected with the existence of two sp^3-hybrid orbitals directed into the vacuum is renewed. Nevertheless, there is a considerable difference between different faces of the same crystal. This difference is caused however not only by the different orientations of the atoms in the surface (different geometry of the surface layer of atoms) but also by the different electronic structures of different surfaces.

F. Chemisorption of Atoms on the Surface of a Crystal

(1) *Chemisorption of an Atom on an Infinite Two-Dimensional Lattice*

Let us consider a two-dimensional lattice of atoms of one kind. For each atom let us consider only one orbital. The resonance integrals are different from zero only for nearest neighbours which are lying in the direction of the elementary translations. The resonance integrals between nearest neighbours have a single value, which is used as the unit for the energy scale the zero of which is again the Coulomb integral. The energy of a non-localized state is given by the relation

$$E = 2(\cos \xi_1 + \cos \xi_2) \tag{177}$$

An atom denoted by the index 00 is bound to the C atom lying above the atom 00. The parameters characterizing the bond, the electronegativity of the chemisorbed atom and the change in the Coulomb integral of the orbital χ_{00}, will be denoted in the same way as in Section II-B-(1). The equation for the calculation of the energy is obtained in the usual way, according to Koster and Slater (cf. refs. 47, 49, 50):

$$D(W) = (L_{00})^{-1} \tag{178}$$

where[63]

$$L_{00} = \int \chi_{00} \mathscr{L} \chi_{00} d\tau = \frac{2}{\pi W} K\left(\frac{4}{W}\right) \tag{179}$$

and $K(k)$ is the complete elliptical integral of the first kind. The right-hand side of Eq. (178) is equal to zero for $W = 4$, i.e. for the boundary of the energy band of the delocalized states. For $W < -4$, the right-hand side of Eq. (178) increases from $-\infty$ to zero and for $W > 4$, from zero to $+\infty$. From these properties of Eq. (178) it follows that the case where not a single localized state exists is excluded. Existence conditions for the occurrence of different numbers of localized states can be obtained from Table I, where it is necessary to put:

$$\alpha = \alpha' = \alpha'', \quad \rho' = \rho - 4, \quad \rho'' = \rho + 4 \tag{180}$$

(2) *Chemisorption of an Atom on the Surface* (1, 0, 0) *of a Semi-infinite Cubic Crystal*[29,47,51]

For the cubic crystal analogous suppositions are made as for the two-dimensional crystal, so that the energies of the volume states of the infinite crystal in question are given by the equation

$$E = 2 \sum_{j=1}^{3} \cos \xi_j \tag{181}$$

The surface of the semi-infinite crystal is created by the plane $m_3 = 0$. The Coulomb integrals of the surface atoms are generally changed by the quantity c (cf. Eq. (183)).

To an atom of the surface denoted by the index 000 is bound a C atom which is lying above this surface atom. We start from a clean semi-infinite crystal as a sub-system (according to the definitions of Section II-A) and use Koster–Slater's method. With the same notation as before we obtain for the calculation of the energy W of a localized state the following equation:[51]

$$D(W) = (L_{000})^{-1} \tag{182}$$

where

$$L_{000} = L_A + L_B = \int \chi_{000} \mathscr{L} \chi_{000} d\tau$$

$$L_A = \pi^{-2} \int_0^{2\pi} \frac{\sin^2 y}{(1 + c^2 - 2c \cos y)} \frac{4}{(W - 2c \cos y)} \times K \left(\frac{4}{W - 2 \cos y} \right) dy$$

$$|c| < 1: \quad L_B = 0$$

$$|c| > 1: \quad L_B = \frac{1 - c^{-2}}{2\pi W} \frac{4}{(W - c - c^{-1})} K \left(\frac{4}{W - c - c^{-1}} \right) \tag{183}$$

$$c = \int a_0(\xi_1, \xi_2) \mathscr{H}_0 a_0(\xi_1, \xi_2) d\tau - \int a_p(\xi_1, \xi_2) \mathscr{H}_0 a_p(\xi_1, \xi_2) d\tau$$

In these relations we have used (cf. Eq. (18)) the ($p \geqslant 1$) extension coefficients for a semi-infinite crystal according to Artmann.[8] The two forms of the function L_B are related to the existence of surface states of Tamm's type for $|c| > 1$. For large values of W the right-hand side of Eq. (182) increases beyond all limits. At the boundaries of the bands of the surface states the function L_B increases beyond all limits so that the right-hand side of Eq. (182) is equal to zero at the boundaries of the surface states of a finite crystal before chemisorption. At the boundaries of the volume states, however, the right-hand side of Eq. (182) has a finite value.

If surface states do not exist before chemisorption then the conditions summarized in Table I for the existence of localized states exhibited by the chemisorption are valid. We have only to substitute[63]

$$\alpha' = \alpha - \lambda(c), \quad \alpha'' = \alpha + \lambda(-c), \quad \rho' = \rho - 6, \quad \rho'' = \rho + 6 \tag{184}$$

where

$$\lambda(c) = (L_A(W, c))_{W=6} \tag{185}$$

are the values of the function L_A for the energy boundary of the volume states. From a closer analysis it follows that when $|c| < 1$ in general no localized state exhibited by the chemisorption need appear.

More complicated conditions occur if surface states exist in the three-dimensional crystal before chemisorption for $|c| > 1$. It can be seen however that the probability of the formation of localized states exhibited by the chemisorption is considerably increased. This is connected with the fact that at the boundaries of the energy bands of the localized surface states of Tamm's type the right-hand side of Eq. (182) becomes zero. This is also readily understood as we saw that chemisorption localized states must always appear in a two-dimensional crystal, and the wave function corresponding to Tamm's surface state is mainly localized in the "two-dimensional crystal" consisting of surface atoms.

(3) *Interaction between Two Atoms Chemisorbed on the Surface of a Crystal*[51]

Let us consider the semi-infinite cubic crystal of the previous section limited by the plane $m_3 = 0$. With two surface atoms in the cells (000) and (p00) are connected by chemical bonds two atoms from the gaseous phase. For each of them one atomic orbital is taken into account. The resulting system can therefore be considered as composed of two subsystems, one of them consisting of two chemisorbed atoms and the other being a semi-infinite crystal with a clean surface. The two energy levels of the "molecule" creating the first subsystem will be denoted by F_1 and F_2 respectively for the bonding and antibonding states resulting from the mutual interaction of both atoms of this "molecule". It is further assumed that only the resonance integrals between the first chemisorbed atom and the surface atom denoted by (000) and between the second chemisorbed atom and the surface atom denoted by (p00) are non-zero. Both integrals are assumed to have the same value $\beta\sigma$. Finally, denoting the changes in the Coulomb integrals of the chemisorbed atoms by $\beta\nu$ and those in the Coulomb integrals of the neighbouring atoms by $\alpha\beta$, the equations suitable for the estimation of the energy in our energy scale read:

$$[L_{000} + L_{p00}]^{-1} = \alpha + \frac{\sigma^2}{W - \nu - F_1}$$

$$[L_{000} - L_{p00}]^{-1} = \alpha + \frac{\sigma^2}{W - \nu - F_2} \tag{186}$$

$$L_{p00} = \int \chi_{000} \mathscr{L} \chi_{p00} d\tau = \langle \chi_{000} | \mathscr{L} | \chi_{000} \rangle$$

The first equation belongs to the state symmetric to reflection in a plane perpendicular to the surface of the crystal and to the line connecting the two chemisorbed atoms and bisecting this line. The second equation belongs to the antisymmetric state. The quantities L_{000} and L_{p00} are defined as usual according to Eq. (18), taking into consideration the fact that one of the subsystems is a semi-infinite crystal. From a comparison of Eq. (186) with Eq. (178) it follows that the energy of the symmetric state is decreased when compared with the energy of the two isolated chemisorbed atoms. This effect occurs even if the direct mutual interaction of the atoms is negligible ($F_1 = F_2 = \rho - \nu$). Both chemisorbed atoms can mutually interact owing to the electron

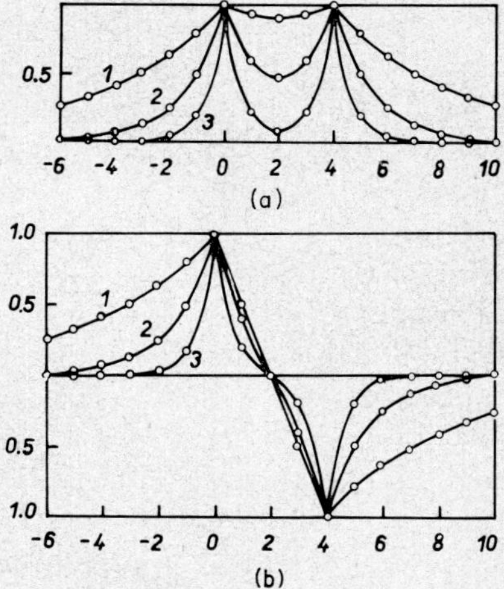

Fig. 21. The bonding (a) and antibonding (b) states in the chemisorption of two atoms on a linear infinite chain. The amplitudes of the wave functions at the atomic orbitals in the chain are plotted for $p = 4$. 1: $z_1 = 0.8$; 2: $z_1 = 0.5$; 3: $z_1 = 0.2$ (cf. Eqs. (30) and (187)).

cloud connecting them through the crystal. A prerequisite for this interaction is, naturally, a certain delocalization of the electron cloud inside the crystal. It is probable that the corresponding wave function is less localized in the neighbourhood of atoms where chemisorption takes place than in the neighbourhood of atoms chemisorbed outside the crystal.

A concrete calculation can easily be performed for the simplified model of the chemisorption of two atoms on an infinite one-dimensional chain of equal atoms using the tight-binding approximation. The equation for the calculation of the energy for $F_1 = F_2$ is

$$D(W) = [z_2(W) - z_1(W)]/[1 \pm z_1^{(p)}(W)] \tag{187}$$

where $z_1(W)$ and $z_2(W)$ are defined in Eq. (30). From the diagram of the extension coefficients of the corresponding wave function in Fig. 21, the bonding and antibonding character of the localized states is evident.

G. Generalized Model of Kronig and Penney

(1) *General Relations Concerning the Equivalence of the MO LCAO Method and of Kronig–Penney's Model*

Kronig–Penney's model[68] in the broadest sense of the term will be called that model in which along the bonds of a crystal we assume a one-dimensional constant potential and in those places where atoms are found the potential has the form of a Dirac function.[16,41] In the study of a finite crystal, a vacuum is represented by a constant (or infinite) potential in the direction given by the broken bond of the surface atom. Thus, we obtain a network composed of sections of constant potential with discontinuities at the joint points.

The potential has the form

$$V = -\sum_j g_j \delta(j) + \sum_{(j-p)} V_{jp} \tag{188}$$

where V_{jp} is the potential which is constant along the $(j - p)$-th branch of the network and is equal to zero outside this branch,

$\delta(j)$ is the Dirac delta function at the j-th joint point, and g_j is the "strength" of the Dirac function, which has a negative value. The branch $j - p$ leads to the j-th atom and is limited by the p-th atom. If at the end of some of the branches an atom is lacking, we put there for simplicity a "pseudo-atom" with $g = 0$.

The solution of Schrödinger's equation for any one of these branches has the form (in atomic units)

$$\psi = \psi_0 e^{\pm[2(V_{jp} - E)]^{\frac{1}{2}} x} \tag{189}$$

where E is the energy and V_{jp} the constant potential of the $(j - p)$-th branch. This energy can be determined from the condition requiring that the wave function be a continuous one[16,23,70,71] (it is natural that at the points of discontinuity it cannot have a derivative). For us however it is more advantageous to use the equivalent method of Frost and Leland,[24] which is formally analogous to the MO LCAO method. According to Frost–Leland's method a one-electron wave function can be written in the form of a linear combination:

$$\psi = \sum_j c_j \psi_j \tag{190}$$

where ψ_j is the DAO (delta atomic orbital) function of the j-th atom. This function is defined on the $(j - p)$-th branch as:

$$\psi_j = \frac{\sinh(c_{jp} x_{jp})}{\sinh(c_{jp} R_{jp})} \tag{191}$$

where x_{jp} is the coordinate on the $(j - p)$-th branch measured from the atom p, R_{jp} is the length of the $(j - p)$-th branch and c_{jp} is defined by the relation:

$$c_{jp} = [2(V_{jp} - E)]^{\frac{1}{2}} \tag{192}$$

Outside the branches arising on the atom p

$$\psi_j = 0 \tag{193}$$

It is obvious that at the atom j this function is equal to unity.

As the function of the form (190) is obviously an exact solution of Schrödinger's equation for suitable E, it is possible to insert it into the characteristic equation of the Hamiltonian operator. The equation thus obtained can be multiplied by the function ψ_j and integrated over the whole network, when we obtain equations of the form[24, 64]

$$c_j H_{jj} + \sum_{\substack{k \\ \text{(neighbours)}}} c_k H_{jk} = 0 \tag{194}$$

where

$$H_{jj} = \int \psi_j (\mathscr{H} - E) \psi_j d\tau = -m_j g_j + \tfrac{1}{2} \sum_{(j-p)} c_{jp} \coth (c_{jp} R_{jp})$$

$$H_{jk} = \int \psi_j (\mathscr{H} - E) \psi_k d\tau = -\frac{1}{2} \frac{c_{jk}}{\sinh (c_{jk} R_{jk})} \tag{195}$$

and m_j is the number of branches arising on the atom j. The relations (194) are formally fully analogous to the equations used in the simple MO LCAO method in the tight-binding approximation. The topology of the bonds determines the points at which non-diagonal elements appear. The character of the bonds determines which of these non-diagonal elements are equal to each other. Diagonal elements which belong to joint points occupied by the same atoms (the same g) and have the same surroundings will be the same.

Contrary to the simple MO LCAO method, the elements considered are not independent of the energy, as can be seen from Eq. (195). But as long as the relations estimating which of the elements H_{jj} or H_{jk} respectively are equal to each other are the same in both methods, it is possible to use the results obtained for Hückel's MO LCAO method directly for the derivation of the relation between these elements. This makes the calculation of the energy possible. The interrelation of both methods follows directly from the form of the matrix elements of the Hamiltonian:

MO LCAO method	Kronig–Penney's model
$\alpha_j - E$	$-m_j g_j + \frac{1}{2} \sum_{(j-p)} c_{jp} \coth (c_{jp} R_{jp})$ (196)
β_{jk}	$-\dfrac{1}{2} \dfrac{c_{jk}}{\sinh (c_{jk} R_{jk})}$

For the surface and chemisorption localized states, the energy of these states and their corresponding existence conditions in Kronig–Penney's model, expressed by means of H_{ij}, have the same form as in Hückel's MO LCAO method.

(2) *A Discussion of Some Models*

(a) *Linear semi-infinite chain with different atoms and different bonds terminating at an atom*[1,31,64,83,97]

Let us number the atoms from the beginning of the chain. The zero of the coordinates is put at atom zero and "atom" -1 is put at $-\infty$. We suppose further (cf. Fig. 22) that

$$\begin{aligned} &R_{2k,2k+1} = x_1, && g_{2k} = g_1\,(k \neq 0), && V_{2k,2k+1} = 0 \\ &R_{2k+1,2k+2} = x_2, && g_{2k+1} = g_2, && V_{2k+1,2k+2} = V_0 \\ &R_{-1,0} = \infty, && g_{-1} = 0, && V_{-1,0} = V_1 \end{aligned} \tag{197}$$

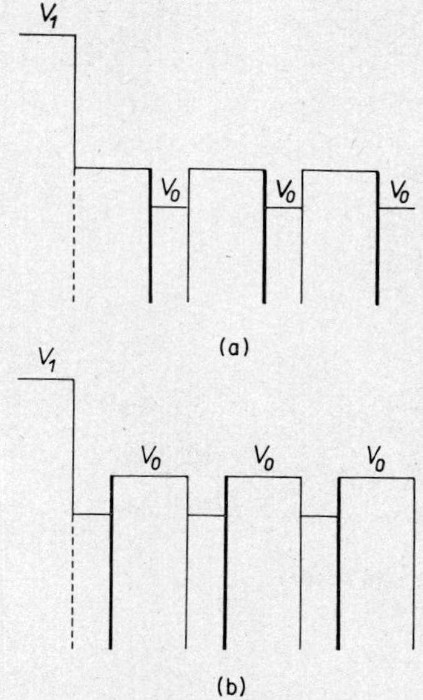

Fig. 22. Diagrammatic representation of the potential in a semi-infinite one-dimensional chain. The chain ends with (a) a weaker, (b) a stronger bond.

According to the general relations (191):

$$
\begin{aligned}
H_{00} &= -g + \varepsilon + \alpha \\
H_{2k,2k} &= -(g + \Delta g) + \varepsilon, \quad H_{2k+1,2k+1} = -(g - \Delta g) + \varepsilon \\
H_{2k,2k+1} &= \frac{-c_1}{2 \sinh (c_1 x_1)} = \beta_2, \quad H_{2k-1,2k} = \frac{-c_2}{2 \sinh (c_2 x_2)} = \beta_1 \\
\varepsilon &= \tfrac{1}{2}(\coth (c_1 x_1) + \coth (c_2 x_2)) \\
\alpha &= (g - g_0) + \beta_2 \cosh (c_2 x_2) + c_3/2 \\
g &= (g_1 + g_2)/2, \quad \Delta g = (g_1 - g_2)/2 \\
c_1 &= (-2E)^{\frac{1}{2}}, \quad c_2 = [2(V_0 - E)]^{\frac{1}{2}}, \quad c_3 = [2(V_1 - E)]^{\frac{1}{2}}
\end{aligned}
\tag{198}
$$

Following the general procedure suggested in the previous section, we immediately obtain from the results of Section II-B-(2a) the following relations:

The energy of the volume states is given by:

$$g = \varepsilon \pm [(\Delta g)^2 + \beta_1^2 + \beta_2^2 + 2\beta_1\beta_2 \cos \xi]^{\frac{1}{2}} \tag{199}$$

The energy of the surface states is given by:

$$g - \varepsilon = (2\alpha)^{-1} \cdot \{\alpha^2 + \beta_2^2 \pm \operatorname{Sign} (\beta_1\beta_2) \cdot [(\beta_2^2 - \alpha^2 + 2\alpha\Delta g)^2 + 4\beta_1^2\alpha^2]^{\frac{1}{2}}\} \tag{200}$$

For positive energies of the electron the quantities β_1 and β_2 can naturally have different signs, which is impossible in the tight-binding method. For the surface states given by Eq. (200) with the upper sign, the existence condition is

$$\alpha^2 - \beta_2^2 > \frac{2\alpha\Delta g}{1 + |\beta_1/\beta_2|} \tag{201}$$

This condition, when $\Delta g = 0$, changes into:

$$\alpha^2 > \beta_2^2 \tag{202}$$

For the surface states given by Eq. (200) with the lower sign, the existence condition has the form:

$$|\beta_2/\beta_1| < 1: \quad \alpha^2 - \beta_2^2 > \frac{2\alpha\Delta g}{|\beta_1/\beta_2| - 1} \tag{203}$$

or

$$|\beta_2/\beta_1| > 1: \quad \alpha^2 - \beta_2^2 < \frac{2\alpha\Delta g}{1 - |\beta_1/\beta_2|} \tag{204}$$

For $\Delta g = 0$ these relations are simplified to:

$$|\beta_1| > |\beta_2|, \quad |\alpha| > |\beta_2| \tag{205}$$

or

$$|\beta_1| < |\beta_2|, \quad |\alpha| < |\beta_2| \tag{206}$$

It is evident that a complete analogy is obtained with Tamm's and Shockley's surface states and Shockley's subsurface states in the MO LCAO method. From this analogy it directly follows that the behaviour of wave functions of particular kinds of states is the same as that of the analogous states in the simple MO LCAO method.[64]

If the energies of an electron are lower than the constant potential which characterizes the lower of the barriers between the atoms, which means that

$$E < \min\,(0, V_0) \tag{207}$$

The inequalities between the quantities $|\beta_1|$ and $|\beta_2|$ are given by the character of the barriers between the atoms, i.e. by the character of the bonds between these atoms. If the "bond" is stronger, the length of the barrier is smaller and its height is lower, and therefore the quantity β is larger irrespective of the energy of the electron as long as condition (207) holds.

The parameter α depends upon the energy in a complicated way and therefore it is impossible to expect that in the whole region of negative energies one of the inequalities (201), (203) or (204) will hold. Nevertheless, for bonding states for which the inequality (207) holds and where we can expect the analogy with the results of simple MO LCAO method to be valid, the fundamental interpretation of the surface states based on the character of the bond broken is still maintained.[57]

For energies higher than the energy of the barriers between atoms, all kinds of surface states can appear. Because of the periodic behaviour of the quantities α and $\beta_{1,2}$, in some regions of the spectrum the conditions for the occurrence of Tamm's states are fulfilled and in other regions the conditions for Shockley's states.

(b) *Linear semi-infinite chain with different atoms and bonds terminating in the region of the barrier between atoms*[2]

The cases solved by the simple MO LCAO method can also be taken over for this model. If we denote the distance between the interruption of the crystal and the border atom by x_3, the potential between the border atom and the interruption mentioned, which is chosen to be the zero of the coordinate scale, by V_2, and finally the potential for negative values of x by V_1, we select the DAO function for the points $x = 0$ and $x = x_3$ in the following way

$$x < 0: \ \psi_0 = e^{-c_3 x}, \quad 0 < x < x_3: \ \psi_0 = \frac{\sinh [c'(x_3 - x)]}{\sinh (c'x_3)}$$

$$0 < x < x_3: \ \psi_1 = \frac{\sinh (c'x)}{\sinh (c'x_3)}, \quad x_3 < x < x_3 + x_j:$$

$$\psi_1 = \frac{\sinh [c_j(x_j - x)]}{\sinh (c_j x_j]} \qquad (208)$$

where $c' = [2(V_2 - E)]^{\frac{1}{2}}$ and j is equal to 1 or 2 if the potential in the interval $x_3 < x < x_3 + x_j$ is 0 or V_0 respectively. The DAO function for the remaining atoms of the semi-infinite chain will have the same form as in the previous section. Only the matrix elements H_{00}, H_{11}, H_{01} of the Hamiltonians of the new DAO are different from the matrix elements occurring in the original infinite chain.

Putting

$$H_{00} = (\rho - \bar{\alpha} - W)\beta, \quad H_{11} = (\gamma - \bar{\alpha} - W)\beta, \quad H_{01} = \sigma\beta \quad (209)$$

we obtain a complete analogy with the chemisorption case of Section II-B-(3b). In an analogous way we could directly obtain the results for the "chemisorption" at the end of the chain studied in the previous section.

We can conclude that even in these cases the character of the localized states with energies lower than the heights of the barriers between atoms (i.e. for strongly bonded electrons) is the same in Kronig–Penney's model as in the MO LCAO method.

(c) *Shockley's surface states in graphite*[67]

For a finite graphite lattice limited by the straight line $a = 0$ two models will be chosen. Both form networks, model A according to Fig. 10 and model B according to Fig. 23. In this second

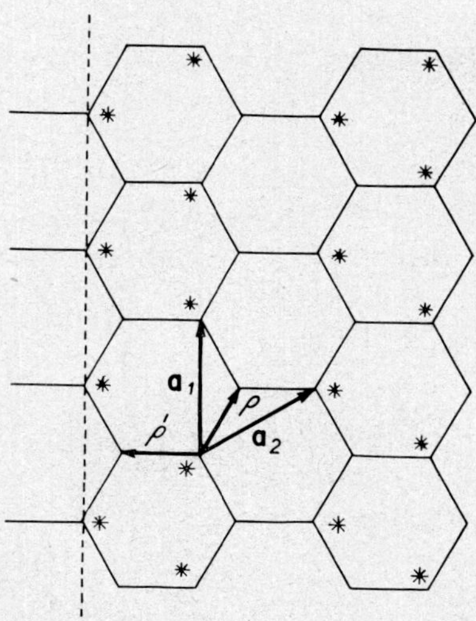

Fig. 23. Model B for the surface of a graphite lattice. The surface is perpendicular to the translation $\mathbf{a}_1$.

model, branches perpendicular to the surface and of length equal to the distance R between two atoms inside the lattice are connected to the surface atoms. The same constant potential is assumed on all branches of both models. In model B all joint points are equivalent so that $\alpha = 0$. In both models $\beta_1 = \beta_2|a|$ (cf. Section II-C-(2b)).

From these considerations for the Shockley's states with negative energy it follows in the usual way that $H_{ii} = 0$, i.e.

$$\tanh (cR) = c/g \tag{210}$$

From the existence condition (206) for Shockley's surface states follows the necessity of fulfilling Eq. (87).

In model A, apart from fulfilling condition (87), it is necessary for the second relation of Eq. (206) to be valid as the m_j for surface atoms is different from that for atoms inside the lattice and therefore $\alpha \neq 0$.

By a simple discussion of the equation for the calculation of the energy, Eq. (200), where the left-hand side is multiplied by $\frac{2}{3}$, we can easily establish that this condition cannot be satisfied at the same time as condition (87). As in model B, when the surface atoms differ from those inside the lattice a considerable disturbance occurs inside the lattice. This disturbance prevents the existence of Shockley's surface states in model A. However, subsurface states occur if the following condition is fulfilled

$$|a| = 2|\cos(\xi_1/2)| > 1 \tag{211}$$

III. CALCULATION OF CHEMISORPTION ENERGY

A. General Formulae[60]

When passing from an infinite crystal to a semi-infinite or a finite one, or equally to a crystal with chemisorption on its surface, the quasicontinuous energy spectrum of the volume states is at first sight not changed. In reality, however, each of the separate levels of the energy band of the volume states is subject to small changes which may, with regard to the great number of levels, lead to changes in the total energy of an electron. Shifts due to these changes are comparable with the width of the occupied bands considered.

There are two possible ways of calculating this energy shift. The first way is to calculate the energies of the volume states for a great number of atoms in the crystal and to obtain the total energy of an electron in the system by a convenient limiting transition and by integration.[10,42,44,81] The second way is to calculate the difference between the electron energies in the two systems directly. For this calculation, the integral formulae derived by Coulson and Longuet-Higgins[15–18] are, after some rearrangements, very useful.

The difference between the electronic energies of two systems can be written in Hückel's MO LCAO approximation in the following way:[15–18]

$$\Delta\varepsilon = \sum_j \sum_{k=1}^{n_j} \Delta\alpha_{jk} + \frac{\beta}{\pi}\int_{-\infty}^{+\infty} \ln\,[\Delta(iy)/\Delta_0(iy)]dy \tag{212}$$

where n_j stands for the number of atomic orbitals $|j, \xi_k\rangle$ forming a basis of P_0^j.

In Eq. (212) β is the energy unit chosen and $\Delta\alpha_{jk}$ means the difference between the Coulomb integrals:

$$\Delta\alpha_{jk} = \langle j, \xi_k|\mathscr{V}|j, \xi_k\rangle \tag{213}$$

$\Delta(W)$, $\Delta_0(W)$ are the determinants:

$$\begin{aligned}\Delta(W) &= |\langle j, \xi_k|\mathscr{H} - W\mathscr{I}|l, \xi_m\rangle| \\ \Delta_0(W) &= |\langle j, \xi_k|\mathscr{H}_0 - W\mathscr{I}|l, \xi_m\rangle|\end{aligned} \tag{214}$$

$\Delta(W)$ has the form

$$|\langle j, \xi_k|\mathscr{P}(\mathscr{H} - W\mathscr{I})\mathscr{P}|l, \xi_m\rangle| \times |\langle j, \xi_k|\mathscr{S}(\mathscr{H} - W\mathscr{I})\mathscr{S}|l, \xi_m\rangle| \tag{215}$$

Let us introduce the determinant

$$\Delta'(W) = |\langle j, \xi_k|\mathscr{S}|\mathscr{H}_0 - W\mathscr{I})\mathscr{S} + \mathscr{P}(\mathscr{H} - W\mathscr{I})\mathscr{P} + \mathscr{Q}\mathscr{H}_0\mathscr{S}|l, \xi_m\rangle| \tag{216}$$

in which the addition of the operator $\mathscr{Q}\mathscr{H}_0\mathscr{S}$ does not change the value of $\Delta(W)$ and the addition of the operator $\mathscr{S}\mathscr{V}\mathscr{S}$ changes its value in the ratio $D(W)/(-W)^s$, where

$$D(W) = |\langle j, \xi_k|\mathscr{S}(\mathscr{H}_0 - W\mathscr{I})\mathscr{S}|l, \xi_m\rangle| \tag{217}$$

and s is the number of atomic orbitals in S:

$$\Delta(W) = \Delta'(W)\,\frac{(-W)^s}{D(W)} \tag{218}$$

We can write

$$\begin{aligned}\Delta'(W)/\Delta_0(W) &= |\langle j, \xi_k| - \mathscr{L}(W)[\mathscr{S}(\mathscr{H}_0 - W\mathscr{I})\mathscr{S} \\ &\quad + \mathscr{P}(\mathscr{H}_0 - W\mathscr{I})\mathscr{P} + \mathscr{Q}\mathscr{H}_0\mathscr{S} + \mathscr{Q}\mathscr{V}\mathscr{Q}]|l, \xi_m\rangle| \\ &= |\langle j, \xi_k|\mathscr{P} + \mathscr{S} + \mathscr{L}\mathscr{P}\mathscr{H}_0\mathscr{S} + \mathscr{L}\mathscr{S}\mathscr{H}_0\mathscr{P} \\ &\quad - \mathscr{L}\mathscr{Q}\mathscr{V}\mathscr{Q} - \mathscr{L}\mathscr{Q}\mathscr{H}_0\mathscr{S}|l, \xi_m\rangle|\end{aligned} \tag{219}$$

and by means of Eqs. (17), (18) and (23) we finally obtain:

$$\Delta'(W)/\Delta_0(W) = |\langle j, \xi_k|\mathscr{Q}(\mathscr{I} + \mathscr{L}\mathscr{S}\mathscr{H}_0 - \mathscr{L}\mathscr{Q}\mathscr{V})\mathscr{Q}|l, \xi_m\rangle| \tag{220}$$

In the special case $\mathscr{Q}\mathscr{V}\mathscr{Q} = \mathscr{O}$, we can use the relationship (cf. the definition of the operator $\mathscr{L}$):

$$\mathscr{L}(W)\mathscr{H}_0 = -\mathscr{I} + W\mathscr{L} \tag{221}$$

in the equation

$$\begin{aligned}\Delta(W)/\Delta_0(W) = |\langle j, \xi_k| - \mathscr{L}(\mathscr{H}_0 - W\mathscr{I} - \mathscr{H}_0\mathscr{S} \\ - \mathscr{S}\mathscr{H}_0\mathscr{P} + \mathscr{S}\mathscr{H}_0\mathscr{P})|l, \xi_m\rangle| = |\langle j, \xi_k|\mathscr{I} \\ + \mathscr{L}\mathscr{H}_0\mathscr{S}|l, \xi_m\rangle| = (W)^s|\langle j, \xi_k|\mathscr{S}\mathscr{L}\mathscr{S}|l, \xi_m\rangle|\end{aligned} \tag{222}$$

Using the relations mentioned we can calculate the energy of chemisorption in two steps: we can calculate the difference in energy between the finite or semi-infinite crystal and an infinite crystal and then the difference in energy between the finite crystal in interaction with the chemisorbed substance and the energy of an infinite crystal. The difference between these two quantities is equal to the electronic contribution to the chemisorption energy.

B. Simple Examples of the Calculation of Chemisorption Energy

(1) *Chemisorption at the End of a Semi-infinite Atomic Chain*

The chain has the same properties as that in Section II-B-(1). The symbols are also the same as those used in this section.

In the first step we calculate the electronic contribution to the energy necessary to break the cyclic infinite chain by removing one of the atoms of type A:

$$\begin{aligned}\beta^{-1}A_{\mathrm{A}} &= -\gamma - \frac{1}{\pi}\int_{-\infty}^{+\infty} \ln\,[iyL_{\mathrm{AA}}^{00}(iy)]dy \\ &= -\gamma - \frac{1}{\pi}\int_{-\infty}^{+\infty} \ln\left[\frac{\gamma - iy}{[(y^2 + 1 + \delta^2 + \gamma^2)^2 - 4\delta^2]^{\frac{1}{2}}}\right] dy\end{aligned} \tag{223}$$

Performing the integration in Eq. (223) we obtain

$$\beta^{-1}A_{\mathrm{A}} = -\gamma - |\gamma| + [\gamma^2 + (\delta - 1)^2]^{\frac{1}{2}} + [\gamma^2 + (\delta + 1)^2]^{\frac{1}{2}} \tag{224}$$

In particular, when $\gamma = 0$, $\delta \neq 1$ we have

$$A_{\mathrm{A}} = 2\beta_m, \quad \beta_m = \mathrm{Max}\,(\beta, \beta') \tag{225}$$

The energy necessary to remove a B atom is obtained by interchanging α_A and α_B. We then calculate the difference between the total electronic energy of a semi-infinite chain built up of atoms of the same kind joined by the same bonds ($\delta = 1$, $\gamma = 0$), with an atom chemisorbed at one end, and the total electronic energy of an infinite chain. Again the symbols are the same as before. If we create the semi-infinite chain by breaking one bond, this energy difference is given by (cf. Eqs. (220) and (27)):

$$\Delta\varepsilon' = \alpha + \frac{1}{\pi}\int_{-\infty}^{+\infty} \ln \Lambda(iy)dy \tag{226}$$

where:

$$\Lambda(iy) = \begin{vmatrix} 1 + L_1(iy) - \left(\alpha + \dfrac{\sigma^2}{iy - \rho}\right) L_0(iy), & L_0(iy) \\ L_0(iy) - (\alpha + \left(\dfrac{\sigma^2}{iy - \rho}\right) L_1(iy), & 1 + L_1(iy) \end{vmatrix} \tag{227}$$

$$L_0(iy) = \frac{\text{Sign}\ (y)}{i(y^2 + 4)^{\frac{1}{2}}} \quad L_1(iy) = \frac{|y| - \sqrt{y^2 + 4}}{2\sqrt{y^2 + 4}}$$

After a simple rearrangement we have

$$\Delta\varepsilon' = \alpha + \frac{1}{\pi}\int_0^{\infty} \ln \{[\![[(y^2 + 4)^{\frac{1}{2}} + y][y^2 + \rho^2] + 2y\sigma^2]\!]^2 + 4[\alpha(y^2 + \rho^2) - \rho\sigma^2]^2\}dy - \frac{1}{\pi}\int_0^{\infty} \ln \{4(y^2 + 4)(y^2 + \rho^2)^2\}dy \tag{228}$$

If $\rho = \alpha = 0$ (cf. Section II-B-(3a)) it is easy to perform the integrations giving

$\sigma \leqslant 1$:

$$\Delta\varepsilon' = -2 + \frac{4}{\pi}\left\{1 + \frac{\sigma^2}{2(1 - \sigma^2)^{\frac{1}{2}}} \ln \left[\frac{1 + (1 - \sigma^2)^{\frac{1}{2}}}{1 - (1 - \sigma^2)^{\frac{1}{2}}}\right]\right\} \tag{229}$$

$\sigma \geqslant 1$:

$$\Delta\varepsilon' = -2 + \frac{4}{\pi} + \frac{2\sigma^2}{(\sigma^2 - 1)^{\frac{1}{2}}}\left[1 - \frac{2}{\pi} \text{arc cot } (\sigma^2 - 1)^{\frac{1}{2}}\right]$$

From the relations derived above we have for the energy of chemisorption for our simple model (cf. Eq. (225)):

$$\varepsilon = \Delta\varepsilon' + A - R + |\rho| = \Delta\varepsilon' + 2 - \frac{4}{\pi} + |\rho| \tag{230}$$

where R is the difference between the energy of a chain of $N + 1$ atoms and the energy of a chain of N atoms if $N \to \infty$.[44] For the case when $\sigma = 1$, $\rho = 0$, $\alpha \neq 0$, we obtain

$$\Delta\varepsilon' = \alpha + \frac{2}{\pi} + (|\alpha| + |\alpha|^{-1})\left[1 - \frac{2}{\pi}\text{ arc tan }(|\alpha|^{-1})\right] - 2 \tag{231}$$

The difference between the energies of two semi-infinite chains whose end atoms have Coulomb integrals α and 0, respectively, is (cf. ref. 10)

$$\begin{aligned}\varepsilon &= \Delta\varepsilon' + 2 - \frac{4}{\pi} \\ &= \alpha - \frac{2}{\pi} + (|\alpha| + |\alpha|^{-1})\left[1 - \frac{2}{\pi}\text{ arc tan }(|\alpha|^{-1})\right]\end{aligned} \tag{232}$$

(2) *Chemisorption on a One-Dimensional Infinite Chain* ($\delta = 1$, $\gamma = 0$)

Using the symbols of Section II-B-(4) the difference between the electronic energy of the system in which the interaction between the chemisorbed atom and the chain is considered and the electronic energy of the system in which this interaction is not considered is:

$$\Delta\varepsilon = \alpha + |\rho| + \frac{1}{\pi}\int_{-\infty}^{+\infty} \ln \left[\begin{vmatrix} 1 - \alpha L_0(iy), & -\sigma L_0(iy) \\ -\dfrac{\sigma}{iy - \rho}, & 1 \end{vmatrix}\right] dy \tag{233}$$

After a simple rearrangement this equation takes the form:

$$\begin{aligned}\Delta\varepsilon &= \alpha + |\rho| - \frac{1}{\pi}\int_0^\infty \ln\,[(\rho^2 + y^2)^2(y^2 + 4)]dy \\ &+ \frac{1}{\pi}\int_0^\infty \ln\,\{[(\rho^2 + y^2)(y^2 + 4)^{\frac{1}{2}} + \sigma^2 y]^2 + [\alpha(\rho^2 + y^2) - \sigma^2\rho]^2\}dy\end{aligned} \tag{234}$$

If $\rho = \alpha = 0$, $\sigma \neq 0$, it is easy to perform the integrations to give

$$\Delta\varepsilon = -2\left\{1 + \frac{1}{\pi}(\nu_1 - \nu_2)\ln\frac{1+\nu_1}{1-\nu_1} - (\nu_1 + \nu_2)(1 - \frac{2}{\pi}\operatorname{arcot}\nu_2)\right\} \tag{235}$$

where

$$\nu_{1,2}^2 = (1 + \sigma^4/4)^{\frac{1}{2}} \mp \sigma^2/2 \tag{236}$$

The energy divided by the number of chain atoms of the infinite chain to each atom of which one "chemisorbed" atom is bound ($\sigma \neq 0$, $\rho = \alpha = 0$) is given by

$$\varepsilon' = \frac{4}{\pi}\left[(\sigma^2 + 1)^{\frac{1}{2}}E((1 + \sigma^2)^{-\frac{1}{2}}) - 1\right] \tag{237}$$

where $E(k)$ is a complete elliptic integral of the second kind.

It is clear that the following inequality holds for $\rho = \alpha = 0$

$$\varepsilon' > \Delta\varepsilon \tag{238}$$

If the Coulomb integral of the chemisorbed atom is very large an opposite effect can occur. In the limiting case, if $|\rho| \rightarrow \infty$

$$\Delta\varepsilon = |\rho|, \text{ whilst } \varepsilon' = |\rho| - R = |\rho| - 4/\pi.$$

IV. CONCLUDING REMARKS

In this survey the main emphasis has been placed on those results of the quantum chemistry of surfaces which can be used as starting points for the qualitative description of a surface in terms of quantum chemistry. Thus, the methods to be reviewed were determined with this in mind. Papers using methods by which no results permitting a clear and illustrative interpretation have so far been obtained were not mentioned in further detail. Nevertheless, it can be expected that in the future development these more complicated treatments will be of great importance. Methods using orthogonalized planar waves[3,7] and spin-polarized orbitals,[5] methods used for the solution of many-electron problems in the theory of solids[13,80] and also methods used in crystal field theory[19] have been applied to surface problems. Equally the method of nearly free electrons and consideration of Kronig–

Penney's model with Dirac delta functions, describing the barrier between elementary cells, were of historical importance in the development of the theory of surface states.

As far as the study of the occurrence and interpretation of localized states at a surface is concerned the possibilities of the simplest methods seem to be exhausted. Tests to study these states in crystals of real substances must mainly be performed by more sophisticated methods.

The interpretation of the localized states is to a large extent independent of the method used. It can be expected that the main features of surface and chemisorption states will be preserved at least in the one-electron approximation and especially using the MO LCAO method. This is the reason why judgement concerning the existence and properties of these states can be made on the basis of the chemical properties of solids, provided electrons in the crystal do not have too high an energy.

When interpreting surface states with loosely bound electrons (i.e. nearly free electrons), the situation becomes less clear and illustrative. The interpretation of these surface states in chemical terms is not simple, which, considering the nature of these terms, is natural. Nevertheless, the general classification of surface states remains valid even for higher electron energies in a crystal, as shown by the generalized Kronig–Penney's model. In this case the probability of the existence of surface states is higher than with more strongly bound electrons.

Relatively little work has been done on the calculation of the total electronic energy of interaction between the surface and atoms or molecules. The question arises in this respect as to whether simpler methods could supply qualitative results which would be interesting from a general point of view.

The use of more complicated methods, including the MO LCAO method, using a considerable number of resonance integrals and assuming several atomic orbitals on one atom could contribute towards the problems of the geometry of surface layers.[20,36,37,89] Changes in the hybridization[35] and bond strength of surface atoms can be used to estimate changes in geometry in these layers.

It would also be necessary to check in more detail the reasoning concerning the influence of the existence of states localized at the surface upon chemisorption and catalytic processes. Attempts which have been made in this direction did not take into account the specific properties of surface and chemisorption states of different kinds when applying the usual methods of semiconductor theory. It is nevertheless necessary to respect the fact that the existence of free valencies on the surface manifested by Shockley's surface states can lead to the formation of surface compounds. It is therefore questionable whether the quantum chemistry of clean surfaces is not nearer to the chemistry of these compounds than to chemisorption and catalysis. But if these compounds were important in chemisorption and catalysis,[86] the theory of these two topics would have to start with the study of chemisorption compounds. A quantum mechanical basis for this study is given even by the quantum chemistry of solid surfaces.

References

1. Aerts, E., *Physica* **26,** 1047, 1057 (1960).
2. Aerts, E., *Physica* **26,** 1063 (1960).
3. Agrain, P. and Dugles, C., *Z. Elektrochem.* **56,** 363 (1952).
4. Amos, A. T., and Davison, S. G., *Physica* **30,** 905 (1964).
5. Amos, A. T., and Davison, S. G., unpublished results.
6. Antončík, E., *Proc. Intern. Conf. Semicond. Phys. Prague,* 1960, 1961, p. 491.
7. Antončík, E., *Phys. Chem. Solids* **21,** 137 (1961).
8. Artmann, K., *Z. Physik* **131,** 244 (1952).
9. Baldock, G. R., *Proc. Cambridge Phil. Soc.* **48,** 457 (1952).
10. Baldock, G. R., *Proc. Phys. Soc. London* **A66,** 2 (1953).
11. Bayliss, N. S., *Quart. Rev. London* **56,** 319 (1952).
12. Bonch-Bruevich, V. L., *Zh. Fiz. Khim.* **25,** 1033 (1951).
13. Bonch-Bruevich, V. L., and Glasko, V. B., *Vest. Mosk. Univ. Ser. Mat. Mekh. Astron. Fiz. i Khim.* **9,** 1 (1958).
14. Center, E. M., *Zh. Eksperim. i Teor. Fiz.* **8,** 682 (1938).
15. Coulson, C. A., *Proc. Cambridge Phil. Soc.* **36,** 201 (1940).
16. Coulson, C. A., *Proc. Phys. Soc. London* **A68,** 1129 (1955).
17. Coulson, C. A., and Baldock, G. R., *Discussions Faraday Soc.* **8,** 27 (1950).
18. Coulson, C. A., and Longuet-Higgins, H. C., *Proc. Roy. Soc. London* **A191,** 39 (1947).
19. Dowden, D. A., and Wells, D., *Actes du II-ème congrès international de catalyse,* 1960, 2nd Ed., Technip, Paris, 1961, p. 1499.
20. Farnworth, R. E., Schlier, R. E., and Dillon, J. A., *Phys. Chem. Solids* **8,** 116 (1959).
21. Feuer, P., *Phys. Rev.* **88,** 92 (1952).

22. Fowler, R. H., *Proc. Roy. Soc. London* **A141,** 56 (1933).
23. Frost, A. A., *J. Chem. Phys.* **23,** 985 (1955).
24. Frost, A. A., and Leland, F. E., *J. Chem. Phys.* **25,** 1154 (1956).
25. Fukui, K., Nagata, Ch., Yonezawa, T., Kato, H., and Morokuma, K., *J. Chem. Phys.* **31,** 287 (1959).
26. Goodwin, E. T., *Proc. Cambridge Phil. Soc.* **35,** 205 (1939).
27. Goodwin, E. T., *Proc. Cambridge Phil. Soc.* **35,** 221 (1939).
28. Goodwin, E. T., *Proc. Cambridge Phil. Soc.* **35,** 232 (1939).
29. Grimley, T. B., *Proc. Phys. Soc. London* **B72,** 103 (1958).
30. Grimley, T. B., *Advan. Catalysis* **12,** 1 (1960).
31. Grimley, T. B., and Holland, B. W., *Proc. Phys. Soc. London* **78,** 217 (1961).
32. Hall, G. G., *Proc. Roy. Soc. London* **A205,** 541 (1951).
33. Hall, G. G., *Phil. Mag.* **43,** 338 (1952).
34. Hall, G. G., *Phys. Rev.* **90,** 317 (1953).
35. Handler, P., in *Semiconductor Surface Physics*, University of Pennsylvania Press, Philadelphia, p. 23; *Semiconductor Surfaces*, Pergamon Press, 1960, p. 1.
36. Haneman, D., *Phys. Chem. Solids* **14,** 162 (1960).
37. Haneman, D., *Phys. Rev.* **121,** 1093 (1961).
38. Hauffe, K., *Advan. Catalysis* **7,** 213 (1955).
39. Hauffe, K., and Engell, H. J., *Z. Elektrochem.* **56,** 366 (1952).
40. Heilbronner, E., *Helv. Chim. Acta* **36,** 921 (1954).
41. Hoerni, A. J., *J. Chem. Phys.* **34,** 508 (1961).
42. Hoffmann, T. A., *J. Chem. Phys.* **18,** 989 (1950).
43. Hoffmann, T. A., *Acta Phys. Acad. Sci. Hung.* **3,** 195 (1953).
44. Hoffmann, T. A., and Kónya, A., *J. Chem. Phys.* **16,** 1172 (1948).
45. Kimball, G. E., *J. Chem. Phys.* **3,** 560 (1935).
46. Koster, G. F., and Slater, J. C., *Phys. Rev.* **95,** 1167 (1954).
47. Koutecký, J., *Z. Elektrochem.* **60,** 835 (1956).
48. Koutecký, J., *Phys. Rev.* **108,** 13 (1957).
49. Koutecký, J., *Collection Czech. Chem. Commun.* **22,** 683 (1957).
50. Koutecký, J., *Collection Czech. Chem. Commun.* **22,** 669 (1957).
51. Koutecký, J., *Trans. Faraday Soc.* **54,** 1038 (1958).
52. Koutecký, J., *Czech. J. Phys.* **8,** 148 (1958).
53. Koutecký, J., *Phys. Chem. Solids* **14,** 233 (1960).
54. Koutecký, J., *Czech. J. Phys.* **B11,** 565 (1961).
55. Koutecký, J., *Physica Status Solidi* **1,** 554 (1961).
56. Koutecký, J., *Czech. J. Phys.* **B12,** 184 (1962).
57. Koutecký, J., *Czech. J. Phys.* **B12,** 177 (1962).
58. Koutecký, J., *Surface Science* 1280 (1964).
59. Koutecký, J., and Davison, S. G., unpublished results.
60. Koutecký, J., and Coulson, C. A., unpublished results.
61. Koutecký, J., and Čížek, J., *Collection Czech. Chem. Commun.* **28,** 2808 (1963).
62. Koutecký, J., and Fingerland, A., *Dokl. Akad. Nauk SSSR* **125,** 841 (1959).

63. Koutecký, J., and Fingerland, A., *Collection Czech. Chem. Commun.* **25,** 1 (1960).
64. Koutecký, J., and Tapilin, V. M., unpublished results.
65. Koutecký, J., and Tomášek, M., *Phys. Rev.* **120,** 1212 (1960).
66. Koutecký, J., and Tomášek, M., *Phys. Chem. Solids* **14,** 241 (1960).
67. Koutecký, J., and Tomášek, M., *Czech. J. Phys.* **B12,** 48 (1962).
68. Kronig, R. L., and Penney, W. G., *Proc. Roy. Soc. London* **A130,** 499 (1951).
69. Kuhn, H., *Experientia* **9,** 41 (1953).
70. Kuhn, H., *J. Chem. Phys.* **22,** 2098 (1954).
71. Kuhn, H., *Z. Naturforsch.* **9a,** 989 (1954).
72. Lax, M., *Proc. Intern. Conf. Semicond. Phys. Prague,* 1960, 1961, p. 484.
73. Lifshitz, I. M., *Zh. Eksperim. i Teor. Fiz.* **17,** 1017, 1076 (1947).
74. Lifshitz, I. M., and Pekar, I. J., *Usp. Fiz. Nauk* **56,** 531 (1955).
75. Longuet-Higgins, H. C., *J. Chem. Phys.* **18,** 265 (1950).
76. Longuet-Higgins, H. C., and Salem, L., *Proc. Roy. Soc. London* **A251** 127 (1959).
77. Löwdin, Per-Olov, *J. Math. Phys.* **3,** 969 (1962).
78. Maue, A. W., *Naturwissenschaften* **38,** 648 (1934).
79. Maue, A. W., *Z. Physik* **94,** 717 (1935).
80. Nagaev, Z. L., *Zh. Fiz. Khim.* **35,** 327 (1961).
81. Osherov, V. I., *Dokl. Akad. Nauk SSSR* **135,** 1168 (1960).
82. Peierls, R., *Ann. Phys.* **4,** 121 (1930).
83. Phariseau, P., *Physica* **26,** 737 (1960).
84. Phariseau, P., *Physica* **26,** 1192 (1960).
85. Pople, J. A., and Wahnsley, S. H., *Mol. Phys.* **5,** 15 (1962).
86. Roginskii, S. Z., private communication.
87. Ryzhanov, S., *Z. Physik* **89,** 806 (1934).
88. Sandorfy, G., *Can. J. Chem.* **33,** 1337 (1955).
89. Schlier, R. E., and Farnsworth, H. E., *J. Chem. Phys.* **30,** 917 (1959).
90. Shockley, W., *Phys. Rev.* **56,** 317 (1939).
91. Slater, J. C., *Phys. Rev.* **76,** 1952 (1949).
92. Slater, J. C., and Koster, G. F., *Phys. Rev.* **94,** 1498 (1954).
93. Sokolov, A. A., *Z. Physik* **90,** 520 (1934).
94. Statz, H., *Z. Naturforsch.* **5a,** 534 (1950).
95. Tamm, I., *Z. Physik* **76,** 849 (1932).
96. Tapilin, V. M., *Kinetika i Kataliz* **3,** 709 (1962).
97. Tomášek, M., *Czech. J. Phys.* **B12,** 159 (1962).
98. Tomášek, M., and Koutecký, J., *Czech. J. Phys.* **B10,** 268 (1960).
99. Tsuji, M., Huzinaga, S., and Hasino, T., *Rev. Mod. Phys.* **32,** 425 (1960).
100. Volkenstein, F. F., *Zh. Fiz. Khim.* **21,** 1317 (1947).
101. Volkenstein, F. F., *Zh. Fiz. Khim.* **22,** 163 (1947).
102. Volkenstein, F. F., *Advan. Catalysis* **9,** 807 (1957).
103. Wallace, P. R., *Phys. Rev.* **71,** 622 (1947).
104. Wannier, G. H., *Phys. Rev.* **52,** 191 (1937).

VIBRATIONAL PROPERTIES OF HEXAFLUORIDE MOLECULES*

BERNARD WEINSTOCK, *Scientific Laboratory, Ford Motor Company, Dearborn, Michigan*

and

GORDON L. GOODMAN, *Argonne National Laboratory, Argonne, Illinois*

CONTENTS

* Based in part on work performed under the auspices of the U.S. Atomic Energy Commission.

I. INTRODUCTION

The vibrational properties of fifteen hexafluoride molecules are reviewed in detail here. Four of these molecules (ReF_6, OsF_6, TcF_6, and RuF_6) show interesting and unusual complications in their infrared and Raman spectra. These features arise because the Jahn–Teller theorem applies to the lowest electronic states in each of these four molecules. The other eleven molecules have infrared and Raman spectra that can be understood simply in terms of normal vibrations about a highly symmetrical nuclear configuration. In this configuration, the six fluorine nuclei are located at the vertices of a regular octahedron and the heavier nucleus is at the center of the octahedron.

The Born–Oppenheimer[1] procedure can be used for defining a potential energy function for each of the eleven, normal hexafluoride molecules. The observed octahedral structure for each of these normal molecules gives a minimum value for the molecular potential energy and thus represents an equilibrium nuclear geometry. The vibrational spectra of the normal molecules can be interpreted in terms of nearly harmonic vibrations of the nuclei about their equilibrium locations.

Spectral features that cannot be explained in the harmonic approximation were discussed by Herzberg and Teller in 1933.[2] They noted that electronic and nuclear motions can be coupled together for polyatomic molecules in much the same way that rotational motions and electronic motions are coupled together for diatomic molecules. In 1934 Renner examined in detail the case of a linear triatomic molecule in a π-electronic state.[3] Renner predicted unusual features for the energy levels in this type of molecule. Twenty-three years later, Dressler and Ramsay[4] identified as Renner effects certain anomalous features in the electronic spectra of the NH_2 free radical and its isotopic modifications. Pople and Longuet-Higgins[5] extended Renner's work to obtain quantitative agreement between theory and experiment for these spectra. Fully resolved rotational features were classified in the work of Dressler and Ramsay;[6] and the successful interpretation of these complete spectra is a beautiful verification of Renner's theory.[7]

One consequence of Renner's work is that for linear molecules with orbitally degenerate electronic states, nuclear vibrations and electronic motions cannot be separated in the usual way, but must be considered simultaneously. The word, vibronic, is used to designate the concomitant coupling of vibrational and electronic motions. The work of Moffitt, Liehr, and Thorson[8,9] and of Longuet-Higgins, Öpik, Pryce, Sack, and Child[10,11] has now made it clear that similar vibronic effects may be expected in situations to which the Jahn–Teller theorem applies.

In 1937 Jahn and Teller[12,13] investigated "the conditions under which a polyatomic molecule can have a stable equilibrium configuration when its electronic state has orbital degeneracy". Jahn and Teller showed that "whenever orbital electronic degeneracy is not accidental in a non-linear molecule, the electronic energy may be expected to depend on the first power of at least one nuclear displacement that distorts its symmetry".

It seems quite difficult to calculate *a priori* the size of the forces that Jahn and Teller discussed, and as a result there has grown up a rather large body of literature speculating on the possible consequences of the Jahn–Teller theorem. These consequences may be divided roughly into two categories: static effects and dynamic effects. A static Jahn–Teller effect arises if the distortion forces are large and lead to a new, distorted equilibrium configuration. A dynamic Jahn–Teller effect arises if the distortion forces are small and it is impossible to specify any equilibrium configuration.

Most of the experimental evidence concerning Jahn–Teller effects has been obtained from studies of solids. In solids there are many forces at work and what could be a dynamic Jahn–Teller effect for an isolated molecule in the vapor phase can be frozen into a static distortion in the solid. Moreover, as observed many years ago by Van Vleck,[14] there are many reasons other than Jahn–Teller forces why distortions occur in solids.

The hexafluoride molecules offer several unusual opportunities to investigate possible consequences of the Jahn–Teller theorem. In the first place, their high volatility makes feasible studies of their spectra in the vapor phase, where the observed transitions arise from isolated molecules. Since spectral studies in the solid phase are also possible with these molecules, comparisons between these observations and those in the vapor phase should eventually make it possible to distinguish Jahn–Teller effects from other effects that arise in solids.

Secondly, the study of many electronic energy levels for an extensive series of homologous compounds makes possible a thorough test of the qualitative aspects of the Jahn–Teller theorem. Each time there is an orbital electronic degeneracy, some qualitative evidence for a Jahn–Teller effect must appear; when this degeneracy is absent, these effects must also be absent.

Lastly, ligand-field theory provides a detailed classification of the electronic states of lowest energy. For the third transition series, Moffitt, Goodman, Fred, and Weinstock[15] have assigned values for the two adjustable parameters on which the ligand-field theory depends. These assignments permit the identification of the degeneracy and other characteristics of each of seventeen energy levels for ReF_6, OsF_6, IrF_6, and PtF_6. We have now extended their analysis to apply to the second transition series.

In this article we are concerned mainly with infrared and Raman spectra of hexafluoride molecules. These spectra show transitions associated with the lowest electronic energy level for each molecule. No rotational features can be resolved in these spectra because of the high moment of inertia for a hexafluoride molecule. As long as the interaction between the internal motions and the rotational motions of a molecule is small, this lack of resolution is not a serious drawback. No inconsistencies seem to arise in the interpretations of the spectra of either normal or abnormal hexafluoride molecules because of our neglect of these interactions. But, the energy-level analyses given here are less complete than those for the light free radical, NH_2, for which a full rotational analysis was possible.

The specific dynamic Jahn–Teller effect exhibited by gaseous metal hexafluorides and examined in detail here is the occurrence of several spectral transitions spaced around the frequency at which there would be just one transition in the absence of vibronic coupling. The energy levels leading to this effect were first reported by Moffitt and his co-workers,[8,9] but only a matter of months before the appearance of independent work by Longuet-Higgins and his co-workers.[10] Indications of this effect in spectra of metal hexafluorides were first noted by Weinstock and Claassen.[19]

Because of the late William Moffitt's interest in problems of the metal hexafluoride molecules and because of the priority of his work on the dynamical consequences of the Jahn–Teller theorem, we feel that it is appropriate to dedicate this review of vibrational properties to his memory.

The outline of this present article is as follows. After this introductory section, we summarize the published vibrational spectra for the normal hexafluoride molecules and give the arguments leading to assignments of their fundamental vibrational frequencies. In Section III a survey is given of the theory of vibronic spectra, drawing heavily on the work of Moffitt and of Longuet-Higgins and their co-workers. Section IV presents our interpretation of the unusual spectra of ReF_6, TcF_6, OsF_6, and RuF_6. In this section we assign magnitudes to the one-electron integrals that govern the coupling of electronic and nuclear motions for hexafluoride molecules of elements in the second and third transition series. Finally, in the fifth section, there is a discussion of force constants and thermodynamic functions for hexafluoride molecules.

II. VIBRATIONAL SPECTRA OF NORMAL HEXAFLUORIDE MOLECULES

The eighteen elements that form hexafluoride molecules are listed in Table I, arranged in their periodic order. The nonbonding electronic configurations for the $4d$, $5d$, and $5f$ transition series compounds are also indicated. Where the symmetry of the electronic ground state is known, it is given in parentheses beside each formula. Three elements are listed additionally; these are the most likely candidates to form hexafluoride molecules, although no evidence for such compounds of these elements currently exists.

TABLE I. The Hexafluoride Molecules: Number and Type of Nonbonding Valence Electrons and Ground-State Symmetries

Nonbonding valence electrons					
Type	Number				
	0	1	2	3	4
	$SF_6(A_{1g})$				
	CrF_6[a]				
	$SeF_6(A_{1g})$				
$4d$	$MoF_6(A_{1g})$	$TcF_6(G_{\frac{3}{2}g})$	$RuF_6(F_{2g})$	$RhF_6(G_{\frac{3}{2}g})$	Pd[b]
	$TeF_6(A_{1g})$	I[b]	XeF_6[a]		
$5d$	$WF_6(A_{1g})$	$ReF_6(G_{\frac{3}{2}g})$	$OsF_6(E_g, F_{2g})$	$IrF_6(G_{\frac{3}{2}g})$	$PtF_6(A_{1g})$
	PoF_6[a]				
$5f$	$UF_6(A_{1g})$	$NpF_6(E_{\frac{5}{2}u})$	$PuF_6(A_{1g})$	Am[b]	

[a] The molecular structure has not been determined for these compounds.
[b] Hexafluorides have not been formed for these elements.

Three compounds, PoF_6,[16] XeF_6,[17] and CrF_6,[18] have only recently been prepared and their structures have either not been studied or are incompletely worked out. Vibrational spectra have been reported for the other fifteen molecules in the table and have been interpreted in terms of a regular octahedral structure, point group O_h. Eleven of these molecules either have nondegenerate electronic ground states or have degenerate ground states with predominantly spin character. For these compounds, vibronic effects are not important and their vibrational spectra are normal. The spectra of these molecules are considered in detail in this section.

The remaining four molecules are $TcF_6(4d^1)$, $RuF_6(4d^2)$, and their third transition series opposite numbers, $ReF_6(5d^1)$, and $OsF_6(5d^2)$. All of these molecules show anomalies in their vibrational spectra that have been ascribed to a dynamic Jahn–Teller effect.[19–22] The vibrational spectra of these four compounds and the interpretation of the observed vibronic effects will be given in Section IV.

A. General Discussion

Herzberg[23] has given an introduction to the use of group theory for the vibrational properties of molecules with octahedrally symmetric equilibrium configurations and his analysis is used in this section. The symmetry characteristics of the displacement coordinates are given in Mulliken's notation. For a seven-atom molecule, there are fifteen internal degrees of freedom that are characterized in O_h symmetry by six vibrational frequencies. Three are Raman-active and customarily designated as: ν_1, which is nondegenerate, symmetry a_{1g}; ν_2, doubly degenerate, symmetry e_g; and ν_5, triply degenerate, symmetry f_{2g}. Two are infrared-active, ν_3 and ν_4; both are triply degenerate and symmetry f_{1u}. The sixth, ν_6, is spectrally inactive, triply degenerate, and symmetry f_{2u}.

The nuclear displacements that correspond to normal modes appropriate for NpF_6 are shown in Fig. 1. There is only one vibrational mode for each of the symmetry types, $a_{1g}\,(\nu_1)$, $e_g\,(\nu_2)$, $f_{2g}\,(\nu_5)$, and $f_{2u}\,(\nu_6)$, and the normal displacements shown here for NpF_6 are also appropriate for all of the other hexafluoride molecules discussed in this article. However, there are two infrared-active f_{1u} vibrations, ν_3 and ν_4. The normal displacements for these modes are not the same for all the hexafluoride molecules, but depend on the force constants of the particular molecule. The normal displacements of ν_3 and ν_4 shown in Fig. 1 for NpF_6 were derived from force constants for NpF_6 given in Section V. Force constants for the other hexafluoride molecules are also given in Section V.

Spectroscopic proof of an octahedral structure for a hexafluoride molecule thus requires the observation of three Raman shifts and two infrared-active fundamentals. Additionally, for a molecule with a center of symmetry, no Raman-active fundamental can be active in the infrared. This proof has been obtained directly for five of the hexafluoride molecules. For the other six normal molecules, some active fundamentals have not been observed because of experimental difficulties. In three cases, ν_4 was not observed because it was outside the range of the spectrometer used. In four instances, the Raman spectrum has not been obtained because the molecules are colored and absorb too strongly or because of photochemical decomposition. However, failure to observe all of the active fundamentals cannot be said to imply deviation from octahedral symmetry, because all symmetries lower than O_h require more active fundamentals rather than fewer.

For the molecules in which all of the active fundamentals have not been observed directly, the missing frequencies are derived from infrared-active binary sum and difference bands. To be

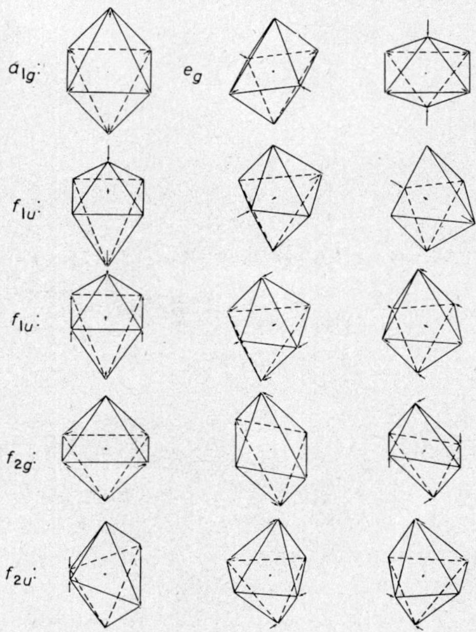

Fig. 1. Normal displacement coordinates for NpF_6, based on the assumption that $2(f_{d\alpha} - \rho)/(f_\alpha - \gamma + 2\kappa - 2\varepsilon) = 4m/(M + 2m) = 0.2763$. (The definitions of force constants and atomic masses are those used by Claassen[85] and they are summarized in Section V.)

infrared-active, these bands must contain the symmetry species f_{1u}. The inactive fundamental, ν_6, is also evaluated from these binary combination bands. The eight binary sum bands and the corresponding difference bands containing f_{1u} are listed in Table II together with their entire symmetry species. When binary bands are referred to, the symbol ν is omitted. It is worth noting that binary overtones are not allowed in O_h symmetry, since they are g vibrations. In no case has a binary overtone been observed

TABLE II. Infrared-Active Binary Combination Bands in O_h Symmetry

Band	Symmetry
(1 + 3), (1 − 3)	f_{1u}
(1 + 4), (1 − 4)	f_{1u}
(2 + 3), (2 − 3)	$f_{1u} + f_{2u}$
(2 + 4), (2 − 4)	$f_{1u} + f_{2u}$
(2 + 6), (2 − 6)	$f_{1u} + f_{2u}$
(3 + 5), (3 − 5)	$a_{2u} + e_u + f_{1u} + f_{2u}$
(4 + 5), (4 − 5)	$a_{2u} + e_u + f_{1u} + f_{2u}$
(5 + 6), (5 − 6)	$a_{1u} + e_u + f_{1u} + f_{2u}$

for a hexafluoride molecule. The infrared spectrum of NpF_6 reported by Malm, Weinstock, and Claassen[24] is shown in Fig. 2 as representative. For this molecule, all of the allowed binary sum bands were observed and three of the binary difference bands; ν_4 was not observed because it was below the range of observation.

A great deal of confidence can be placed in the values of the fundamental vibrational frequencies derived from the binary bands. In most instances where the observed frequency of a binary band can be compared with the corresponding value calculated from the directly observed fundamentals, there is agreement to within an experimental uncertainty of ± 2 cm^{-1}. This result also suggests that anharmonicity is not important in the lower vibrational states of these molecules.

As will be seen in Section IV, a knowledge of the quantitative variation of vibrational frequency with number of nonbonding electrons for the metal hexafluorides plays an essential role in our understanding of the spectra of the abnormal molecules. We have, therefore, made a critical re-examination of the available vibrational spectra of the normal hexafluoride molecules, in order to obtain the most reliable values for their fundamental vibrational frequencies. The nonmetal hexafluoride molecules are included mainly for completeness.

In order to make the intercomparisons among the molecules as meaningful as possible, a consistent method of analysis for all of the data was developed. In general, where the fundamental vibrational frequencies were directly observed for the vapor, these numbers were used in preference to numbers obtained from combination bands. More recent work, done with high-precision instruments and with higher purity samples, was given more weight than earlier measurements. Where combination-band

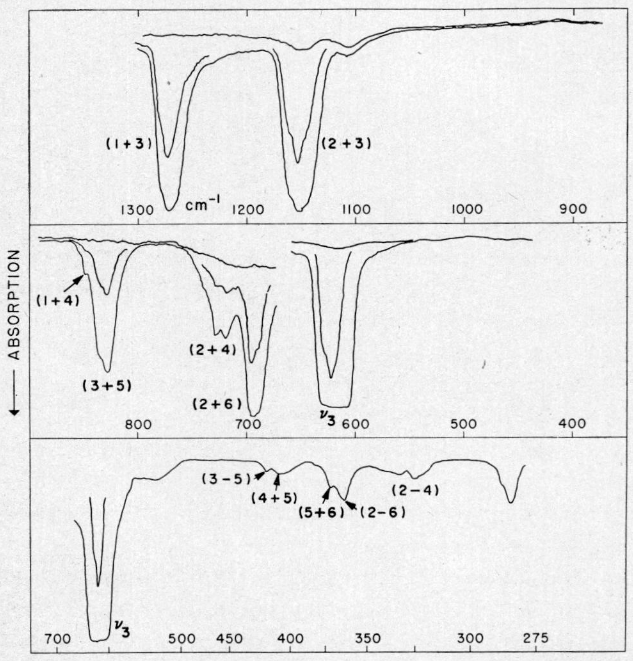

Fig. 2. Infrared spectrum of NpF_6 vapor, using a NaCl prism from 1300 to 670 cm^{-1}, KBr from 670 to 400 cm^{-1}, and CsBr from 650 to 275 cm^{-1} (Malm, Weinstock, and Claassen[24]).

data were used to evaluate fundamentals, stronger bands were given more weight than weaker bands and an average of all the pertinent data was usually taken. The combination bands were also used to resolve discrepancies in the values of the fundamentals obtained by different investigators. Although liquid–vapor shifts were found to be small, better agreement among the binary sum bands usually resulted if the liquid Raman data were not used to assign frequencies for the molecules in the vapor phase.

This spectral re-analysis has uncovered a number of minor discrepancies, including some instances where assignments have been made for impurity bands. The most significant changes, however, are in the assignment of bands involving ν_6 and in the evaluation of frequencies for this inactive fundamental. Generally, low intensity and near degeneracy with more intense bands make these assignments ambiguous. In the present analysis, intercomparisons among the spectra of all the hexafluoride molecules helped to reduce this difficulty. As a consequence, new assignments have been made for the binary bands involving ν_6 for MoF_6 and WF_6, and the values of ν_6 derived from these assignments are markedly different from the values previously reported for these molecules.

In all cases, including the vibronic molecules, the experimental investigators had concluded that the structure is octahedral, i.e., that there is no indication of a static distortion from octahedral symmetry. No evidence to alter these conclusions has been found in this investigation. For several hexafluorides low temperature heat capacity data are available and a Third Law entropy comparison can be made. A convincing overall confirmation of the molecular structure and the vibrational frequency assignments is provided in every case.

Details of the data analysis for the individual molecules are given next, as well as some brief historical background.

The experimental data obtained with these molecules are summarized in Tables III–X. The entries are given in cm^{-1}. The values for the six fundamental vibrational frequencies derived from the present analysis are given in the second column of each table. Also included in this column are the values calculated for the binary bands based on these assignments. Where the entry is given in parentheses, this frequency has not been observed directly for the vapor; all of the entries for ν_6, the inactive fundamental, are therefore in parentheses. Values obtained for the liquid (l), solid (s), and solution (sln) are given on separate lines. Where the individual experimentalists did not obtain the fundamental vibrational frequencies directly, the values they assigned are also given in parentheses. Included as the last column in these tables is a qualitative comparison of the relative intensity of the combination bands. The abbreviations used have the following meanings: s, strong; m, moderate; w, weak; v, very; a d, near or accidental degeneracy; n o, not observed.

B. Nonmetal Hexafluoride Molecules

The nonmetal hexafluoride molecules, SF_6, SeF_6, and TeF_6, differ from hexafluoride molecules containing metallic elements. (PoF_6, the fourth member of this series, is not considered here: only inferential evidence for its existence has been obtained owing to the short half-life of the available Po isotopes.[16]) The nonmetal hexafluorides are characterized by an unusual chemical inertness and are very stable to dissociation into fluorine and a lower fluoride. In contrast, the metal hexafluoride molecules are all very reactive compounds that hydrolyze rapidly and fluorinate most common materials of construction. Consequently, special techniques are used for spectral studies with the metal hexafluorides, while more usual methods are adequate with the nonmetal compounds. The nonmetal hexafluorides are also much more volatile than the metal hexafluorides. The boiling points of the nonmetal hexafluorides lie in the range —64° to —38°C; the range of boiling or sublimation points for the metal hexafluorides is 17° to about 69°C.

For the nonmetal hexafluoride molecules the six valence electrons of the central atom are all used in the formation of the fluorine bonds; the electronic ground state is thus totally symmetric (A_{1g}).

Values for the six fundamental vibrational frequencies for some or all of these molecules have been derived by the following authors. Yost, Steffens, and Gross[25] (YSG) measured the Raman spectra of liquid and gaseous SF_6, SeF_6, and TeF_6 from which they assigned values for ν_1, ν_2, and ν_5. They estimated the values of the other frequencies using central-force and valence-bond models. Eucken, Ahrens, Bartholomé, and Bewilogua[26] (EABB) measured the Raman spectra of liquid and gaseous SF_6 and the infrared spectra of SF_6 vapor from 5 to 25 μ. From these measurements they obtained the five active fundamentals directly; they derived ν_6 using these data and measurements of the heat capacity of SF_6 vapor. These experiments of Yost *et al.* and Eucken *et al.* were the first vibrational spectral measurements with hexafluoride molecules. They found these spectra to be consistent with the regular octahedral structure for hexafluoride molecules that had been originally deduced from electron-diffraction measurements.[27,28] The first suggestion of an octahedral structure had been made on the basis of Raman measurements with the hexachloride ions, $SbCl_6^-$ and $SnCl_6^{2-}$.[29] Sachsse and Bartholomé[30] (SB) measured the infrared spectra of SeF_6 and TeF_6 from 3 to 25 μ. For SeF_6 both ν_3 and ν_4 were observed; a value for ν_6 was derived from two binary and eighteen ternary bands. For TeF_6, they observed only ν_3 directly because ν_4 was outside the range of their spectrometer; values for ν_4 and ν_6 were derived using heat-capacity measurements with TeF_6 vapor, the Raman data of Yost *et al.*, and a large number of binary and ternary combination bands. Lagemann and Jones[31] (LJ) measured the infrared spectrum of SF_6 vapor from 2 to 25 μ with improved accuracy; they obtained ν_3 and ν_4 but did not derive a value of ν_6 from their data. Gaunt[32] (G), with similarly improved accuracy, measured the infrared spectra of gaseous SF_6, SeF_6, and TeF_6 from 2 to 25 μ and of TeF_6 vapor from 25 to 40 μ.[33] He observed ν_3 and ν_4 for all of these molecules and also derived values of ν_6 from his infrared data. The results of these measurements are summarized in Tables III–V, as well as the assignments of the fundamental vibrational frequencies derived from our analysis of the data.

(1) *Sulfur Hexafluoride*

For SF_6 the infrared data of Lagemann and Jones and of Gaunt are in general quantitative agreement and their values were given equal weight in the frequency evaluations. The infrared data of Eucken *et al.* were not used in the averaging. About half of their frequency assignments agree well with the more recent data, but the rest differ by more than the uncertainty of the later measurements. The value of 984 cm^{-1} assigned by Lagemann and Jones to $(2 + 6)$ has been changed to 990; the band shape of $(2 + 6)$ shows structure and 990 is the weighted average of the values of 984 and 995 reported by these authors for the two peaks. Eucken *et al.* did not resolve $(2 + 6)$ from ν_3 and the high value of 965 cm^{-1} that they report for ν_3 may be a consequence of this. The strong band at 830 cm^{-1} that Eucken *et al.* assign as $(2 + 6)$ was probably due to an impurity.

Eucken *et al.* calculated the value of 363 cm^{-1} for ν_6 from their measurements of the heat capacity of SF_6 vapor and their other five vibrational assignments. This value is in fair agreement with the value of 349 cm^{-1} derived here. However, their values for the heat capacity of SF_6 are probably too low, possibly because of a small systematic error in their measurements. This conclusion is reached because the frequency assignments of Eucken *et al.* are all higher than the values derived here. If the current frequencies were used to derive ν_6 from their heat-capacity measurements, a value higher than 363 cm^{-1} would be obtained, which would be in poorer agreement with the assigned value of 349 cm^{-1}. The value given by Gaunt for ν_6 was 344 cm^{-1}. The estimates

TABLE III. SF_6

Mode	Assigned value	(YSG)[25]	(EABB)[26]	(LJ)[31]	(G)[32]	(GNS)[86]	Intensity
1	770	772.4	775	(775)	(775)	769.4	
1(l)	775	776	775				
1(s)	772					772	
2	640	(642)	(645)	(644)	(644)	639.5	
2(l)	643	642	645				
2(s)	639					639	
3	939		965[a]	940	932		
4	614		617	615	613		
5	522	(522)	(525)	(524)	(524)	522	
5(l)	523	522	525				
5(s)	522					522	
6	(349)		(363)	(363)	(344)		
1 + 3	1709		1703	1710	1712		m
2 + 3	1579		1578	1580	1580		m
3 + 5	1461			1455	1455		w
1 + 4	1384		1380	1380	1388		m
2 + 4	1254		1262	1250	1255		m
4 + 5	1136		1163	1132	1136		w
2 + 6	989		(830)[a]	990[a]	987		s[a]
5 + 6	871		885	875	869		m
3 − 5	(417)						n o

[a] These assignments are discussed in the text.

given by Yost *et al.* for ν_3, ν_4, and ν_6 were: by the central-forces model 1510, 620, and 280; by the valence-bond model 970, 540, and 370 cm^{-1}.

Although we originally overlooked the most recent study of SF_6 by Gullikson, Nielsen, and Stair[86](GNS), their data are included in Table III and have been used for our own assignments. They studied the Raman spectra of SF_6 in the gaseous and solid states. For the gas they observed: $\nu_1 = 769.4$, $\nu_2 = 639.5$, and $\nu_5 = 522$ cm^{-1}. In addition, a very weak sharp band was found at 690 cm^{-1} which they interpreted as the overtone, $2\nu_6$. For the solid, they observed: $\nu_1 = 772$, $\nu_2 = 639$, and $\nu_5 = 522$ cm^{-1}. Two very weak bands at 505 and 543 cm^{-1} were also observed with the solid which they explained as probably arising from combinations of ν_5 with low vibrational lattice modes.

TABLE IV. SeF_6

Mode	Assigned value	(YSG)[25]	(SB)[30]	(G)[32]	Intensity
1	708	708	(708)	(708)	
1(l)	710	710			
2	(661)	(662)	(662)	(662)	
2(l)	662	662			
3	780		787	780	
4	437		461	437	
5	(403)	(405)	(405)	(405)	
5(l)	405	405			
6	(262)		(245)	(260)	
1 + 3	1488		1476	1484	m
2 + 3	1441		1440	1438[a]	m
3 + 5	1183		1209	1182	v w
1 + 4	1145		1170	1148	v v w
2 + 4	1098		1114	1104[a]	v w
4 + 5	840			841	m
2 + 6	923		934	924	m
5 + 6	665		620	664	m
2 − 6	(399)				n o

[a] These assignments are discussed in the text.

(2) *Selenium Hexafluoride*

For SeF_6, the infrared data of Sachsse and Bartholomé were not used in the analysis. Although their data are in qualitative agreement with those of Gaunt, they are less reliable. The values of ν_2 derived from Gaunt's values for (2 + 3) and (2 + 4) are not in close agreement, so that the value of ν_2 obtained by Yost *et al.* for the liquid was also considered. Gaunt's infrared spectrum for SeF_6 shows absorption in the region 1200–1300 cm^{-1}, which probably arises from impurities; however, these bands do not appear to affect the reliability of the frequency assignments. The values estimated by Yost *et al.* for ν_3, ν_4, and ν_6 are: central-forces, 960, 570, and 230; valence-bond, 650, 430, and 286 cm^{-1}.

TABLE V. TeF_6

Mode	Assigned value	(YSG)[25]	(SB)[30]	(G)[32,33]	Intensity
1	701	701.2	(701)	(701)	
1(l)	697	697			
2	674	674.4	(674)	(674)	
2(l)	672	672			
3	752		752	752	
4	325		(370)	325	
5	313	313	(313)	(313)	
5(l)	316	316			
6	(195)		(165)	(197)	
1 + 3	1453		1446	1451	m
2 + 3	1426		1425	1425	m
3 + 5	1065		1094[a]	1063	v w
1 + 4	1026		1046[a]	1025	v w
2 + 4	999		990[a]		v v w
4 + 5	638			639	v v w
2 + 6	869		832	871	m
5 + 6	508		496[a]	509	w
3 − 5	439		410	439	w
2 − 6	479		478[a]	481	w

[a] These assignments are discussed in the text.

(3) *Tellurium Hexafluoride*

For TeF_6 all five of the spectrally active fundamentals have been observed directly. Very close agreement is found between the infrared sum bands observed by Gaunt and the values calculated from these directly measured fundamentals. This demonstrates the reliability of the procedure of deriving the Raman-active fundamentals from the infrared sum bands when Raman data are available only for the liquid. Two binary difference bands have been observed for TeF_6. The data of Sachsse and Bartholomé were not used in the evaluation of ν_6 for TeF_6 because they appear less reliable than those of Gaunt. These authors (SB) also made a number of incorrect band assignments. They assign 1094 cm^{-1} to $(1 + 4)$ rather than $(3 + 5)$; 1046 cm^{-1} is assigned as $(3 + 5)$ and $(2 + 4)$ rather than $(1 + 4)$; 990 cm^{-1} is assigned as $(2\nu_5 + \nu_4)$ rather than $(2 + 4)$; 496 cm^{-1} is called $(2 - 6)$ instead of $(5 + 6)$ and for 478 cm^{-1} the reverse assignment was made. Yost *et al.* estimated for ν_3, ν_4, and ν_6: central-forces, 850, 420, and 190; valence-bond, 610, 310, and 222 cm^{-1}.

C. Actinide Hexafluoride Molecules

For these molecules, UF_6, NpF_6, and PuF_6, of the actinide transition series, the nonbonding valence electrons are $5f$ electrons. The other metal hexafluorides form transition series involving d nonbonding valence electrons. UF_6 $(5f^0)$ and PuF_6 $(5f^2)$[34] have totally symmetric electronic ground states, (A_{1g}). NpF_6 $(5f^1)$ has a Kramers doublet $(E_{\frac{5}{2}u})$ for its lowest electronic state.[35] Since none of these molecules have Jahn–Teller effects they are being considered before the other transition-metal hexafloride molecules.

The actinide hexafluorides offer experimental difficulties to the study of their vibrational spectra that are not present for the nonmetal hexafluorides. UF_6, although quite stable to thermal decomposition, is readily hydrolyzed by traces of water. This difficulty is common to all of the metal hexafluorides but can be overcome by careful technique. NpF_6 and PuF_6 are both sensitive to photochemical decomposition with the result that Raman data have not been obtained for them. NpF_6 and PuF_6 are both radioactive gases and their study presents a severe biological hazard; this is particularly true for PuF_6, which is one of the most toxic gases known. Stability with respect to dissociation decreases for the actinide hexafluorides as one goes across the series, with the consequence that the preparation of AmF_6, the next member of the series, appears unlikely.[36]

More attention has been focussed on UF_6 than on any other hexafluoride because of the important use of this compound as the process gas for the separation of the uranium isotopes by gaseous diffusion. The Raman spectra of liquid UF_6 and of UF_6 dissolved in C_7F_{16} and the infrared spectrum of UF_6 vapor from 2 to 17 μ were measured by Bigeleisen, Mayer, Stevenson, and Turkevich[37] (BMST). This was the pioneering effort in the measurement of the vibrational spectra of the metal hexafluorides and these authors concluded that their data were consistent with a symmetrical octahedral structure. Earlier electron-diffraction studies with the metal hexafluorides UF_6, MoF_6 and WF_6[38] had suggested unsymmetrical structures for these molecules. A careful reinvestigation of UF_6 by electron diffraction[39] confirmed the possibility of an unsymmetrical structure for this molecule. The discrepancy between the vibrational-spectra and the electron-diffraction results was resolved when Schomaker and Glauber[40] showed the difficulty to lie with the Born approximation as customarily used in electron-diffraction analysis. They were able to explain the complex diffraction patterns obtained with symmetrical molecules by the inclusion of an angle-dependent phase factor in the scattering amplitude.

Burke, Smith, and Nielsen[41] (BSN) remeasured the infrared spectrum of UF_6 vapor from 2–38 μ, and confirmed the earlier results. Gaunt[32] (G) has also reported the infrared spectrum of UF_6 vapor in the range 2–25 μ. Hawkins, Mattraw, and Carpenter[42] (HMC) measured the infrared spectrum of UF_6 from 2–23 μ at temperatures up to 100°C. Claassen, Weinstock, and Malm[43] (CWM) measured the Raman spectrum of UF_6 vapor and observed shifts for ν_1 and ν_2. Claassen[44] (C) has measured the Raman spectra of liquid and solid UF_6. These results and our assignments for the fundamental frequencies are given in Table VI.

Much less work has been done with NpF_6 and PuF_6, owing to the special experimental difficulties mentioned earlier. Malm, Weinstock, and Claassen[24] (MWC) measured the infrared spectra of NpF_6 and PuF_6 vapors from 2–38 μ; they also unsuccessfully attempted to observe the Raman spectrum of PuF_6 vapor. Hawkins, Mattraw, and Sabol[45] (HMS) studied PuF_6 vapor from 2–23 μ. These results and our assignments for the fundamental frequencies are given in Table VII.

(1) *Uranium Hexafluoride*

The infrared data of Bigeleisen *et al.* are seen to be in excellent agreement with those of the later workers. However, several discrepancies exist between their assignments of the fundamental frequencies and the current values. Two factors contribute to this disparity. Bigeleisen *et al.* assigned ν_1 and ν_2 on the basis of their measurements with liquid UF_6, which values are seen to be much lower than the values assigned for the vapor. Additionally, they suggested that ν_3 and $(2 + 6)$ are accidentally degenerate at about 640 cm^{-1} and that the transitions actually observed, above and below this frequency, are the result of a Fermi resonance.[46] They were led to this interpretation by their observation that $(2 + 6)$ is relatively more intense than any other binary band, presumably because it borrows intensity from the fundamental, ν_3.

Burke *et al.* found their data to be in agreement with those of Bigeleisen *et al.* and suggested no changes in interpretation or assignment. Gaunt's data were in agreement with those of the others, but he suggested a different evaluation of the fundamental frequencies. He pointed out that, starting with ν_3, the other five fundamentals for UF_6 could be derived from the binary infrared bands without using the Raman data. His assignments for all but ν_3 and ν_5 are significantly different from those of the earlier authors. The data reported by Hawkins *et al.* are also in agreement with the others; they accepted Gaunt's method of analysis and derived values for the fundamentals similar to his. Gaunt's method of evaluation of the fundamental frequencies for UF_6 was shown to be correct by Claassen, Weinstock, and Malm,[43] who observed the Raman shifts for ν_1 and ν_2 with UF_6 vapor and found them to agree with the values deduced from the infrared data alone. Claassen's observations of the Raman shifts for liquid UF_6[44] are higher than the values obtained by Bigeleisen *et al.*; his value for ν_1 is identical with the vapor value, although

ν_2 is still significantly lower. Bigeleisen *et al.* reported that extensive decomposition had occurred during their Raman exposures with liquid UF_6; this probably contributed to the uncertainty of their measurements. It is of interest to note that the Raman shifts obtained by Bigeleisen *et al.* for ν_1 and ν_5 with UF_6 solutions in C_7F_{16} agree closely with the values derived from the vapor data and that the liquid–vapor shifts for ν_2 and ν_5 are larger than those observed for the nonmetal hexafluorides.

Two other assignments made by the original authors have also been changed. Burke *et al.* assigned a band at 392 cm^{-1} as $(4 + 5)$; Gaunt also assigned his band at 394 as $(4 + 5)$. In the present analysis this band is assigned as $(2 - 6)$ because this assignment gives better agreement between the observed and calculated frequencies and because $(2 - 6)$ is found to be more intense than $(4 + 5)$ for the other actinide hexafluorides. Burke *et al.* assigned their band at 350 cm^{-1} as $(4 + 6)$, which is not infrared-active; a better assignment for this band is $(2 - 4)$, for which there is good agreement between calculated and observed values.

There seems to be no necessity to assume that a Fermi resonance influences the observed frequencies of ν_3 and $(2 + 6)$, but some uncertainty on this point must remain. The most cogent argument against a strong interaction between ν_3 and $(2 + 6)$ can be made in the case of MoF_6. There the bands ν_3 and $(2 + 6)$ are separated by only 23 cm^{-1} rather than by 50 cm^{-1} as in UF_6. No apparent shift of these frequencies is observed for MoF_6 either. (MoF_6 is discussed in detail later in this section.) On the other hand, when the relative intensities of $(2 + 6)$ and $(1 + 3)$ are compared for all of the hexafluorides, the intensity of $(2 + 6)$ is found to increase as its frequency difference from ν_3 decreases. This correlation then suggests that $(2 + 6)$ borrows intensity from ν_3, and that the corresponding normal coordinates mix.

The assignment of a frequency value for ν_6 has proven a vexing problem for the hexafluorides. As mentioned earlier, Eucken *et al.* and Sachsse and Bartholomé used measurements of the heat capacity of the vapor in their evaluation of ν_6 for the nonmetal hexafluorides. Bigeleisen *et al.* used the Third-Law entropy procedure to distinguish between a symmetrical or unsymmetrical structure for UF_6. In this procedure the entropy of the gas is calculated by the methods of statistical mechanics from molecular data, including the vibrational frequencies, and compared with the value derived from low-temperature heat-capacity measurements and the entropy of vaporization. From this evaluation Bigeleisen *et al.* were able to confirm the symmetrical structure for UF_6, but the agreement was not precise enough to verify their vibrational-frequency assignments. A more recent calculation,[47] using more reliable heat of vaporization data and vibrational assignments close to those given here, gave excellent agreement: S^0, calculated,

TABLE VI. UF_6

Mode	Assigned value	(BMST)[37]	(BSN)[41]	(G)[32]	(HMC)[42]	(CWM)[43]	(C)[44]	Intensity
1	667	(656)	(656)	(668)	(665)	666.6		
1(sln)	666	666						
1(l)	666	656					666	
1(s)	663						663	
2	535	(511)	(511)	(532)	(536)	535		
2(l)	524	511					524	
2(s)	518						518	
3	624	623[a]	623[a]	626	623			
4	(184)	(200)	(200)	(189)	(186)			
5	(201)	(200)	(200)	(202)	(202)			
5(sln)	202	202						
5(l)	207						207	
5(s)	211						211	
6	(140)	(130)	(130)	(144)	(136)			
1 + 3	1291	1295	1288	1294	1288			m
2 + 3	1159	1163	1159	1158	1159			m
3 + 5	825	825	825	825	825			m
1 + 4	851	850	850	854	852			w
2 + 4	719	716	715	717	715			w
4 + 5	(385)							n o
2 + 6	675	675[a]	674[a]	676	672			s
5 + 6	(341)							n o
2 − 6	395		392[a]	394[a]				w
2 − 4	351		350[a]					w
1 − 4	483				479			v w
3 − 5	(423)							n o

[a] These assignments are discussed in the text.

at 298°K was 90.21 cal deg^{-1} mole^{-1} compared with the experimental value of 90.24. Bigeleisen *et al.* calculated a corresponding value of 90.34 and Gaunt obtained 89.86. The good agreement of the value of S^0 obtained by Bigeleisen *et al.* at 298°K is somewhat fortuitous; comparisons at other temperatures show larger deviations.

(2) *Neptunium Hexafluoride*

There has been only one study of NpF_6, by Malm, Weinstock, and Claassen,[24] and only ν_3 has been directly observed. (Their spectrum of NpF_6 is shown in Fig. 2.) However, in this work, all of the allowed binary sum bands were observed and three of the binary difference bands, so that each of the fundamentals is involved at least twice in the binary bands. The values of the fundamentals given in Table VII differ slightly from those given by Malm *et al.*; the changes were made on the basis of consistency with the method of analysis used for the other molecules.

Low-temperature heat-capacity measurements have been done with NpF_6[48] and there is excellent agreement between the Third-Law entropy and the value calculated using these assignments. At 340°K, the value for S^0, experimental, is 94.09 cal deg^{-1} mole^{-1}, and 94.17 for S^0, calculated; there is similar agreement at other temperatures.

(3) *Plutonium Hexafluoride*

For PuF_6, Malm *et al.*[24] reported three more binary bands than Hawkins, Mattraw, and Sabol;[45] otherwise, the two spectra are in agreement and were weighted equally. As with NpF_6, only ν_3 was observed directly. Three fewer binary bands were observed for PuF_6 than for NpF_6, but all of the fundamentals still occur at least twice in the binary bands to lend confidence to the assignments. Reliable low-temperature heat-capacity measurements are not possible with PuF_6 owing to the radioactivity of the principally available plutonium isotope, ^{239}Pu (half-life, 25,000 years), so that an overall check of the assignments is not presently available. However, comparison of the frequencies derived for PuF_6 with the corresponding values for UF_6 and NpF_6 reveals no irregularities.

TABLE VII

	NpF_6			PuF_6			
Mode	Assigned value	(MWC)[24]	Intensity	Assigned value	(MWC)[24]	(HMS)[45]	Intensity
1	(648)	(648)		(628)	(628)	(625)	
2	(528)	(528)		(523)	(523)	(522)	
3	624	624		616	615	617	
4	(198)	(200)		(203)	(203)	(205)	
5	(205)	(206)		(211)	(211)	(211)	
6	(165)	(164)		(173)	(171)	(176)	
1 + 3	1272	1272	s	1244	1246	1242	m
2 + 3	1152	1153	s	1139	1138	1139	m
3 + 5	829	827	m	827	822	828	w
1 + 4	846	849	v w	831	828		v w
2 + 4	726	724	m	726	725	727	m
4 + 5	403	407	v v w	414	418		v w
2 + 6	693	693	s	696	696	698	s
5 + 6	370	371	v w	(384)			n o
2 − 6	363	364	v w	350	353		w
2 − 4	330	323	v v w	(320)			n o
3 − 5	419	418	v v w	(405)			n o
1 − 4	(450)		n o	(425)			n o

D. Third Transition Series Hexafluoride Molecules

The hexafluorides in the third or 5*d* transition series discussed here are: WF_6, IrF_6, and PtF_6. The other two, ReF_6 and OsF_6, are discussed in Section IV because of their vibronic anomalies. WF_6 ($5d^0$) and PtF_6 ($5d^4$)[15] have totally symmetric electronic ground states (A_{1g}). The electronic ground state for IrF_6 ($5d^3$) is fourfold degenerate, symmetry $G_{\frac{3}{2}g}$.[15] Vibronic effects in this molecule, although allowed by symmetry, are unimportant because the electronic ground state is essentially a spin quartet with little orbital character.[49, 50]

WF_6 offers relatively little difficulty in its study although care must be taken to prevent contamination by a volatile oxyfluoride. IrF_6, which forms a bright yellow vapor and a brown liquid, offers difficulty in the observation of Raman shifts because of its color.

PtF_6, which is one of the most corrosive and least stable of the hexafluoride molecules, is very difficult to study; its dark red color also prevents observation of the Raman shifts.

Tanner and Duncan[51] (TD) did the first spectroscopic studies with WF_6 and measured the Raman shifts for ν_1, ν_2, and ν_5 with the liquid. They also measured depolarization ratios, from which they could distinguish between ν_1, which has a low ratio, and ν_2 and ν_5, which have high ratios in O_h symmetry. Burke, Smith, and Nielsen[41] (BSN) obtained the infrared spectrum of WF_6 from 2–40 μ and the Raman spectrum of the liquid. Gaunt[32] (G) studied the infrared spectrum of WF_6 from 2–25 μ. These results for WF_6 are summarized in Table VIII as well as the present assignments. Also included in Table VIII are unpublished data for the infrared spectrum of WF_6 by Weinstock and Claassen[52] (WC) in the range 600–1800 cm^{-1} obtained with a very pure sample of WF_6.

Mattraw, Hawkins, Carpenter, and Sabol[53] (MHCS) measured the infrared spectrum of IrF_6 from 2–40 μ. They listed their band assignments, but did not publish a representative spectral trace. Claassen and Weinstock[49] (CW) reinvestigated the vibrational spectra of IrF_6 for evidence of a dynamic Jahn–Teller effect. They repeated the infrared measurements of Mattraw *et al.* over the more limited range of 600–1800 cm^{-1} and also obtained a Raman spectrum with a 20 mole% solution of IrF_6 in n-C_7F_{16}. Only ν_2 was observed in this Raman spectrum; ν_1 coincides with the 5677 A Hg line and ν_5 could not be observed because of absorption and scattering background. The Raman spectrum could not be obtained with liquid IrF_6 because of its dark color.

Weinstock, Claassen, and Malm[20] (WCM) studied the infrared absorption spectrum of PtF_6 vapor from 6–50 μ. The results for IrF_6 and PtF_6 are given in Table IX.

(1) *Tungsten Hexafluoride*

For WF_6, there is a large difference between the value of 134 cm^{-1} for ν_6 assigned here and the values of 215 and 216 derived by Burke *et al.* and by Gaunt, respectively. Burke *et al.* assigned their band at 796 cm^{-1} as $3\nu_4$ instead of $(2 + 6)$ as assigned here. Gaunt assigned the equivalent band, at 808 cm^{-1} in his spectrum, as $(1 + 3 - 2)$. These assignments as ternary bands are surprising in view of the fact that $(2 + 6)$ is usually much more intense than any other band, barring a fundamental. There is an apparent misprint in the paper of Burke *et al.*: the intensity of the 796 cm^{-1} band is listed as weak instead of strong. Burke *et al.* assign a very weak band at 871 cm^{-1} as $(2 + 6)$; Gaunt assigns a similar band at 886 cm^{-1} as $(2 + 6)$. Neither of these bands appear in the spectrum of Weinstock and Claassen[52] and are probably due to impurities. Burke *et al.* assign a band at 448 cm^{-1} as $(2 - 6)$; Gaunt observes this band at 450 cm^{-1} and also assigns it as $(2 - 6)$. From the analysis given here, this band is assigned instead as $(5 + 6)$; $(2 - 6)$ cannot be seen in their spectra of WF_6. These changes in assignments lead to the large discrepancy

TABLE VIII. WF_6

Mode	Assigned value	(TD)[51]	(BSN)[41]	(G)[32]	(WC)[52]	Intensity
1	(771)		(772)	(769)		
1(l)	771	769	772			
2	(673)		(672)	(670)		
2(l)	671	670	672			
3	711		712	712	710	
4	258		258	(256)		
5	(315)		(316)	(322)		
5(l)	319	322	316			
6	(134)		(215)	(216)		
1 + 3	1482		1481	1483	1480	m
2 + 3	1384		1383	1384	1387	m
3 + 5	1026		1025[a]	1025[a]	1027	w
1 + 4	1029					a d
2 + 4	931		928	930	928	m
4 + 5	573		573	575		v w
2 + 6	807		796[a]	808[a]	808	s[a]
5 + 6	449		448[a]	450[a]		w
2 − 6	(539)					n o
2 − 4	(415)					n o
1 − 4	513			514		v v w
3 − 5	(396)					n o

[a] These assignments are discussed in the text.

between the value of ν_6 derived here and the values derived by Burke *et al.* and by Gaunt.

There is another assignment change made here which does not affect the values derived for the fundamental vibrations involved. The band at 1025 cm^{-1} is assigned by Burke *et al.* and by Gaunt as $(1 + 4)$, but is reassigned here as $(3 + 5)$. These bands are accidentally degenerate, but $(3 + 5)$ is much more intense than $(1 + 4)$ in spectra of metal hexafluorides for which the two bands are resolved.

The disagreement between the value of 134 cm^{-1} derived here for ν_6 and the much higher values derived by Burke *et al.* and by Gaunt is resolved in favor of our assignment by consideration of the entropy of WF_6 gas. Using the fundamental vibrational frequencies given here, S^0 for WF_6 at 300°K is 84.49 cal deg^{-1} $mole^{-1}$. This is to be contrasted with the value of 81.95 obtained using Gaunt's assignments. A preliminary evaluation of S^0 for WF_6 at 300°K, using the low-temperature heat-capacity measurements of Westrum and Weinstock,[54] gives 84.29, in agreement with the value calculated from our assignments.

(2) *Iridium Hexafluoride*

Because of the wider frequency range covered by Mattraw *et al.* they observed four more infrared-active bands than Claassen and Weinstock. Both papers report frequency values for the bands $(1 + 3)$ and $(1 + 4)$, but the values are significantly different from one another. Since there is no basis for choosing between these values, the two values have been averaged in the current analysis. As with WF_6, $(1 + 4)$ and $(3 + 5)$ are nearly degenerate at 977 cm^{-1}; the assignment as $(3 + 5)$ is again preferred because of its probable greater intensity. Consequently, in the present analysis, ν_1 was derived using $(1 + 3)$ only; ν_5 was derived using $(3 + 5)$ and $(5 + 6)$. The band assigned by Mattraw *et al.* as $(3 - 5)$ differs enough from the calculated value for $(3 - 5)$ to make this assignment questionable.

Claassen and Weinstock[49] reported that the vibrational spectra of IrF_6 do not show evidence of vibronic anomalies. They had previously attributed two features of the vibrational spectra of OsF_6 to a vibronic anomaly: an abnormal band profile for $(2 + 3)$ and a failure to observe ν_2 in the Raman effect under favorable conditions.[19,20] They had also pointed out that Gaunt's spectra for ReF_6[55] showed similar vibronic anomalies: the band profile for (2 + 3) was abnormal and ν_2, although observed in the Raman spectrum, was diffuse. In contrast, in their spectra for IrF_6 the band shape of (2 + 3) is normal (see Fig. 11, Section IV) and ν_2 is readily observed in the Raman effect, albeit in solution. From these observations, they concluded that IrF_6 is different from ReF_6 and OsF_6 because its ground electronic state is a spin quartet, for which vibronic coupling is not important.

(3) *Platinum Hexafluoride*

The data for PtF_6 are relatively meager, but the values of the fundamentals assigned by Weinstock, Claassen, and Malm[20]

TABLE IX

Mode	IrF_6				PtF_6		
	Assigned value	(MHCS)[53]	(CW)[49]	Intensity	Assigned value	(WCM)[20]	Intensity
1	(701)	(696)	(705)		(655)	(655)	
2	(646)	(643)	(644)		(600)	(601)	
2(sln)	651		651				
3	719	718	720		705	705	
4	276	276	(276)		273	273	
5	(258)	(260)			(242)	(242)	
6	(206)	(205)	(208)		(211)	(211)	
1 + 3	1420	1414	1425	m	1360	1360	m
2 + 3	1365	1361	1364	m	1305	1306	m
3 + 5	977	972	979[a]	w	947	947	w
1 + 4	977	972[a]		a d	928	928[a]	v w[a]
2 + 4	922	918	921	w	873	869	w
4 + 5	(534)			n o	(515)		n o
2 + 6	852	850	852	m w	811	812	w
5 + 6	464	465		v w	(453)		n o
2 − 6	440	440		v w	389	390	v w
2 − 4	(370)			n o	(327)		n o
1 − 4	(425)			n o	(382)		n o
3 − 5	461	450[a]		v w	(463)		n o

[a] These assignments are discussed in the text.

appear to be quite reliable. (1 + 4) was not reported, but can be seen in their spectra as a shoulder on (3 + 5) at about the calculated value of 928 cm^{-1}. The band shape of (2 + 3) for PtF_6 is normal, in agreement with its assigned nondegenerate ground state.

E. Second Transition Series Hexafluoride Molecules

The compounds MoF_6 and RhF_6 are two of the four hexafluoride molecules in the second or $4d$ transition series. The other two, TcF_6 and RuF_6, will be discussed in Section IV together with ReF_6 and OsF_6. The prospects for the preparation of the missing member of this series, PdF_6, are not favorable.[56]

MoF_6 is chemically similar to WF_6 and offers relatively little difficulty in its study. RhF_6 is quite difficult to study. It is strongly colored (reddish brown) and comparable to PtF_6 in instability and reactivity.

MoF_6 ($4d^0$) has a totally symmetric electronic ground state (A_{1g}). The electronic ground state for RhF_6 ($4d^3$) is essentially a spin quartet of symmetry $G_{\frac{3}{2}g}$, closely analogous to that of IrF_6 ($5d^3$). The electronic structure of RhF_6 is discussed in more detail in Section III.

Tanner and Duncan[51] (TD) made the initial spectroscopic measurements with MoF_6. As they had done with WF_6, they measured the three Raman shifts for liquid MoF_6 and determined the depolarization ratios for these bands. The depolarization ratios they obtained were: $\rho_1 = 0.09$, $\rho_2 = 0.94$, and $\rho_5 = 0.98$. These compare favorably with the respective theoretical values in O_h symmetry of 0.00, 0.87 and 0.87. The values they had obtained with WF_6 were respectively 0.04, 0.76, and 0.66. These depolarization measurements provided important information confirming the octahedral symmetry of these molecules. Before the studies of Tanner and Duncan the only structural evidence for these molecules was the electron-diffraction data of Braune and Pinnow.[38] The electron-diffraction data suggested an unsymmetrical structure rather than a regular octahedron. The resolution of this dilemma in favor of the symmetrical structure was previously discussed in the UF_6 section.

Burke, Smith, and Nielsen[41] (BSN) also measured the Raman

spectrum of liquid MoF_6 and, additionally, reported the infrared spectrum of the vapor from 2–40 μ. Gaunt[32] (G) measured the infrared spectrum of MoF_6 vapor from 2–25 μ. Claassen, Selig, and Malm[21] (CSM) measured the Raman shifts for ν_1 and ν_2 with MoF_6 vapor and also reported its infrared spectrum from 6–43 μ. The infrared spectrum of RhF_6 was measured by Weinstock, Claassen, and Chernick[22] (WCC). The MoF_6 and RhF_6 results as well as the current assignments are summarized in Table X.

(1) *Molybdenum Hexafluoride*

As was the case with WF_6, the current analysis gives an interpretation of the infrared spectral bands involving ν_6 for MoF_6 which is different from that given in the three experimental papers. Burke *et al.* assign a very weak band at 882 cm^{-1} to $(2 + 6)$; Gaunt also assigns the band at 882 cm^{-1} in his spectrum to $(2 + 6)$. Claassen, Selig, and Malm suggest that this band at 882 cm^{-1} is due to an impurity since they find it relatively much less intense in their spectra than in that of the others. Instead, they assign a very weak shoulder on ν_3 at 832 cm^{-1} to $(2 + 6)$; Burke *et al.* and Gaunt both assign this band as a ternary, $(1 + 5 + 6)$. In the current interpretation the shoulder on ν_3 near 765 cm^{-1} has been assigned as $(2 + 6)$. Burke *et al.* assign this band as $(2\nu_4 + \nu_6)$ and Gaunt calls it $(2\nu_6 + \nu_4)$. Claassen *et al.* do not report an absorption near 765 cm^{-1} although it has the same appearance in their spectra as in that of the other investigators. Burke *et al.* designate the intensity of their 764 cm^{-1} band as moderate. On the basis of intensity, it seems preferable to assign this band of moderate intensity at 764 cm^{-1} as $(2 + 6)$ rather than as a ternary, and, by the same token, it seems unlikely that the very weak bands at 882 or 832 cm^{-1} can possibly be $(2 + 6)$. This situation is analogous to that previously discussed for WF_6, although in that case the $(2 + 6)$ band was completely resolved.

This reassignment for $(2 + 6)$ affects other assignments involving ν_6. The band reported at 435 cm^{-1} in the three experimental papers is reassigned here as $(5 + 6)$. Burke *et al.* called it $(3 - 5)$ or $(2 - 6)$; Gaunt and Claassen *et al.* both assigned it as $(3 - 5)$. The value calculated for $(3 - 5)$ for the current assignments is 429 cm^{-1}, which would appear as a shoulder on $(5 + 6)$. Gaunt and Claassen *et al.* report a shoulder on the high-frequency side of (5 + 6) at 456 and 453 cm^{-1} respectively, which they assign as (2 — 6); Burke *et al.* report it at 455 cm^{-1} and assign it as a forbidden frequency, $2\nu_6$. From the current analysis this band is not assigned; (2 — 6), calculated, is 521 cm^{-1}.

Two other points are worth noting for MoF_6: a rather large liquid–vapor shift of 8 cm^{-1} is found for ν_5, and (2 + 4), calculated, is about 8 cm^{-1} displaced from (2 + 4), observed. A similar discrepancy involving (2 + 4) is observed for SeF_6.

Recent low-temperature heat-capacity measurements with MoF_6[57] provide evidence in favor of the present assignments and against those presented in the three experimental papers. The major discrepancy among the several assignments is in the value for ν_6. Burke *et al.* derived 234 cm^{-1}; Gaunt, 240 cm^{-1}; Claassen *et al.*, 190 cm^{-1}; this paper gives 122 cm^{-1}. The value of S^0 at 298.15°K using Gaunt's frequencies is 80.02 cal deg^{-1} $mole^{-1}$; Claassen *et al.* derive 81.30; using the frequencies given here the value is 83.77. The value of S^0 derived from the heat-capacity measurements is 83.75, in excellent agreement with the present assignments and in poor agreement with the alternative assignments for ν_6.

(2) *Rhodium Hexafluoride*

The infrared spectrum obtained by Weinstock, Claassen, and Chernick with RhF_6 shows one more sum band than that obtained with PtF_6. Again, all of the fundamental vibrational frequencies for RhF_6 have been derived with a measure of confidence. The values assigned to ν_2 and ν_6 in this paper differ slightly from the values given in the experimental paper. The value for ν_2 is the average of values derived from (2 + 3), (2 + 4), and (2 — 4), with the (2 + 3) value weighted twice. Using this value (592 cm^{-1}) for ν_2, the observed (2 + 6) and (2 — 6) bands then give identical values for ν_6.

Weinstock *et al.* point out that there is no evidence of vibronic effects in the infrared spectrum of RhF_6. For vibronic molecules, the binary bands involving ν_2 are either greatly broadened or missing from the infrared spectra. For RhF_6, the profile of the (2 + 3) band is similar to that for the normal hexafluoride molecules (see Fig. 10, Section IV), and four other binary bands

TABLE X

Mode	MoF_6						RhF_6		
	Assigned value	(TD)[51]	(BSN)[41]	(G)[32]	(CSM)[21]	Intensity	Assigned value	(WCC)[22]	Intensity
1	741		(741)	(736)	741		(634)	(634)	
1(l)	739	736	741						
2	643		(645)	(641)	643		(592)[a]	(595)	
2(l)	643	641	645						
3	741		741	742	741		724	724	
4	262		260	(269)	264		283	283	
5	(312)		(322)	(319)	(306)		(269)	(269)	
5(l)	320	319	322						
6	(122)		(234)	(240)	(190)		(189)[a]	(192)	
1 + 3	1482		1480	1481	1479	m	1358	1358	m
2 + 3	1384		1385	1384	1390	m	1316	1314	m
3 + 5	1053		1052	1052	1054	w	993	993	v w
1 + 4	1003		1002	1004	1005	v w	(917)		n o
2 + 4	905		913	914	914	w	875	878	v w
4 + 5	(574)					n o	(552)		n o
2 + 6	765		764[a]	763[a]		m	781	781	s
5 + 6	434		435[a]	434[a]	435[a]	v w	(458)		n o
2 − 6	(521)					n o	403	403	v w
2 − 4	(381)					n o	309	312	v w
1 − 4	479				480	v v w	(351)		n o
3 − 5	(429)					n o	(455)		n o

[a] These assignments are discussed in the text.

involving ν_2 are observed. In view of the extreme experimental difficulties in studying RhF_6, the observation of five bands containing ν_2 would be unlikely if the ν_2 transitions were vibronic (*cp.* RuF_6, Section IV). The nature of the infrared spectrum of RhF_6 thus strongly supports the analogy between this molecule and IrF_6.

III. THEORY

In this section, we discuss briefly electronic assignments for gross spectral features of transition metal hexafluoride molecules. Then, we review the influence of nuclear motions on details of the lowest energy levels for these molecules.

A. Electronic Spectra and Assignments

Weak absorption bands at wavelengths longer than 4000 A occur for hexafluoride molecules formed by transition metals with more than six valence electrons. Moffitt, Goodman, Fred, and Weinstock[15] reported that for the third transition series these absorptions come between 3000 and 20,000 cm^{-1}. They assigned these bands to primarily electronic transitions within the ligand-field energy levels for an octahedrally symmetric nuclear configuration.

An octahedral ligand field separates the five components of atomic *d*-orbitals into a group of two (e_g) at higher energy and another group of three (f_{2g}) at lower energy. The energy that separates the two e_g-orbitals from the three f_{2g}-orbitals was assigned by Moffitt *et al.*[15] to be some 30,000 cm^{-1}, nearly 4 eV, for hexafluoride molecules of elements in the third transition series.

The possibility of mixing together e_g-orbitals and f_{2g}-orbitals may thus be expected to be unimportant for a qualitative, roughly quantitative understanding of the lowest energy levels in these molecules. The preceding statement is supported, for example, by the results of a calculation for IrF_6 by Runciman and Schroeder,[58] who took full account of the possible mixing of e_g- and f_{2g}-states induced by spin–orbit and electron-correlation energies. They found that each of the lower energy levels of IrF_6 is more than 90% composed of states arising from the configuration f_{2g}^3. We shall, therefore, follow Moffitt *et al.* and consider in detail only configurations of the type f_{2g}^n, where n is the number of valence electrons in excess of six for the neutral metal atom of interest.

The configuration f_{2g}^n yields, in general, several energy levels under the combined influence of spin–orbit coupling and Coulomb repulsions of the nonbonding valence electrons. Approximate energies separating these levels may be conveniently expressed in

TABLE XI. The Coefficient Matrices for Calculating the Ligand-Field Energy Levels of the Configurations f_{2g}^n, $n = 1, 2, 3, 4$, or 5. Only elements on or above the principal diagonal are shown

$n = 1$

$$
\begin{array}{c}
J' \\ \tfrac{3}{2} \\ \tfrac{1}{2} \\ J'
\end{array}
\begin{array}{c}
\mathbf{A}^{(1)} \\
\begin{bmatrix} 0 & 0 \\ & 0 \end{bmatrix} \\
\begin{array}{cc} \tfrac{3}{2} & \tfrac{1}{2} \end{array}
\end{array}
\qquad\qquad
\begin{array}{c}
\mathbf{B}^{(1)} \\
\begin{bmatrix} -\tfrac{1}{2} & 0 \\ & 1 \end{bmatrix} \\
\begin{array}{cc} \tfrac{3}{2} & \tfrac{1}{2} \end{array}
\end{array}
$$

$n = 2$

$$
\begin{array}{c|ccccc|ccccc}
J' & & & \mathbf{A}^{(2)} & & & & & \mathbf{B}^{(2)} & & \\
0 & 4 & 0 & 0 & 0 & 0 & 0 & \sqrt{2} & 0 & 0 & 0 \\
0 & & -1 & 0 & 0 & 0 & & 1 & 0 & 0 & 0 \\
1 & & & -1 & 0 & 0 & & & \tfrac{1}{2} & 0 & 0 \\
2 & & & & -1 & 0 & & & & -\tfrac{1}{2} & -\sqrt{\tfrac{1}{2}} \\
2 & & & & & 1 & & & & & 0 \\
J' & 0 & 0 & 1 & 2 & 2 & 0 & 0 & 1 & 2 & 2
\end{array}
$$

$n = 3$

$$
\begin{array}{c|ccccc|ccccc}
J' & & & \mathbf{A}^{(3)} & & & & & \mathbf{B}^{(3)} & & \\
\tfrac{3}{2} & -3 & 0 & 0 & 0 & 0 & 0 & 0 & 1 & 0 & 0 \\
\tfrac{1}{2} & & 2 & 0 & 0 & 0 & & 0 & 0 & 0 & 0 \\
\tfrac{3}{2} & & & 2 & 0 & 0 & & & 0 & \sqrt{\tfrac{5}{4}} & 0 \\
\tfrac{3}{2} & & & & 0 & 0 & & & & 0 & 0 \\
\tfrac{5}{2} & & & & & 0 & & & & & 0 \\
J' & \tfrac{3}{2} & \tfrac{1}{2} & \tfrac{3}{2} & \tfrac{3}{2} & \tfrac{5}{2} & \tfrac{3}{2} & \tfrac{1}{2} & \tfrac{3}{2} & \tfrac{3}{2} & \tfrac{5}{2}
\end{array}
$$

$\mathbf{A}^{(4)} = \mathbf{A}^{(2)}$; $\mathbf{B}^{(4)} = -\mathbf{B}^{(2)}$ and $\mathbf{A}^{(5)} = \mathbf{A}^{(1)}$; $\mathbf{B}^{(5)} = -\mathbf{B}^{(1)}$

TABLE XII. Ligand-Field Energy Levels in cm^{-1} Units for Configurations f_{2g}^{n}

Case I: $\zeta = 3400$ cm^{-1} and $G = 2400$ cm^{-1}

	J'	$\frac{1}{2}$		$\frac{3}{2}$		
ReF_6	$W^{(1)}$	3400		−1700		
	$(W^{(1)} + 1700)$	5100		0		
	J'	0	2	1	0	2
OsF_6	$W^{(2)}$	11,751	3193	−700	−1151	−4893
	$(W^{(2)} + 4893)$	16,644	8086	4193	3742	0
	J'	$\frac{3}{2}$	$\frac{1}{2}$	$\frac{5}{2}$	$\frac{3}{2}$	$\frac{3}{2}$
IrF_6	$W^{(3)}$	7510	4800	0	−1684	−8226
	$(W^{(3)} + 8226)$	15,736	13,026	8226	6542	0
	J'	0	2	2	1	0
PtF_6	$W^{(4)}$	10,978	3711	−2011	−4100	−7178
	$(W^{(4)} + 7178)$	18,156	10,889	5167	3078	0

Case II: $\zeta = 1281$ cm^{-1} and $G = 2989$ cm^{-1}

	J'	$\frac{1}{2}$	$\frac{3}{2}$			
TcF_6	$W^{(1)}$	1281	−640			
	$(W^{(1)} + 640)$	1921	0			
	J'	0	2	0	1	2
RuF_6	$W^{(2)}$	12,192	3111	−1944	−2348	−3751
	$(W^{(2)} + 3751)$	15,943	6862	1807	1403	0
	J'	$\frac{3}{2}$	$\frac{1}{2}$	$\frac{5}{2}$	$\frac{3}{2}$	$\frac{3}{2}$
RhF_6	$W^{(3)}$	6405	5978	0	−316	−9078
	$(W^{(3)} + 9078)$	15,483	15,056	9078	8762	0
	J'	0	2	2	1	0
PdF_6	$W^{(4)}$	12,156	3139	−2498	−3630	−4470
	$(W^{(4)} + 4470)$	16,626	7609	1972	840	0

terms of two parameters: the spin–orbit energy ζ and a Coulomb energy G. Diagonalization of the matrix

$$\mathbf{W}^{(n)}(G, \zeta) = G\mathbf{A}^{(n)} + \zeta\mathbf{B}^{(n)}$$

yields the energy levels of interest with the energetic center of gravity for the configuration f_{2g}^{n} as zero. The coefficient matrices $\mathbf{A}^{(n)}$ and $\mathbf{B}^{(n)}$ may be evaluated as explained by Moffitt *et al.*[15] or by Griffith[59] through an analogy between p- and f_{2g}-orbital properties. The resulting coefficient matrices are collected in Table XI. Only elements on or above the principal diagonal are shown for these symmetric matrices.

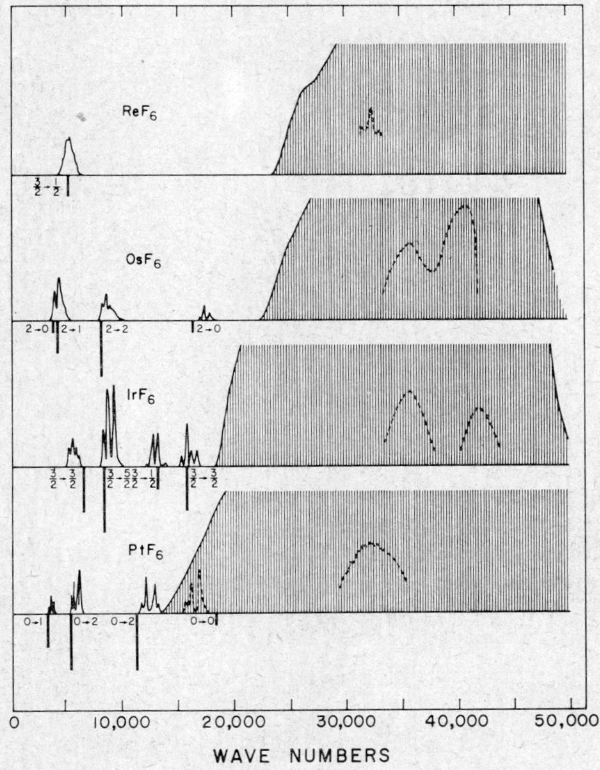

Fig. 3. A comparison of observed and calculated spectra for hexafluoride molecules formed by metals from the third transition series (Moffitt, Goodman, Fred, and Weinstock[15]).

Using these coefficient matrices and the values $G = 2400\ \text{cm}^{-1}$ and $\zeta = 3400\ \text{cm}^{-1}$, Moffitt *et al.* obtained the energy levels listed in Table XII. The agreement between theory and experiment is shown in Fig. 3.

The value of J', the total effective angular momentum, for each energy level indicates characteristics of the electronic wave functions. For example, $d_{J'} = (2J' + 1)$ is the number of independent states that make up an energy level designated by J'. A summary of the relationships between these values of J' and the irreducible representations of O_h^*, the octahedral double-group, is given in the following table. The notation used here for the representations of the octahedral double-group is similar to that used by Moffitt and Thorson,[9] with the exceptions that F is used instead of T to indicate a triply degenerate representation and fractional subscripts are used instead of primes to indicate the double-valued representations.

TABLE XIII. The Correspondence between Values of J', the Effective Electronic Angular Momentum, and Irreducible Representations of O_h^*

J'	0	$\frac{1}{2}$	1	$\frac{3}{2}$	2	$\frac{5}{2}$
$\Gamma(J')$	A_{1g}	$E_{\frac{5}{2}g}$	F_{1g}	$G_{\frac{3}{2}g}$	E_g, F_{2g}	$E_{\frac{1}{2}g}$, $G_{\frac{3}{2}g}$

For the purposes of this article, we have extended the work of Moffitt *et al.*[15] to make assignments of near-infrared and visible absorption spectra that have been observed for recently synthesized hexafluorides of elements from the second transition series: TcF_6 (Selig, Chernick, and Malm[60]) and RuF_6 (Claassen, Selig, Malm, Chernick, and Weinstock[61]). The first member of this series, MoF_6, has been known for some time and is transparent in the spectral region of interest here. The last known member of this series, RhF_6, prepared by Chernick, Claassen, and Weinstock,[56] has not yet been studied in this spectral range because of its high chemical reactivity. It turns out, however, that preliminary and previously unpublished observations for TcF_6 and RuF_6 are just sufficient to permit the estimation of ζ and G

for this series of compounds and to provide crude predictions about the absorption features to be expected for RhF_6 and for the still unknown molecule PdF_6.

A broad, weak absorption band occurs for TcF_6 in the infrared, centered at approximately 1900 cm^{-1}. If this energy is identified with the transition from $J' = \frac{3}{2}$ to $J' = \frac{1}{2}$ within f_{2g}, then the value of ζ for $4d$ hexafluoride molecules is approximately $\frac{2}{3}$ of 1900 cm^{-1}. The calculated transition energy for TcF_6 listed in Table XII is 1921 cm^{-1}. This value comes from a preliminary set of energy parameters that seem to fit the spectra of both TcF_6 and RuF_6 reasonably well.

There are five weak absorption bands observed in the near-infrared for RuF_6. Two of these bands, centered at 1400 cm^{-1} and at 1950 cm^{-1}, are mentioned by Weinstock, Claassen, and Chernick.[22] The remaining ones, centered at 5900 cm^{-1}, at 7800 cm^{-1}, and at 11,900 cm^{-1}, can be identified in unpublished spectra obtained by Claassen.[44] A value of G in the neighborhood of 3000 cm^{-1} yields what seem the most reasonable assignments for these features of RuF_6. For $(G + \zeta) = 4270$ cm^{-1} and $x = \zeta/(G + \zeta) = 0.3$, the energy levels given in Table XII are calculated for RuF_6. Thus, we assign the band at 1400 cm^{-1} to a transition from the ground state of this molecule to an electronically excited, $J' = 1$ level; the band at 1950 cm^{-1} corresponds to an excited $J' = 0$ state; and the band at 11,900 cm^{-1} we assign to a transition from the ground state to another excited $J' = 0$ state, the one calculated to be at 15,943 cm^{-1}. The main reason for this discrepancy of some 4000 cm^{-1} for the spectral position of the higher $J' = 0$ level is probably the increasing importance of the mixing of the f_{2g}^2 configuration with the $f_{2g}e_g$ and the e_g^2 configurations as the energy of the levels becomes higher. More evidence for the importance of such mixing of configurations can be seen in the fact that two separate bands occur in the region of 6800 cm^{-1}, where one fivefold degenerate energy level is expected in the strong-ligand-field approximation. The bands at 7800 cm^{-1} and at 5900 cm^{-1} can be assigned to an E_g and an F_{2g} excited state for the RuF_6 molecule; but there is no information at present about which is which. We must expect a corresponding, probably smaller splitting of the lower quintet, $J' = 2$, to influence the vibronic properties of RuF_6; however, no transition between these two lowest electronic energy levels can be identified in the presently available spectra of RuF_6.

The predicted strong-ligand-field energy levels for RhF_6 based on $(\zeta + G) = 4270$ cm^{-1} and $x = 0.3$ are also given in Table XII. Of course, these predictions must be viewed with caution. For example, the $J' = \frac{5}{2}$ sextet will no doubt be split noticeably into a doublet and a quartet. Nevertheless, the prediction that the lowest energy level of RhF_6 is rather well removed from the remaining ones seems quite reliable. Indeed, the spectrum reported by Weinstock *et al.*[22] for RhF_6 seems to bear out this prediction at least up to energies of 1400 cm^{-1}.

Corresponding, approximate predictions for the still unknown molecule PdF_6 are also tabulated for $(G + \zeta) = 4270$ cm^{-1} and $x = 0.3$. The only features of these predictions that deserve comment at this time are that the lowest energy level for PdF_6 should be nondegenerate and that this molecule should have an electronic transition in the spectral region commonly associated with the fundamental absorption for fluorine–metal stretching vibrations, that is, just above 700 cm^{-1}.

According to the assignments of Moffitt *et al.*,[15] the lowest electronic states for ReF_6 and OsF_6 are orbitally degenerate for an octahedrally symmetric nuclear arrangement. Thus, the theorem of Jahn and Teller[12,13] applies to these two molecules, predicting that their symmetric nuclear configurations cannot be stable with respect to all nuclear displacements. However, several qualitative arguments lead to the inference that this expected instability does not produce a static distortion of ReF_6 or of OsF_6 in the vapor phase.

First, the gross features of the electronic absorption spectra for these molecules, ReF_6 and OsF_6, agree with the simple ligand-field theory based on an octahedrally symmetric field just as well as do those of the neighboring molecules, IrF_6 and PtF_6, for which, according to the assignments of Moffitt *et al.*, the Jahn–Teller theorem predicts no instability of an octahedral nuclear configuration.* If ReF_6 or OsF_6 were statically distorted, there

* For IrF_6, the lowest electronic energy level is fourfold degenerate, but this degeneracy arises principally from the spin portion of the electronic wave function and is thus relatively insensitive to vibronic coupling. For PtF_6, the electronic state of lowest energy is nondegenerate, and therefore the Jahn–Teller theorem is not applicable.

should be more electronic band-systems for these two molecules than the number expected on the basis of simple ligand-field theory.

Secondly, electron-diffraction studies have been made on several of these molecules, WF_6, OsF_6, and IrF_6, by Schomaker, Weinstock, Smith, and Kimura.[62] Their diffraction studies yield results that are similar for all these molecules and compatible with a regular octahedral structure for each of them.

Thirdly, infrared absorption spectra of ReF_6 and OsF_6 are only slightly different from the spectra of the certainly octahedral molecules WF_6, IrF_6, and PtF_6. In each case, for example, only one fundamental absorption occurs in the infrared region associated with fluorine–metal stretching and only one fundamental absorption in the region associated with bending of fluorine–metal–fluorine angles. If either ReF_6 or OsF_6 were distorted, force constants associated with its longer bonds (or wider angles) should be different from those associated with its shorter bonds (or smaller angles); and a splitting of the infrared fundamental absorption bands should be observed.

Thus, Claassen and Weinstock[19,49] were led to conclude quite correctly (as we show in some detail in Section IV) that certain of the rather subtle differences in spectral profiles for these molecules are manifestations of dynamic coupling between nuclear and electronic motions. The term vibronic is used generally to indicate this type of intimate mixture of vibrational and electronic properties.

Fewer data are available for hexafluoride molecules in the palladium series than for those in the platinum series. But the qualitative evidence for vibronic coupling in TcF_6 and in RuF_6 is similar to that for ReF_6 and for OsF_6, respectively. That RhF_6 shows no vibronic peculiarities in its infrared spectra is not surprising because the quartet of states at lowest energy for RhF_6 has almost entirely spin character to its degeneracy. This spin character is more pronounced for RhF_6 with $x \sim 0.3$ than for IrF_6 with $x \sim 0.6$. The qualitative similarities between the second and the third transition series were mentioned by Claassen *et al.*[61] and by Claassen, Selig, and Malm[21] and stressed by Weinstock *et al.*[22] in their respective discussions of the vibrational spectra of RuF_6, of TcF_6, and of RuF_6 and RhF_6. A summary of such systematic evidence about vibronic coupling is given at the beginning of Section IV.

B. Theory of Vibronic Coupling

Moffitt and Thorson[9,63] first reported vibronic energy levels for octahedral molecules. Energy levels that they calculated for E_g, F_{2g}, and $G_{\frac{3}{2}g}$ electronic states are most pertinent here. Molecular properties that are sensitive to vibronic coupling for octahedral molecules have also been described in detail by Thorson[64] and by Child.[50,65] A more complete review than that given here of the literature on the theory of vibronic coupling can be found in an essay by Liehr.[66]

The phenomenological approach to vibronic coupling starts from an observation that the Hamiltonian operator or energy for a molecule is invariant under all permutations of the identical nuclei in a molecule. (Recently, the importance of this invariance with respect to nuclear permutations has been emphasized in incisive discussions of molecular symmetry properties by Hougen[67] and by Longuet-Higgins.[68]) In the phenomenological analysis of vibronic coupling, the wave function for each steady state of the molecule is factored into an "electronic part" and a "nuclear part". Viewed in terms of their distribution on a scale of energy, the various possible "electronic parts" occur in energetic families of one or more, with wide spacings in energy between different families and very little, if any, energy difference between the members of a particular family. Associated with each family of "electronic parts", there is a whole spectrum of "nuclear parts" for the molecular wave function. These "nuclear parts" describe the possible translational, rotational, and vibrational motions of the molecule, and their separations from each other on the energy scale are much smaller than the separations between families of "electronic parts". This disparity in energy differences forms the basis for the next step in the procedure: the functional form of the "electronic part" of the molecular wave function is taken to be independent of the particular "nuclear part" with which it is combined. This last step is made reasonable energetically by letting the "electronic part" of the molecular wave function depend on the nuclear coordinates. In this way the energy associated with each "electronic part" usually becomes a well behaved, rather mildly varying function of the nuclear arrangement. The Born–Oppenheimer approximation[1] consists of regarding this parametric nuclear dependence of the "electronic parts" as negligible with regard to the nuclear kinetic energy. Even when the Born–Oppenheimer approximation is quite valid, however, the distinction between the "electronic part" and the "nuclear part" of a molecular wave function is by no means as clearcut as these names might suggest, if they were used without some reservation, such as quotation marks.

In particular, permutations of indistinguishable nuclei in a molecule transform certain of the "electronic parts" of a molecule's wave functions into each other by interchanging their parametric dependencies on nuclear coordinates. Because the entire energy of the molecule remains unaltered by these nuclear permutations, only "electronic parts" which belong to the same energetic families are interchanged in this way.*

Often there is a special geometry for the nuclear configuration of the molecule in which all of the "electronic parts" in a particular energetic family have exactly the same energy associated with them. Simple and useful heuristic treatments of some molecular symmetry properties are based on the point group of rotations and reflections that exchange this symmetrical nuclear geometry for an indistinguishable one. One result of the use of the algorism based on a molecular point group is, however, to make difficult the placing of proper emphasis on the significance of nuclear permutations. Thus, we have chosen not to begin our discussion of vibronic properties with a reference to the most highly symmetric nuclear configuration that can arise.

Probably the clearest way to obtain useful results from those transformations of "electronic" and "nuclear" parts in a molecular

* Of course, the "nuclear parts" of a molecule's wave functions are also interchanged selectively by nuclear permutations; and only an entire molecular steady-state wave function, the product in this procedure of "electronic" and "nuclear" parts, transforms in such a way as to keep the molecular energy exactly unchanged. Nevertheless, if there is one (or even a set of) nuclear configuration(s) from which only small displacements yield all of the nuclear configurations of importance for the steady states of interest, then this characterization of "electronic parts" according to energetic families remains an extremely useful concept.

wave function that are induced by nuclear permutations is the way suggested by Longuet-Higgins:[68] namely, to select the largest set of these permutations that satisfy the rules of combination for an abstract group and that produce transformations of molecular properties which are what Longuet-Higgins calls " 'feasible transformations', those which can be achieved without passing over an insuperable energy barrier". This set of permutations must in general be supplemented by additional operations in order to express more fully the symmetry properties of the molecular Hamiltonian. These additional operations are those combinations of the inversion of the coordinate system for the molecule with some permutation of its identical nuclei for which the composite operation induces "feasible transformations" of molecular properties. The molecular Hamiltonian itself is, of course, considered to be unchanged either by inversion of the coordinates of every particle in the molecule or by permutations of identical nuclei, and thus by any composite of these. But, in certain molecules, e.g., octahedral or tetrahedral ones, inversion of the coordinates of every particle induces transformations that are not "feasible" in the sense of Longuet-Higgins, unless combined with a nuclear permutation that returns the nuclei to their original order relative to each other, so that a superposition of similarly labelled nuclei in the initial and the final forms of the molecule can be obtained by a rotation of the molecular axes. The largest possible set of these composite operations is annexed to the already selected set of permutations so that the combined set still satisfies the rules of combination for an abstract group. The group formed in this way is what Longuet-Higgins calls the *molecular symmetry group*,[68] and proves to be most useful for specifying molecular symmetry properties.

The molecular symmetry group for an octahedral MF_6 molecule $O_h(M)$, has 48 elements and 10 classes in it. They are listed in Table XIV as permutations of the labels attached to the six fluorine nuclei in Fig. 4. The standard notation for permutations is used here so that, for example, $(1\bar{1})(2\bar{2})$ indicates that the labels 1 and $\bar{1}$ are interchanged at the same time that the labels 2 and $\bar{2}$ are interchanged, and $(12\bar{1}\bar{2})$ means that the label 1 is substituted for 2, 2 for $\bar{1}$, $\bar{1}$ for $\bar{2}$, and $\bar{2}$ for 1. The operation E^*, as defined by Longuet-Higgins,[68] inverts the spatial coordinates of all electrons and nuclei in the molecule. It is noteworthy that E^* itself is not an element in the molecular symmetry group for an octahedral MF_6 molecule, although various operations involving E^* are elements in this group: an example is $E^*(3\bar{3})$, which is written $(3\bar{3})^*$. It may be verified that this molecular symmetry group has the same structure to its multiplication table as has the point group of rotations and reflections that take a regular octahedron into itself, O_h. Thus, the two groups are

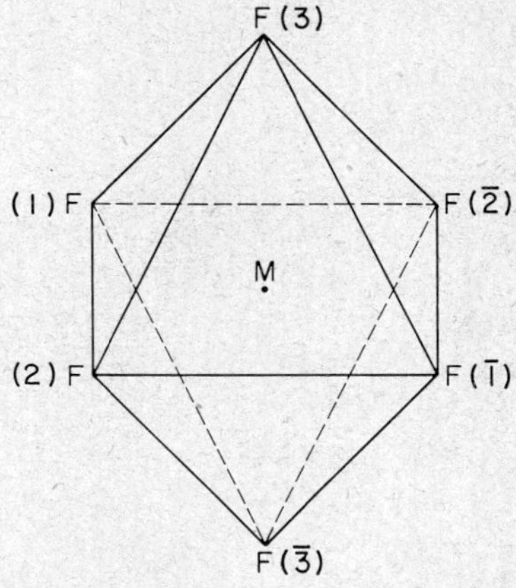

Fig. 4. Nuclear labels associated with an MF_6 molecule that has an octahedrally symmetric nuclear configuration.

said to be isomorphic; that is to say, at least one one-to-one correspondence exists between the individual elements in these two groups. (Details of the theory of abstract groups and of their representations can be found in any of a large number of books, for example, Wigner's *Group Theory*.[69])

Two ways suggest themselves immediately for exhibiting the isomorphism between the point group O_h and the molecular symmetry group $O_h(M)$ for an octahedral MF_6 molecule. Both of these ways rely on the identification of the element $(1\bar{1})(2\bar{2})(3\bar{3})^*$ in the molecular symmetry group with the inversion element I in the point group. The first way of exhibiting the isomorphism is to identify each simple permutation (i.e., each element of $O_h(M)$ that does not involve E^*) with the point-group operation that would accomplish the same relabelling of fluorine nuclei for an octahedrally symmetric configuration as the permutation accomplishes. The second way of exhibiting the isomorphism is to identify each simple permutation with the point-group operation that would accomplish the inverse relabelling of fluorine nuclei and thereby restore the labels to their original locations if both the simple permutation and the point-group operation were applied to an octahedrally symmetric nuclear configuration. [It should be noticed that the combination of a rotation–reflection element of O_h with the corresponding element of $O_h(M)$ (one that now involves E^*) does *not* restore the labels to their original locations. For an example of this statement, one needs only to consider the inversion element I of O_h and the corresponding element $(1\bar{1})(2\bar{2})(3\bar{3})^*$ of $O_h(M)$.] We shall return shortly to the question of which of these two correspondences seems to be the more useful to us.

We note that the character table for the molecular symmetry group $O_h(M)$ is just the same as that for the point group O_h because they are isomorphic groups. This character table may be obtained from any of many sources, such as Herzberg's *Infrared and Raman Spectra*, page 123,[23] and it is reproduced in Table XIV for convenience. The class labels used here for $O_h(M)$, namely E, C_2, etc., are chosen to point up the analogy with O_h, for which such labels are standard.

The molecular symmetry group $O_h(M)$ can be modified to produce another group that is more useful for discussing the internal motions of hexafluoride molecules. Here we mean to distinguish external motions, such as translations and rotations of the molecule as a whole, from internal motions, such as nuclear vibrations and electronic motions relative to the nuclear framework. A symmetry group modified to apply only to these internal motions may be termed *a vibronic symmetry group of the molecule.* A vibronic group is constructed for what Longuet-Higgins[68] calls a rigid molecule along the lines laid down by Hougen.[67] (The general conditions that must be satisfied in order that a vibronic group exists for a nonrigid molecule were discussed by Longuet-Higgins and Goodman.[70]) Two essential features of the elements in a vibronic symmetry group for a rigid molecule are, first, that they form a group, and, secondly, that they leave *the nuclear labels*

TABLE XIV. The Character Table for $O_h(M)$

	E	C_2	C_4	C_3	C_2'	I	σ_h	S_4	S_6	σ_d
	1	3	6	8	6	1	3	6	8	6
	E	$(1\bar{1})(2\bar{2})$ $(2\bar{2})(3\bar{3})$ $(3\bar{3})(1\bar{1})$	$(12\bar{1}\bar{2})$ $(1\bar{2}\bar{1}2)$ $(13\bar{1}\bar{3})$ $(1\bar{3}\bar{1}3)$ $(23\bar{2}\bar{3})$ $(2\bar{3}\bar{2}3)$	$(123)(\bar{3}\bar{1}\bar{2})$ $(132)(\bar{3}\bar{2}\bar{1})$ $(2\bar{1}3)(\bar{3}\bar{2}1)$ $(23\bar{1})(\bar{3}1\bar{2})$ $(12\bar{3})(3\bar{1}\bar{2})$ $(1\bar{3}2)(3\bar{2}\bar{1})$ $(2\bar{1}\bar{3})(13\bar{2})$ $(2\bar{3}\bar{1})(1\bar{2}\bar{3})$	$(\bar{1}\bar{2})(12)(\bar{3}3)$ $(1\bar{2})(2\bar{1})(\bar{3}3)$ $(23)(\bar{2}\bar{3})(1\bar{1})$ $(2\bar{3})(\bar{2}3)(1\bar{1})$ $(1\bar{3})(\bar{1}3)(2\bar{2})$ $(13)(\bar{1}\bar{3})(2\bar{2})$	$(\bar{3}3)(1\bar{1})(2\bar{2})$*	$(\bar{3}3)$* $(1\bar{1})$* $(2\bar{2})$*	$(\bar{3}3)(1\bar{2}\bar{1}2)$* $(\bar{3}3)(12\bar{1}\bar{2})$* $(2\bar{2})(1\bar{3}\bar{1}3)$* $(2\bar{2})(13\bar{1}\bar{3})$* $(1\bar{1})(2\bar{3}\bar{2}3)$* $(1\bar{1})(23\bar{2}\bar{3})$*	$(1\bar{2}3\bar{1}2\bar{3})$* $(1\bar{3}2\bar{1}3\bar{2})$* $(213\bar{2}\bar{1}\bar{3})$* $(2\bar{3}\bar{1}\bar{2}31)$* $(1\bar{2}\bar{3}\bar{1}23)$* $(132\bar{1}\bar{3}\bar{2})$* $(21\bar{3}\bar{2}\bar{1}3)$* $(23\bar{1}\bar{2}\bar{3}1)$*	$(12)(\bar{1}\bar{2})$* $(1\bar{2})(2\bar{1})$* $(23)(\bar{2}\bar{3})$* $(2\bar{3})(\bar{2}3)$* $(1\bar{3})(\bar{1}3)$* $(13)(\bar{1}\bar{3})$*
A_{1g}	1	1	1	1	1	1	1	1	1	1
A_{1u}	1	1	1	1	1	−1	−1	−1	−1	−1
A_{2g}	1	1	−1	1	−1	1	1	−1	1	−1
A_{2u}	1	1	−1	1	−1	−1	−1	1	−1	1
E_g	2	2	0	−1	0	2	2	0	−1	0
E_u	2	2	0	−1	0	−2	−2	0	1	0
F_{1g}	3	−1	1	0	−1	3	−1	1	0	−1
F_{1u}	3	−1	1	0	−1	−3	1	−1	0	1
F_{2g}	3	−1	−1	0	1	3	−1	−1	0	1
F_{2u}	3	−1	−1	0	1	−3	1	1	0	−1

and the nuclear locations unaltered for a particular nuclear configuration: what we call the reference configuration for this vibronic symmetry group.

For example, with an octahedrally symmetric reference configuration for an MF_6 molecule, as shown in Fig. 4, the element $(1\bar{1})(2\bar{2})$ of $O_h(M)$ is combined with an overall rotation through π radians about the $3\bar{3}$-axis of the molecule to yield an element, $C(1\bar{1})(2\bar{2})$, say, of the vibronic symmetry group $O_h(V)$. Two elements of $O_h(M)$, namely E and $(1\bar{1})(2\bar{2})(3\bar{3})^*$, are combined with the identity rotation to yield elements of $O_h(V)$, E, and $C(1\bar{1})(2\bar{2})(3\bar{3})^*$, respectively. To obtain elements of $O_h(V)$ more generally, each simple permutation of $O_h(M)$ is combined with that element of O_h with which it was identified in the second correspondence given above for exhibiting the isomorphism between O_h and $O_h(M)$; and each element of $O_h(M)$ that involves E^* is combined with I times the corresponding element of O_h. Thus, it seems that the second or "inverse" correspondence between elements of O_h and those of $O_h(M)$ is the more useful to us. As one may suspect, $O_h(V)$ is also isomorphic to O_h. But before we go into this correspondence, we take into account the possibility of molecules with an odd number of electron spins.

In order to discuss problems involving the spins of an odd number of electrons, it is useful to extend the definition of the vibronic symmetry group $O_h(V)$ to obtain *a vibronic symmetry double-group*, $O_h^*(V)$. In the vibronic symmetry double-group, a rotation through 2π radians is an element distinct from the identity element and is given the designation R. Thus for example, in $O_h^*(V)$, $C(1\bar{1})(2\bar{2})\cdot C(1\bar{1})(2\bar{2}) = R$, not E as in $O_h(V)$. The reason for introducing this new element R is to permit a simulation of the behavior of wave functions with half-integral angular momenta. Under a rotation through 2π radians these wave functions go into the negative of themselves and only after a further rotation through 2π radians are they restored to their original form. Thus, $R\cdot R = E$, in $O_h^*(V)$. In listing the elements of $O_h^*(V)$, we find it convenient to write $R(1\bar{2}\bar{1}2)$ for $R\cdot[C(12\bar{1}\bar{2})]^3$, and $R(1\bar{3}2\bar{1}3\bar{2})^*$ for $R\cdot[C(1\bar{2}3\bar{1}2\bar{3})^*]^5$, etc.

The vibronic symmetry double-group for octahedral metal hexafluoride molecules has 96 elements and 16 classes. It is isomorphic to the better known double-group O_h^*. The character table for $O_h^*(V)$ can therefore be constructed by consulting any of a number of sources, for example, Ballhausen's *Introduction to Ligand Field Theory*,[71] and is given here in Table XV for convenience.

The internal properties of a molecule form bases for linear substitutions that represent the elements of its vibronic symmetry double-group. Three of these bases that are most important for describing phenomenologically vibronic interactions in a molecule are the following ones: nuclear displacement coordinates measured from the molecule's reference configuration, "electronic parts" of the molecular wave functions, and, finally, matrix elements between these various "electronic parts". Each of these representatives for the vibronic symmetry double-group $O_h^*(V)$ is now discussed in turn, starting with the nuclear displacements.

Figure 1 shows two-dimensional projections of nuclear displacements from an octahedrally symmetric configuration for a MF_6 molecule. The first displacement, shown in the upper left-hand corner, preserves the symmetry of the octahedral arrangement. It simply lengthens each M–F distance by the same amount. This displacement is left completely unchanged by every element in the vibronic symmetry double-group, $O_h^*(V)$. That is to say, it forms a basis for the identity or A_{1g} representation of $O_h^*(V)$.

The second and the third nuclear displacements pictured in the top line of Fig. 1 span a two-dimensional representation of $O_h^*(V)$. This representation is called E_g in Table XV.

The three displacements in each of the other four rows of Fig. 1 span three-dimensional representations of $O_h^*(V)$. The similarity of the pictured distortions in the second and the third row makes it clear that the two corresponding sets of displacements span the same irreducible representation; it is called F_{1u} in Table XV. The displacements in the fourth row span the representation F_{2g}; while those in the fifth row span F_{2u}.

The reader may wish to verify that the sum of the diagonal elements in the matrices which stand for various elements of $O_h^*(V)$ in each of these representations is equal to the character for that element in the proper representation, as given in Table XV. For example, the character of $C(123)(\bar{1}\bar{2}\bar{3})$ acting on the e_g displacements is $(-\frac{1}{2}-\frac{1}{2})=-1$, in agreement with the corresponding entry in Table XV.

TABLE XV. The Chara

	E	R	C_2	C_4	RC_4	C_3	RC_3	C_2'	I
	1	1	6	6	6	8	8	12	1
	E	R	$C(1\bar{1})(2\bar{2})$ $C(2\bar{2})(3\bar{3})$ $C(3\bar{3})(1\bar{1})$ $R(1\bar{1})(2\bar{2})$ $R(2\bar{2})(3\bar{3})$ $R(3\bar{3})(1\bar{1})$	$C(12\bar{1}\bar{2})$ $R(32\bar{3}\bar{2})$ $C(31\bar{3}\bar{1})$ $R(21\bar{2}\bar{1})$ $C(23\bar{2}\bar{3})$ $R(13\bar{1}\bar{3})$	$R(12\bar{1}\bar{2})$ $C(32\bar{3}\bar{2})$ $R(31\bar{3}\bar{1})$ $C(21\bar{2}\bar{1})$ $R(23\bar{2}\bar{3})$ $C(13\bar{1}\bar{3})$	$C(123)(\bar{1}\bar{2}\bar{3})$ $R(132)(\bar{1}\bar{3}\bar{2})$ $C(13\bar{2})(\bar{1}\bar{3}2)$ $R(1\bar{2}3)(\bar{1}2\bar{3})$ $C(1\bar{2}\bar{3})(\bar{1}23)$ $R(1\bar{3}\bar{2})(\bar{1}32)$ $C(1\bar{3}2)(\bar{1}3\bar{2})$ $R(12\bar{3})(\bar{1}\bar{2}3)$	$R(123)(\bar{1}\bar{2}\bar{3})$ $C(132)(\bar{1}\bar{3}\bar{2})$ $R(13\bar{2})(\bar{1}\bar{3}2)$ $C(1\bar{2}3)(\bar{1}2\bar{3})$ $R(1\bar{2}\bar{3})(\bar{1}23)$ $C(1\bar{3}\bar{2})(\bar{1}32)$ $R(1\bar{3}2)(\bar{1}3\bar{2})$ $C(12\bar{3})(\bar{1}\bar{2}3)$	$C(12)(\bar{1}\bar{2})(3\bar{3})$ $C(1\bar{2})(\bar{1}2)(3\bar{3})$ $C(13)(\bar{1}\bar{3})(2\bar{2})$ $C(1\bar{3})(\bar{1}3)(2\bar{2})$ $C(23)(\bar{2}\bar{3})(1\bar{1})$ $C(2\bar{3})(\bar{2}3)(1\bar{1})$ $R(12)(\bar{1}\bar{2})(3\bar{3})$ $R(1\bar{2})(\bar{1}2)(3\bar{3})$ $R(13)(\bar{1}\bar{3})(2\bar{2})$ $R(1\bar{3})(\bar{1}3)(2\bar{2})$ $R(23)(\bar{2}\bar{3})(1\bar{1})$ $R(2\bar{3})(\bar{2}3)(1\bar{1})$	$C(1\bar{1})(2\bar{2})$
A_{1g}	1	1	1	1	1	1	1	1	1
A_{1u}	1	1	1	1	1	1	1	1	−1
A_{2g}	1	1	1	−1	−1	1	1	−1	1
A_{2u}	1	1	1	−1	−1	1	1	−1	−1
E_g	2	2	2	0	0	−1	−1	0	2
E_u	2	2	2	0	0	−1	−1	0	−2
$E_{\frac{1}{2}g}$	2	−2	0	$\sqrt{2}$	$-\sqrt{2}$	1	−1	0	2
$E_{\frac{1}{2}u}$	2	−2	0	$\sqrt{2}$	$-\sqrt{2}$	1	−1	0	−2
$E_{\frac{5}{2}g}$	2	−2	0	$-\sqrt{2}$	$\sqrt{2}$	1	−1	0	2
$E_{\frac{5}{2}u}$	2	−2	0	$-\sqrt{2}$	$\sqrt{2}$	1	−1	0	−2
F_{1g}	3	3	−1	1	1	0	0	−1	3
F_{1u}	3	3	−1	1	1	0	0	−1	−3
F_{2g}	3	3	−1	−1	−1	0	0	1	3
F_{2u}	3	3	−1	−1	−1	0	0	1	−3
$G_{\frac{3}{2}g}$	4	−4	0	0	0	−1	1	0	4
$G_{\frac{3}{2}u}$	4	−4	0	0	0	−1	1	0	−4

In order to discuss the transformation properties of the "electronic parts" of a molecular wave function, we must examine the behavior of the quantities on which each "electronic part" depends. We consider first vectors between a particular electron and the various nuclei in the molecule. We use $D(n, e)$ for the distance between the nucleus labelled n and the electron labelled e. The names $X(n, e)$, $Y(n, e)$, and $Z(n, e)$ are used for the projections of $D(n, e)$ on three mutually orthogonal, molecule-fixed axes. When the particular label for the electron is unimportant, we suppress it to yield $D(n)$ with the projections $X(n)$, $Y(n)$, and $Z(n)$.

For an octahedral MF_6 molecule, the F atoms may conveniently be labelled 1, $\bar{1}$, 2, $\bar{2}$, 3, $\bar{3}$, as shown in Fig. 4, while the M atom is ble for $O_h^*(V)$

RI	σ_h	S_4	RS_4	S_6	RS_6	σ_d
1	6	6	6	8	8	12
$\bar{1})(2\bar{2})(3\bar{3})^*$	$C(3\bar{3})^*$ $C(2\bar{2})^*$ $C(1\bar{1})^*$ $R(3\bar{3})^*$ $R(2\bar{2})^*$ $R(1\bar{1})^*$	$C(3\bar{3})(1\bar{2}\bar{1}2)^*$ $R(3\bar{3})(12\bar{1}\bar{2})^*$ $C(2\bar{2})(13\bar{1}\bar{3})^*$ $R(2\bar{2})(1\bar{3}\bar{1}3)^*$ $C(1\bar{1})(2\bar{3}\bar{2}3)^*$ $R(1\bar{1})(23\bar{2}\bar{3})^*$	$C(3\bar{3})(12\bar{1}\bar{2})^*$ $R(3\bar{3})(1\bar{2}\bar{1}2)^*$ $C(2\bar{2})(1\bar{3}\bar{1}3)^*$ $R(2\bar{2})(13\bar{1}\bar{3})^*$ $C(1\bar{1})(23\bar{2}\bar{3})^*$ $R(1\bar{1})(2\bar{3}\bar{2}3)^*$	$C(1\bar{2}3\bar{1}2\bar{3})^*$ $R(1\bar{3}2\bar{1}3\bar{2})^*$ $C(213\bar{2}\bar{1}\bar{3})^*$ $R(2\bar{3}\bar{1}\bar{2}31)^*$ $C(1\bar{2}\bar{3}\bar{1}23)^*$ $R(132\bar{1}\bar{3}\bar{2})^*$ $C(21\bar{3}\bar{2}\bar{1}3)^*$ $R(23\bar{1}\bar{2}\bar{3}1)^*$	$R(1\bar{2}3\bar{1}2\bar{3})^*$ $C(1\bar{3}2\bar{1}3\bar{2})^*$ $R(213\bar{2}\bar{1}\bar{3})^*$ $C(2\bar{3}\bar{1}\bar{2}31)^*$ $R(1\bar{2}\bar{3}\bar{1}23)^*$ $C(132\bar{1}\bar{3}\bar{2})^*$ $R(21\bar{3}\bar{2}\bar{1}3)^*$ $C(23\bar{1}\bar{2}\bar{3}1)^*$	$C(12)(\bar{1}\bar{2})^*$ $C(1\bar{2})(\bar{1}2)^*$ $C(23)(\bar{2}\bar{3})^*$ $C(2\bar{3})(\bar{2}3)^*$ $C(13)(\bar{1}\bar{3})^*$ $C(1\bar{3})(\bar{1}3)^*$ $R(12)(\bar{1}\bar{2})^*$ $R(1\bar{2})(\bar{1}2)^*$ $R(23)(\bar{2}\bar{3})^*$ $R(2\bar{3})(\bar{2}3)^*$ $R(13)(\bar{1}\bar{3})^*$ $R(1\bar{3})(\bar{1}3)^*$
1	1	1	1	1	1	1
−1	−1	−1	−1	−1	−1	−1
1	1	−1	−1	1	1	−1
−1	−1	1	1	−1	−1	1
2	2	0	0	−1	−1	0
−2	−2	0	0	1	1	0
−2	0	$\sqrt{2}$	$-\sqrt{2}$	1	−1	0
2	0	$-\sqrt{2}$	$\sqrt{2}$	−1	1	0
−2	0	$-\sqrt{2}$	$\sqrt{2}$	1	−1	0
2	0	$\sqrt{2}$	$-\sqrt{2}$	−1	1	0
3	−1	1	1	0	0	−1
−3	1	−1	−1	0	0	1
3	−1	−1	−1	0	0	1
−3	1	1	1	0	0	−1
−4	0	0	0	−1	1	0
4	0	0	0	1	−1	0

labelled simply M. The projections $X(n)$, $Y(n)$, and $Z(n)$ are made on axes called x, y, and z, respectively. These axes are fixed in the molecule according to the positions that the nuclei with particular labels have in the reference configuration which was used to define elements of $O_h^*(V)$. We choose to have the x-axis point from the reference position of M to the reference position of 1. The y-axis points from the reference position of M to the reference position of 2, and the z-axis, from the reference position of M to the reference position of 3. Thus, x, y, and z form a Cartesian, molecule-fixed axis system.

The distance between an electron and the central atom, D(M), is left entirely unchanged by every element in $O_h^*(V)$ and therefore spans the A_{1g} representation of this group. The three projections of D(M) are, however, interchanged with each other by various elements of $O_h^*(V)$. Although these projections are unaffected by any permutations of the fluorine nuclear labels, there are overall molecular rotations that are also parts of the operations that are elements in the vibronic symmetry double-group. These rotations transform the projections X(M), Y(M), and Z(M) according to the F_{1u}-representation. For example, $C(12)(\bar{1}\bar{2})(3\bar{3})$ takes Z(M) into $-Z$(M) and interchanges X(M) with Y(M). Thus, the character for $C(12)(\bar{1}\bar{2})(3\bar{3})$ is -1 in the representation spanned by these projections. This character agrees with the value for F_{1u} given in Table XV.

The six distances between a particular electron and the various fluorine nuclei have their identities interchanged by the nuclear permutations, although they are not affected by molecular rotations. These six distances span the representations A_{1g}, E_g, and F_{1u} of $O_h^*(V)$. There are eighteen projections of the distances between an electron and the various fluorine nuclear positions in an MF_6 molecule. In general, these projections are altered both by the permuting part and by the rotating part of a vibronic symmetry element. They span the following representations of $O_h^*(V)$:

$$A_{1g},\ E_g,\ F_{1g},\ F_{1u}\ \text{(twice)},\ F_{2g},\ \text{and}\ F_{2u}.$$

Consider, for example, the element $C(3\bar{3})^*$. Let P stand for X, Y, or Z; and n stand for 1, $\bar{1}$, 2, $\bar{2}$, 3, or $\bar{3}$. Then $C(3\bar{3})^*$ combines the transformation of $P(n)$ into $-P(n)$ with the interchange of the labels 3 and $\bar{3}$ and also a rotation of π radians about the z-axis. The net result of all these operations is that the six projections involving $n = 3$, or $\bar{3}$, contribute zeros to the sum of diagonal elements in the transformation matrix for $C(3\bar{3})^*$; while for $n = 1$, $\bar{1}$, 2, or $\bar{2}$, the projection $Z(n)$ contributes -1 and the projections $X(n)$ and $Y(n)$ each contribute $+1$ to this sum. Thus, the character for $C(3\bar{3})^*$ is $+4$ in the representation spanned by all eighteen projections of electron–fluorine distances.

It is noteworthy that all of the transformation properties discussed in the last two paragraphs are independent of the particular location for the electron of interest. *These transformation properties are also entirely independent of the specific locations for the various nuclei in the molecule*: they are perfectly valid for a greatly distorted MF_6 molecule, as well as for a highly symmetric one! This feature is of fundamental importance in the phenomenological treatment of vibronic properties.

Now we are ready to discuss a reasonable form for the "electronic part" of the molecular wave function. (In general, this "electronic part" will be the product of an orbital factor and a spin factor; but we ignore any spin factor for the moment.) We are most interested in "electronic parts" that resemble, somewhat at least, the df_{2g}-states that Moffitt *et al.* found so useful for explaining the gross electronic features of the metal hexafluoride molecules. We can easily construct such functions for the electron ε, say, out of its distance from the atom M, $D(\mathrm{M}, \varepsilon)$ and the projections of $D(\mathrm{M}, \varepsilon)$ on the molecular reference axes, $X(\mathrm{M}, \varepsilon)$, $Y(\mathrm{M}, \varepsilon)$, and $Z(\mathrm{M}, \varepsilon)$.

The following are the three f_{2g}-functions:

$$\zeta_\varepsilon = X(\mathrm{M}, \varepsilon)\, Y(\mathrm{M}, \varepsilon)\, S[D(\mathrm{M}, \varepsilon)],$$
$$\eta_\varepsilon = Z(\mathrm{M}, \varepsilon)\, X(\mathrm{M}, \varepsilon)\, S[D(\mathrm{M}, \varepsilon)], \text{ and}$$
$$\xi_\varepsilon = Y(\mathrm{M}, \varepsilon)\, Z(\mathrm{M}, \varepsilon)\, S[D(\mathrm{M}, \varepsilon)].$$

Here $S[x]$ is any useful function of the variable x: for example, one might set

$$S[x] \sim e^{-\alpha x}$$

to simulate closely hydrogenic $3df_{2g}$-orbitals centered on the atom M. The significant point here is that, no matter what form $S[x]$ has or where the nuclei in the molecule are, ζ_ε, η_ε, and ξ_ε transform according to the F_{2g}-representation of $O_h^*(V)$.

Another important type of "electronic part" for the wave function of an MF_6 molecule is a molecular orbital composed of fluorine atomic orbitals but also transforming according to the representation F_{2g}. The simplest functions of this type have the following form:

$$\zeta'_\varepsilon = Y(1, \varepsilon)\, S'[D(1, \varepsilon)] + X(2, \varepsilon)\, S'[D(2, \varepsilon)] - Y(\bar{1}, \varepsilon)\, S'[D(\bar{1}, \varepsilon)] - X(\bar{2}, \varepsilon)\, S'[D(\bar{2}, \varepsilon)],$$
$$\eta'_\varepsilon = Z(1, \varepsilon)\, S'[D(1, \varepsilon)] + X(3, \varepsilon)\, S'[D(3, \varepsilon)] - Z(\bar{1}, \varepsilon)\, S'[D(\bar{1}, \varepsilon)] - X(\bar{3}, \varepsilon)\, S'[D(\bar{3}, \varepsilon)],$$

and

$$\xi'_\varepsilon = Z(2,\,\varepsilon)\,S'[D(2,\,\varepsilon)] + Y(3,\,\varepsilon)\,S'[D(3,\,\varepsilon)]$$
$$-Z(\bar{2},\,\varepsilon)\,S'[D(\bar{2},\,\varepsilon)] - Y(\bar{3},\,\varepsilon)\,S'[D(\bar{3},\,\varepsilon)].$$

Here $S'[x]$ is another arbitrary function that does not influence the transformation properties of ζ'_ε, η'_ε, and ξ'_ε. To make these functions closely resemble molecular orbitals constructed from fluorine $2p$-orbitals, one would set

$$S'[x] \sim e^{-\alpha' x}.$$

The point that is being stressed in each of these examples for "electronic parts" of molecular wave functions is the following: the transformations that are induced by elements of the vibronic symmetry group acting on the "electronic parts" of a molecular wave function have definitions that are entirely independent of where either the electrons or the nuclei in the molecule are instantaneously located. Thus, one can establish conclusions about vibronic properties in an amazingly general form as long as these conclusions are based just on transformation properties of "electronic parts" in a molecular wave function together with transformation properties for nuclear displacement coordinates.

In order to exploit the general validity of those transformation properties that belong to the "electronic parts" of a molecular wave function, we study next the matrix elements that connect these "electronic parts" with each other. These matrix elements themselves have well defined transformation properties. They transform according to just those representations spanned by the product of the representations for the "electronic parts". Here is a pertinent example, based on the group $O_h^*(V)$: the nine matrix elements connecting one set of f_{2g}-"electronic parts" with another set of f_{2g}-"electronic parts" transform according to the representations A_{1g}, E_g, F_{1g}, and F_{2g}. These are representations spanned by the product of two F_{2g}-representations.

For an illustration of how one obtains a representation of a group from the product of two original representations of this group, consider the two nuclear displacements in Fig. 1 that transform according to the E_g-representation of $O_h^*(V)$. These displacements may be called II and III, reading from left to right. Previously, we were interested in the influence of elements of

$O_h^*(V)$ on II and III, themselves; now we wish to know the influence of elements of $O_h^*(V)$ on the four quantities II II, II III, III II, and III III. The character of an element of $O_h^*(V)$ in such a product representation as this can be obtained most simply by taking the product of the characters for this element in the two original representations. In this way, one can verify that the four products of one set of E_g-quantities with another set of E_g-quantities span the representations, A_{1g}, A_{2g}, and E_g. In fact, it is not hard to separate from each other the combinations of these four products that transform within the smallest possible families. These families are said to transform irreducibly under the elements of $O_h^*(V)$. The sum (II II + III III) is the combination that transforms according to A_{1g}. The quantity (II III − III II) transforms according to A_{2g}, while the quantities (II III + III II) and (III III − II II) transform according to E_g.

The line of reasoning that leads one to attribute transformation properties to matrix elements can also be illustrated for the example considered in the preceding paragraph. If the first row and the first column of a two-by-two matrix are identified with the state transforming as II, and the second row and the second column are identified with the state transforming as III, then the matrix

$$\boldsymbol{\alpha}_1 = \begin{bmatrix} 1 & 0 \\ 0 & 1 \end{bmatrix}$$

is to be identified with the quantity (II II + III III) and transforms under the operations of $O_h^*(V)$ according to A_{1g}. The matrix

$$\boldsymbol{\alpha}_2 = \begin{bmatrix} 0 & 1 \\ -1 & 0 \end{bmatrix}$$

is to be identified with (II III − III II) and transforms according to A_{2g}. Finally, the matrices

$$\boldsymbol{\epsilon}_{\mathrm{II}} = \begin{bmatrix} 0 & 1 \\ 1 & 0 \end{bmatrix} \text{ and } \boldsymbol{\epsilon}_{\mathrm{III}} = \begin{bmatrix} 1 & 0 \\ 0 & -1 \end{bmatrix}$$

are to be identified with the quantities (II III + III II) and (II II − III III), respectively, and transform according to E_g.

It is important to notice that we can generate matrices proportional to $\boldsymbol{\alpha}_1$ and $\boldsymbol{\alpha}_2$ by starting with the matrices $\boldsymbol{\epsilon}_{\mathrm{II}}$ and $\boldsymbol{\epsilon}_{\mathrm{III}}$ and by using the standard rules for matrix multiplication together with the same combining coefficients as were used for products of II and III. This observation is true because $\boldsymbol{\epsilon}_{\mathrm{II}}$ transforms like II and $\boldsymbol{\epsilon}_{\mathrm{III}}$ transforms like III. In this way we obtain

$$\boldsymbol{\epsilon}_{\mathrm{II}} \cdot \boldsymbol{\epsilon}_{\mathrm{II}} + \boldsymbol{\epsilon}_{\mathrm{III}} \cdot \boldsymbol{\epsilon}_{\mathrm{III}} = 2 \cdot \boldsymbol{\alpha}_1$$

and

$$\boldsymbol{\epsilon}_{\mathrm{II}} \cdot \boldsymbol{\epsilon}_{\mathrm{III}} - \boldsymbol{\epsilon}_{\mathrm{III}} \cdot \boldsymbol{\epsilon}_{\mathrm{II}} = -2 \cdot \boldsymbol{\alpha}_2.$$

Coincidently, matrices that vanish identically are obtained by applying to products of $\boldsymbol{\epsilon}_{\mathrm{II}}$ and $\boldsymbol{\epsilon}_{\mathrm{III}}$ the same coefficients that were used above to generate $\boldsymbol{\epsilon}_{\mathrm{II}}$ and $\boldsymbol{\epsilon}_{\mathrm{III}}$ from products of II and III. But these two vanishing matrices must be regarded as proportional to $\boldsymbol{\epsilon}_{\mathrm{II}}$ and to $\boldsymbol{\epsilon}_{\mathrm{III}}$ respectively, as follows:

$$\boldsymbol{\epsilon}_{\mathrm{II}} \cdot \boldsymbol{\epsilon}_{\mathrm{III}} + \boldsymbol{\epsilon}_{\mathrm{III}} \cdot \boldsymbol{\epsilon}_{\mathrm{II}} = 0 \cdot \boldsymbol{\epsilon}_{\mathrm{II}}$$

and

$$\boldsymbol{\epsilon}_{\mathrm{II}} \cdot \boldsymbol{\epsilon}_{\mathrm{II}} - \boldsymbol{\epsilon}_{\mathrm{III}} \cdot \boldsymbol{\epsilon}_{\mathrm{III}} = 0 \cdot \boldsymbol{\epsilon}_{\mathrm{III}}.$$

It is no coincidence that only one set of coefficients is required to obtain quantities that transform irreducibly from the products of each of several sets of quantities that transform in the same way under the elements in the relevant group, in this case, $O_h^*(V)$. Except for arbitrary normalizing factors, the values of the reduction coefficients for a specific group are determined by the structure of the multiplication table for this group. Thus, the product representations for each of several isomorphic groups are all reduced by a single set of coefficients. For example, the coefficients given by Tanabe and Sugano[72] or by Griffith[59] for O_h are also useful for $O_h(M)$ and for $O_h(V)$. These important reduction coefficients are frequently called vector-coupling coefficients or Wigner coefficients.

Quantum-mechanical matrix elements for quantities with specific transformation properties are proportional to reduction coefficients for the symmetry group with respect to which these transformation properties are defined. This concept, often called the Wigner–Eckart theorem,[73,74] is the heart of a powerful calculus, known generally as irreducible tensor algebra. The techniques of this calculus are standard in nuclear and atomic spectroscopy and are summarized in Edmonds' book, *Angular Momentum in Quantum Mechanics.*[75] We now apply these techniques to the problem of vibronic coupling in metal hexafluoride molecules.

If the Wigner–Eckart theorem is regarded as the heart of tensor algebra, then angular momentum operators must be its vertebrae. Briefly, these operators are unchanged by space inversion, reverse their signs for time inversion, and are independent of charge conjugation. Usually, they transform according to the three-dimensional, irreducible representation of the rotation group in 3-space, but those angular momenta that are particularly important in discussions of the internal motions of metal hexafluoride molecules are the ones that transform according to the F_{1g}-representation of $O_h^*(V)$.

Each metal hexafluoride molecule can be described as having a characteristic number of open-shell electrons. There are two possible spin states for each electron and, to a first approximation at least, each open-shell electron in these molecules is described by one of three spatial wave functions that transform according to the F_{2g}-representation of $O_h^*(V)$. The two spin states for each electron transform according to the $E_{\frac{1}{2}g}$-representation of this group. As a result of these symmetry properties of the wave functions, there are two different F_{1g} operators (or effective angular momenta) for each open-shell electron of a metal hexafluoride molecule, the orbital angular momentum operator and the spin angular momentum operator. The spin angular momentum, $\mathbf{s}$, arises here because F_{1g} is contained once in the representation spanned by the antisymmetric product of $E_{\frac{1}{2}g}$ with itself. The orbital angular momentum effective for f_{2g}-orbitals, $\mathbf{l}'$, arises in the same way, because F_{1g} is contained once in the representation spanned by the antisymmetric product of F_{2g} with itself.

The effective orbital angular momentum for f_{2g}-orbitals is fictitious in the sense that it is not, in general, equal to the part of the usual orbital angular momentum, $\mathbf{l}$ (defined in terms of position vectors and momentum vectors of the electron), that connects these f_{2g}-states with each other. Both of these orbital angular momenta transform according to F_{1g} and are thus proportional to each other within the f_{2g}-states. The effective angular momentum, $\mathbf{l}'$, is chosen to have just the right magnitude and sign to obey the commutation rules for angular momenta. The spin angular momentum, $\mathbf{s}$, also obeys these rules. It would only be by accident, however, if the usual orbital angular momentum, restricted to a particular set of F_{2g}-states, obeyed these commutation rules. In general the usual angular momentum connects F_{2g}-states with other states and does not satisfy the commutation rules unless all these connections are taken into account. The effective orbital angular momentum, $\mathbf{l}'$, is thus by far the more useful orbital angular momentum for discussing vibronic properties for f_{2g} "electronic parts" of a molecular wave function.

Angular momenta in units of $\hbar$ obey the following commutation rules:

$$[\mathbf{j}_0, \mathbf{j}_\pm] = \pm\, \mathbf{j}_\pm$$

and

$$[\mathbf{j}_+, \mathbf{j}_-] = 2\mathbf{j}_0.$$

Here, $\mathbf{j}_+$, $\mathbf{j}_-$, and $\mathbf{j}_0$ are the three components of an arbitrary angular momentum vector $\mathbf{j}$. In our work, each of these components is usually given as a matrix. The squared magnitude of $\mathbf{j}$,

$$\mathbf{j}^2 = \mathbf{j}_0^2 + (\mathbf{j}_+\mathbf{j}_- + \mathbf{j}_-\mathbf{j}_+)/2,$$

therefore, is also usually a matrix quantity. Within the $(2j + 1)$ states that are connected together by elements of $\mathbf{j}_+$ and $\mathbf{j}_-$ and for which the eigenvalues of $\mathbf{j}_0$, j_0, say, run from $-j$ to j, the matrix $\mathbf{j}^2$ is equal to $j(j + 1)$ times the unit matrix. The numbers j and j_0 are used to identify individual states within such a manifold of $(2j + 1)$ states.

In the two cases of immediate interest to us, j is either $\frac{1}{2}$ or 1; these cases are the $E_{\frac{1}{2}g}$- and F_{2g}- states respectively. The angular momentum matrices for these cases are collected below in the standard form for tensor algebra.

$$\mathbf{s}_+ = \begin{bmatrix} 0 & 1 \\ 0 & 0 \end{bmatrix}, \qquad \mathbf{s}_- = \begin{bmatrix} 0 & 0 \\ 1 & 0 \end{bmatrix},$$

$$\mathbf{s}_0 = \tfrac{1}{2}\begin{bmatrix} 1 & 0 \\ 0 & -1 \end{bmatrix}, \qquad \mathbf{s}^2 = \tfrac{3}{4}\begin{bmatrix} 1 & 0 \\ 0 & 1 \end{bmatrix}.$$

$$\mathbf{l}'_+ = \sqrt{2}\begin{bmatrix} 0 & 1 & 0 \\ 0 & 0 & 1 \\ 0 & 0 & 0 \end{bmatrix}, \qquad \mathbf{l}'_- = \sqrt{2}\begin{bmatrix} 0 & 0 & 0 \\ 1 & 0 & 0 \\ 0 & 1 & 0 \end{bmatrix}.$$

$$\mathbf{l}'_0 = \begin{bmatrix} 1 & 0 & 0 \\ 0 & 0 & 0 \\ 0 & 0 & -1 \end{bmatrix}, \qquad \mathbf{l}'^2 = 2\begin{bmatrix} 1 & 0 & 0 \\ 0 & 1 & 0 \\ 0 & 0 & 1 \end{bmatrix}.$$

Tensor operators of rank k occur in sets of $(2k + 1)$ component operators, where k is some positive integer or zero. The members of each of these sets obey certain standard commutation rules with respect to angular momenta. For $q = -k, -k + 1, \ldots, k$,

$$[\mathbf{j}_0, \mathbf{T}(k, q)] = q\mathbf{T}(k, q)$$

and

$$[\mathbf{j}_\pm, \mathbf{T}(k, q \mp 1)] = \{(k \pm q)(k \mp q + 1)\}^{\frac{1}{2}}\mathbf{T}(k, q).$$

Thus, for example, the tensor operators that correspond to the angular momentum $\mathbf{j}$ are the following quantities:

$$\mathbf{j}(1, \pm 1) = \mp \mathbf{j}_\pm/\sqrt{2}$$

and

$$\mathbf{j}(1, 0) = \mathbf{j}_0.$$

The matrix elements of any tensor operator $\mathbf{T}(k, q)$ can be obtained from the fundamental equation of tensor algebra by substitution of the appropriate reduced matrix element for this set of $(2k + 1)$ quantities,

$$(\bar{\gamma}\bar{j}\|\mathbf{T}(k)\|\gamma j).$$

This fundamental equation written in terms of Wigner 3-j coefficients is

$$(\bar{\gamma}\bar{j}\bar{j}_0|\mathbf{T}(k, q)|\gamma j j_0)$$
$$= (-1)^{j-\bar{j}_0}\begin{pmatrix} \bar{j} & k & j \\ -\bar{j}_0 & q & j_0 \end{pmatrix}(\bar{\gamma}\bar{j}\|\mathbf{T}(k)\|\gamma j).$$

The Wigner 3-j coefficients are simply numbers, for example,

$$\begin{pmatrix} j & j & 0 \\ j_0 & -j_0 & 0 \end{pmatrix} = (-1)^{j-j_0}(2j + 1)^{-\frac{1}{2}}.$$

The 3-j coefficients $\begin{pmatrix} \bar{j} & k & j \\ -\bar{j}_0 & q & j_0 \end{pmatrix}$ are proportional to the reduction (or vector-coupling) coefficients for obtaining a $(2j + 1)$-dimensional representation of the appropriate symmetry group [in this case $O_h^*(V)$] out of the $(2k + 1)(2\bar{j} + 1)$-dimensional representation spanned by products of a $(2k + 1)$-dimensional representation with a $(2\bar{j} + 1)$-dimensional representation. Several tabulations of Wigner 3-j coefficients are now available, for example, *the* 3-j *and* 6-j *symbols.*[76]

There is a simple expression for the reduced matrix element of the special angular momentum $\mathbf{j}$ that is used to specify states within which matrix elements are being calculated, namely,

$$(\bar{\gamma}\bar{j}\|\mathbf{j}(1)\|\gamma j) = \delta(\bar{\gamma},\gamma)\delta(\bar{j}, j)[(2j + 1)(j + 1)j]^{\frac{1}{2}},$$

where $\delta(\alpha, \beta)$ is unity if α is the same as β, and vanishes otherwise. Here $\bar{\gamma}$ and γ refer to other quantum numbers, distinct from j and j_0, the ones that describe how the states transform under the relevant symmetry group [in this case $O_h^*(V)$].

Another particularly useful type of reduced matrix element, introduced by Racah,[77] is that for the "unit" tensor of rank k, $\mathbf{U}^{(n)}(k, q)$, where $-k \leq q \leq k$. Here, $\mathbf{U}^{(n)}(k, q)$ is the sum of "unit" tensors of the same rank, k, and row, q, for each of the n open-shell electrons to be considered. That is to say,

$$\mathbf{U}^{(n)}(k, q) = \sum_{\varepsilon=1}^{n}\mathbf{u}_{\varepsilon}(k, q).$$

The reduced matrix elements of these individual tensors, $\mathbf{u}_{\varepsilon}(k, q)$, seem trivial:

$$(\bar{\gamma}\bar{l}'_{\varepsilon}\|\mathbf{u}_{\varepsilon}(k)\|\gamma l'_{\varepsilon}) = \delta(\bar{\gamma}, \gamma)\delta(\bar{l}'_{\varepsilon}, l'_{\varepsilon}).$$

Here we would identify $\bar{l}'_{\varepsilon}$ and l'_{ε} with the single, possible value for the effective orbital angular momentum of F_{2g}-electrons, namely, $l' = 1 = \bar{l}'_{\varepsilon} = l'_{\varepsilon}$. Thus, in the cases of interest here, the second of the conditions for this reduced matrix element to be nonzero is always satisfied; but the additional label $\bar{\gamma}$ or γ now depends on the relationship between the electron that we are discussing and all the other open-shell electrons in the molecule. As a result, the first condition, $\delta(\bar{\gamma}, \gamma)$, is not always satisfied. Racah has evaluated the reduced matrix elements for unit tensor operators in several cases. His results[77] for equivalent p-electrons can be transferred directly into the present context.

The states for the several open-shell electrons occupying f_{2g}-orbitals in a hexafluoride molecule are classified first according to the values of $\mathbf{S}$ and $\mathbf{L}'$, their total spin angular momentum and their total effective orbital angular momentum, respectively:

$$\mathbf{S} = \sum_{\varepsilon=1}^{n}\mathbf{s}_{\varepsilon}$$

and

$$\mathbf{L}' = \sum_{\varepsilon=1}^{n} \mathbf{l}'_{\varepsilon}.$$

The transformation properties of the various electronic states actually observed for the molecule are finally given in terms of the total effective angular momentum, $\mathbf{J}'$. As used by Moffitt *et al.*,[15]

$$\mathbf{J}' = \mathbf{S} + \mathbf{L}'.$$

Fortunately, the states for one through five open-shell f_{2g}-electrons can adequately be described in terms of just the magnitudes of $\mathbf{S}$, $\mathbf{L}'$, and $\mathbf{J}'$. We can therefore use standard labels of the form

$$^{(2S+1)}L'_{J'}$$

for identifying these states. The arrays of reduced matrix elements

$$(f_{2g}^{n}\,{}^{(2\bar{S}+1)}\bar{L}'_{\bar{J}'}\|\mathbf{U}^{(n)}(2)\|f_{2g}^{n}\,{}^{(2S+1)}L'_{J'})$$

for $n = 1$ through **5** are collected in Table XVI. These arrays are given in terms of the same basis states as those used to specify the matrices $\mathbf{A}^{(n)}$ and $\mathbf{B}^{(n)}$ in Table XI. Thus, the same transformation that diagonalizes the energy

$$\mathbf{W}^{(n)} = G\mathbf{A}^{(n)} + \zeta\mathbf{B}^{(n)}$$

for a particular value of n can be applied to the corresponding array of reduced matrix elements in Table XVI to obtain those reduced matrix elements that are appropriate to the electronic structure dictated for this molecule by the relative sizes of G and ζ. The relative sizes of these parameters are conveniently expressed in terms of $x = \zeta/(G + \zeta)$. The value of x is 0.5862 for the hexafluoride molecules formed by elements in the third transition series; while the value of 0.3 seems reasonable for x in hexafluoride molecules of the second transition series. Transformed arrays of reduced matrix elements for the unit second-rank tensor $\mathbf{U}(2, q)$ are given in Table XVI for $x = 0.5862$ and for $x = 0.3$.

TABLE XVI. Reduced Matrix Elements of the Unit Second-Rank Tensor $\mathbf{U}^{(n)}(2, q)$ for $n = 1, 2, 3, 4$, and 5. Case I: $x = 0.0$, Case II: $x = 0.3$, and Case III: $x = 0.5862$. In each case, the basic states are labelled as $(n; J'; \tilde{w}^{(n)})$, where $\tilde{w}^{(n)}$ is the first one or two significant figures of $W^{(n)}/(G + \zeta)$

Case I: $x = 0.0$

	$(1; \frac{3}{2}; 0)$	$(1; \frac{1}{2}; 0)$			
$(1; \frac{3}{2}; 0)$	$\sqrt{6}/3$	$\sqrt{6}/3$			
$(1; \frac{1}{2}; 0)$	$-\sqrt{6}/3$	0			
	$(2; 0; 4)$	$(2; 0; -1)$	$(2; 1; -1)$	$(2; 2; -1)$	$(2; 2; 1)$
$(2; 0; 4)$	0	0	0	0	$2\sqrt{3}/3$
$(2; 0; -1)$	0	0	0	$-\sqrt{3}/3$	0
$(2; 1; -1)$	0	0	$\frac{1}{2}$	$\sqrt{3}/2$	0
$(2; 2; -1)$	0	$-\sqrt{3}/3$	$-\sqrt{3}/2$	$-\sqrt{21}/6$	0
$(2; 2; 1)$	$2\sqrt{3}/3$	0	0	0	$\sqrt{21}/3$
	$(3; \frac{3}{2}; -3)$	$(3; \frac{1}{2}; 2.0)$	$(3; \frac{3}{2}; 2.0)$	$(3; \frac{3}{2}; 0.0)$	$(3; \frac{5}{2}; 0.0)$
$(3; \frac{3}{2}; -3)$	0	0	0	0	0
$(3; \frac{1}{2}; 2.0)$	0	0	0	$-\sqrt{30}/5$	$2\sqrt{5}/5$
$(3; \frac{3}{2}; 2.0)$	0	0	0	$-\sqrt{30}/5$	$-\sqrt{70}/5$
$(3; \frac{3}{2}; 0.0)$	0	$\sqrt{30}/5$	$-\sqrt{30}/5$	0	0
$(3; \frac{5}{2}; 0.0)$	0	$2\sqrt{5}/5$	$\sqrt{70}/5$	0	0

$$(4; J'; \tilde{w}^{(4)}\|\mathbf{U}^{(4)}(2)\|4; \bar{J}'; \tilde{w}^{(4)}) = -(2; J'; \tilde{w}^{(2)}\|\mathbf{U}^{(2)}(2)\|2; \bar{J}'; \tilde{w}^{(2)})$$

and

$$(5; J'; \tilde{w}^{(5)}\|\mathbf{U}^{(5)}(2)\|5; \bar{J}'; \tilde{w}^{(5)}) = -(1; J'; w^{(1)}\|\mathbf{U}^{(1)}(2)\|1; \bar{J}'; \tilde{w}^{(1)})$$

TABLE XVI (*continued*)

Case II: $x = 0.3$

	(2; 0; 2.8)	(2; 2; 0.7)	(2; 0; −0.4)	(2; 1; −0.5)	(2; 2; −0.8)
(2; 0; 2.8)	0.0	1.1447533	0.0	0.0	−0.0785546
(2; 2; 0.7)	1.1447533	1.4868786	0.0716484	−0.1153462	−0.3024584
(2; 0; −0.4)	0.0	0.0716484	0.0	0.0	−0.5872837
(2; 1; −0.5)	0.0	0.1153462	0.0	0.5000000	0.8583095
(2; 2; −0.8)	−0.0785546	−0.3024584	−0.5872837	−0.8583095	−0.7231160

	(3; $\frac{3}{2}$; 1.5)	(3; $\frac{1}{2}$; 1.4)	(3; $\frac{5}{2}$; 0.0)	(3; $\frac{3}{2}$; −0.07)	(3; $\frac{3}{2}$; −2)
(3; $\frac{3}{2}$; 1.5)	0.4635041	−0.2382592	−1.6276197	−0.9885561	−0.0060376
(3; $\frac{1}{2}$; 1.4)	0.2382592	0.0	0.8944272	−1.0691171	0.0148712
(3; $\frac{5}{2}$; 0.0)	1.6276197	0.8944272	0.0	0.3607217	−0.1439929
(3; $\frac{3}{2}$; −0.07)	−0.9885561	1.0691171	−0.3607217	−0.4609443	0.0952057
(3; $\frac{3}{2}$; −2)	−0.0060376	−0.0148712	0.1439929	0.0952057	−0.0025594

	(4; 0; 2.8)	(4; 2; 0.7)	(4; 2; −0.5)	(4; 1; −0.8)	(4; 0; −1)
(4; 0; 2.8)	0.0	1.1427235	−0.1244925	0.0	0.0
(4; 2; 0.7)	1.1427235	−1.4667403	0.3682131	0.1410551	0.0314244
(4; 2; −0.5)	−0.1244925	0.3682131	0.7029778	0.8544609	−0.5868250
(4; 1; −0.8)	0.0	−0.1410551	−0.8544609	−0.5000000	0.0
(4; 0; −1)	0.0	0.0314244	−0.5868250	0.0	0.0

Case III: $x = 0.5862$

	(2; 0; 2.0)	(2; 2; 0.5)	(2; 1; −0.12)	(2; 0; −0.19)	(2; 2; −0.8)
(2; 0; 2.0)	0.0	1.0748836	0.0	0.0	−0.1061563
(2; 2; 0.5)	1.0748836	1.3029182	0.2711456	0.2827224	−0.6813158
(2; 1; −0.12)	0.0	−0.2711456	0.5000000	0.0	−0.8224841
(2; 0; −0.19)	0.0	0.2827224	0.0	0.0	−0.6481441
(2; 2; −0.8)	−0.1061563	−0.6813158	0.8224841	−0.6481441	−0.5391555

	(3; $\frac{3}{2}$; 1.2)	(3; $\frac{1}{2}$; 0.8)	(3; $\frac{5}{2}$; 0.0)	(3; $\frac{3}{2}$; −0.2)	(3; $\frac{3}{2}$; −1)
(3; $\frac{3}{2}$; 1.2)	0.8467686	−0.4845156	−1.4621973	−0.6587012	−0.0120728
(3; $\frac{1}{2}$; 0.8)	0.4845156	0.0	0.8944272	−0.9717200	0.1449305
(3; $\frac{5}{2}$; 0.0)	1.4621973	0.8944272	0.0	0.6576226	−0.4790735
(3; $\frac{3}{2}$; −0.2)	−0.6587012	0.9717200	−0.6576226	−0.7637810	0.3351629
(3; $\frac{3}{2}$; −1)	−0.0120728	−0.1449305	0.4790735	0.3351629	−0.0829875

	(4; 0; 1.8)	(4; 2; 0.6)	(4; 2; −0.3)	(4; 1; −0.7)	(4; 0; −1)
(4; 0; 1.8)	0.0	1.0507548	−0.3916029	0.0	0.0
(4; 2; 0.6)	1.0507548	−1.0026815	0.9628640	0.4144821	0.0136808
(4; 2; −0.3)	−0.3916029	0.9628640	0.2389189	0.7603976	−0.6395632
(4; 1; −0.7)	0.0	−0.4144821	−0.7603976	−0.5000000	0.0
(4; 0; −1)	0.0	0.0136808	−0.6395632	0.0	0.0

In any case:

	(1; $\frac{1}{2}$; x)	(1; $\frac{3}{2}$; $-x/2$)		(5; $\frac{3}{2}$; $x/2$)	(5; $\frac{1}{2}$; $-x$)
(1; $\frac{1}{2}$; x)	0.0	−0.8165	(5; $\frac{3}{2}$; $x/2$)	−0.8165	−0.8165
(1; $\frac{3}{2}$; $-x/2$)	0.8165	0.8165	(5; $\frac{1}{2}$; $-x$)	0.8165	0.0

The size of the diagonal element in each of these arrays for the electronic state of lowest energy has special significance for the apparent vibrational properties of a molecule. We have therefore presented in Fig. **5** the variation of this quantity, $U^{(n)}(2)_{gr}$, as a function of x.

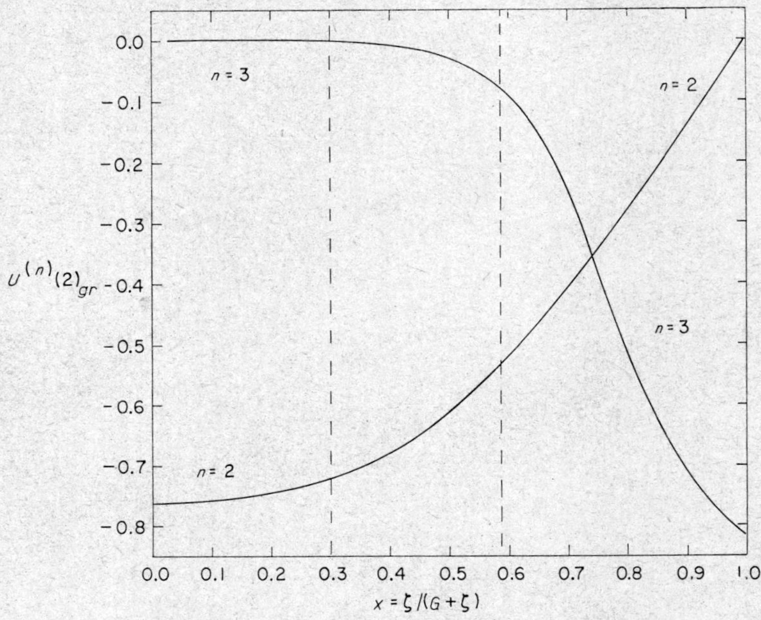

Fig. 5. The variation of $U^{(n)}(2)_{gr}$ with respect to $x = \zeta/(\zeta + G)$ for $n = 2$ and for $n = 3$. $U^{(4)}(2)_{gr} = U^{(5)}(2)_{gr} = 0$ and $U^{(1)}(2)_{gr} = 0.8164966\ldots$

Although we have now seen how a tensor algebra can be used for calculating certain properties of metal hexafluoride molecules, we have not yet examined to what extent the Hamiltonian for this type of molecule is invariant under the operations of $O_h^*(V)$, the vibronic symmetry group on which this tensor algebra is based. If the Hamiltonian were completely unchanged by every element in the group, then steady states for metal hexafluoride molecules could be classified rigorously according to their transformation properties with respect to $O_h^*(V)$: states that transform differently could never mix with each other. This is usually not exactly the case, however, because most of the operations in $O_h^*(V)$ change the orientations of what may be called internal tensors (such as, electron spin and orbital angular momenta, instantaneous dipole, quadrupole, and higher electric moments of the molecule, etc.) relative to what may be called external tensors (such as, the molecule's rotational angular momentum, as determined by spherical-top functions of its Eulerian angles, any electric or magnetic fields arising from charges outside the molecule, etc.). Thus, $O_h^*(V)$ can be regarded as a symmetry group for the molecular Hamiltonian only if we are prepared to neglect all terms in the energy that combine an external tensor with some internal tensor that is not totally symmetric with respect to $O_h^*(V)$. We intend, for several reasons, to make just this approximation. The first reason is that no resolved rotational features can be identified in the presently available vibronic spectra for metal hexafluorides. The second reason is that the energy terms that we neglect here are probably some two orders of magnitude smaller than those that we assign. A final reason is that the inclusion of these neglected terms, which depend on the Eulerian angles, would impossibly complicate the type of calculation that we must make.

The approximation that we have adopted may be described as regarding the internal workings of metal hexafluoride molecules as almost completely isolated from the rest of the universe. The possibility that keeps this isolation from being complete is that any internal property which is invariant with respect to $O_h^*(V)$ may still interact with the external surroundings of the molecules. (A particularly important possibility that we have in mind here is interaction between an external electric field and the average distribution of electrical charge for an octahedrally symmetric nuclear configuration.) In any case, however, the approximation that we have adopted may be stated as follows: only terms that are totally symmetric with respect to $O_h^*(V)$ are to be included in the Hamiltonian operator for a metal hexafluoride molecule. This symmetry condition makes it possible to obtain next explicit expressions for the phenomenological, vibronic Hamiltonian of each of these molecules.

A purely phenomenological molecular Hamiltonian would be one constructed just on the basis of symmetry conditions and having all magnitodes for interactions left to be determined from comparisons with experimental data. Usually, however, any purely phenomenological Hamiltonian contains many possible interactions that are not significant with respect to currently available data. It seems most straightforward in all such cases to set equal to zero any interaction constant that cannot be determined with reasonable precision from the experimental data at hand. We adopt this procedure here to simplify the discussion that follows. Our work is greatly aided in this respect by the fact that the largest terms in the phenomenological Hamiltonian for hexafluorides have already been identified by earlier investigators. These important terms are: the energy of the nonbonding valence electrons for an octahedrally symmetric nuclear configuration, as given by Moffitt *et al.*,[15] the nuclear kinetic energy, and a simple harmonic potential energy, as described by Yost, Steffens, and Gross.[25]

The Hamiltonian used by Moffitt *et al.* to calculate the gross spacing of electronic energy levels for metal hexafluoride molecules consists of four parts: H_O, the kinetic energy of the nonbonding valence electrons and the spherically symmetric part of their potential energy; H_C, the correlation energy that arises from Coulomb repulsions between these electrons; H_S, their energy due to spin–orbit coupling; and H_F, their stereospecific potential energy arising from the assumed octahedrally symmetric disposition of ligands and bonding electrons. We call the sum of these four terms H_E, the effective, electronic part of the molecular Hamiltonian:

$$H_E = H_O + H_C + H_S + H_F.$$

Each of the four terms in H_E is totally symmetric with respect to $O_h^*(V)$. Furthermore, both H_O and H_F represent constant energies for all states arising from the configurations f_{2g}^n, $0 \leq n \leq 6$. The expressions obtained by Moffitt *et al.* for H_C and H_S within these configurations are the following ones:

$$H_C = -G \sum_{\varepsilon < \varepsilon'} (1 + \mathbf{l}'_\varepsilon \cdot \mathbf{l}'_{\varepsilon'} + 4\mathbf{s}_\varepsilon \cdot \mathbf{s}_{\varepsilon'})$$

and

$$H_S = -\zeta_d \sum_\varepsilon \mathbf{l}'_\varepsilon \cdot \mathbf{s}_\varepsilon.$$

We turn now to H_V, the vibrational part of the molecular Hamiltonian. It is the sum of T_N, the nuclear kinetic energy, and of V, a portion of the molecular potential energy as a function of nuclear displacement coordinates:

$$H_V = T_N + V.$$

In particular, V vanishes if the nuclei are in their reference configuration, and it depends on only those combinations of nuclear displacement coordinates that are totally symmetric with respect to $O_h^*(V)$.

Because we have taken a reference configuration that is octahedrally symmetric, the fifteen internal displacement coordinates for a seven-atom hexafluoride transform in six separate families. Each one of these families corresponds to a distinct mode of motion with a particular frequency of vibration. The Jahn–Teller theorem applies only to the a_{1g}, the e_g, and the f_{2g} modes. Thus, these are the only ones that we discuss in detail in the rest of this section. Specifically, we now write

$$T_N = \tfrac{1}{2}\sum_j \sum_r (p_{jr})^2/m_j$$

and

$$V = \bar{l}_1 q_{1a} + \tfrac{1}{2}\sum_j \sum_r \bar{k}_j (q_{jr})^2$$

where for the a_{1g} mode, $j = 1$ and $r = a$; for the e_g mode, $j = 2$ and $r = a$ or b; and for the f_{2g} mode, $j = 5$ and $r = a, b$, or c. The commutation rules obeyed by the momenta p_{jr} and the coordinates $q_{j'r'}$ can be collected in the following equation:

$$[p_{jr}, q_{j'r'}] = -i\hbar\delta(j, j')\ \delta(r, r').$$

The quantum of energy associated with the j-th mode then becomes

$$\nu_j = \hbar(\bar{k}_j/m_j)^{\frac{1}{2}}.$$

The energy levels of H_V are specified by the three quantum numbers, n_1, n_2, and n_5. There are $d(n_1, n_2, n_5)$ independent vibrational states associated with the level at energy $E(n_1, n_2, n_5)$. Here,

$$d(n_1, n_2, n_5) = (n_2 + 1)(n_5 + 1)(n_5 + 2)/2$$

and

$$E(n_1, n_2, n_5) = (n_1 + \tfrac{1}{2})\nu_1 + (n_2 + 1)\nu_2 + (n_5 + \tfrac{3}{2})\nu_5.$$

These several degenerate states can be conveniently distinguished from each other by vibrational angular momenta: M_2, M_{5a}, M_{5b}, and M_{5c}, defined as follows:

$$M_2 = (q_{2a}p_{2b} - p_{2a}q_{2b})/\hbar,$$
$$M_{5a} = (q_{5b}p_{5c} - q_{5c}p_{5b})/\hbar,$$
$$M_{5b} = (q_{5c}p_{5a} - q_{5a}p_{5c})/\hbar,$$

and

$$M_{5c} = (q_{5a}p_{5b} - q_{5b}p_{5a})/\hbar.$$

The eigenvalues of M_2 go from $-n_2$ to n_2 in steps of 2; while those of $\mathbf{M}_5^2 = M_{5a}^2 + M_{5b}^2 + M_{5c}^2$ are $M_5(M_5 + 1)$, where $M_5 = n_5, n_5 - 2, \ldots, 1$ or 0.

Before we finish discussing H_V, the vibrational part of the molecular Hamiltonian, we must comment on the quantities $\bar{l}_j$, $\bar{k}_j$ and m_j. The effective mass for the j-th vibrational mode, m_j, is quite properly regarded as a constant, i.e., independent of the electronic state of the molecule. On the other hand, the force constant, $\bar{k}_j$, for the j-th mode may be rather sensitive to the details of the electronic motion. In particular $\bar{k}_j$ may depend on any combination of electronic angular momenta that is itself totally symmetric with respect to $O_h^*(V)$. For example, we may have

$$\bar{k}_j = \alpha(k_j) + \beta(k_j)\sum_\varepsilon \mathbf{l}'_\varepsilon\cdot\mathbf{s}_\varepsilon + \gamma(k_j)\Big(\sum_{\varepsilon<\varepsilon'} \mathbf{l}'_\varepsilon.\mathbf{l}'_{\varepsilon'} + 4\mathbf{s}_\varepsilon.\mathbf{s}_{\varepsilon'}\Big),$$

if both the spin–orbit interactions and the Coulomb repulsions of the nonbonding valence electrons are sensitive to the nuclear displacements for the j-th mode. In short, according to this line of reasoning, we expect to see a variation of apparent force constants from one electronic energy level to another for a particular molecule.

Only the totally symmetric displacement, q_{1a}, can appear linearly in H_V. The coefficient with which it appears, $\bar{l}_1$, may depend on electronic parameters in the same way that $\bar{k}_j$ may. The variation in the effective value of $\bar{l}_1$ from one gross electronic energy level to another determines the change in apparent size of a molecule when it makes a transition from one of these electronic energy levels to another one. (If the ratio $\gamma(l_1)/\beta(l_1)$ is different from the ratio G/ζ_d then a nonzero value of $\bar{l}_1$ can also lead to changes in the apparent force constant of the a_{1g} mode from one molecule to another, as well as to changes from one electronic energy level to another for a particular molecule.)

At the present time, we only have detailed evidence about the force constants appropriate to the lowest energy levels of each metal hexafluoride molecule and we use the symbol k_j to stand for this effective, or so-called "ground-state", force constant.

The part of the molecular Hamiltonian that gives rise to Jahn–Teller forces in metal hexafluoride molecules we call $H_{\mathrm{J-T}}$. This part of the Hamiltonian depends essentially on electronic matrix elements that are not totally symmetric with respect to $O_h^*(V)$. For f_{2g}-orbitals, these matrix elements transform according to the E_g, the F_{2g} or the F_{1g} representation of $O_h^*(V)$.

The possibility of contributions to $H_{\mathrm{J-T}}$ arising from F_{1g}-matrix elements has been considered by Goodman.[78] These matrix elements are odd with respect to time-reversal and thus do not lead to forces tending to distort the molecule. Instead, they describe direct coupling of electronic orbital angular momentum with vibrational angular momentum and with rotational angular momentum. These terms may become important when rotational features of a molecule's energy levels are discussed, but they are properly neglected in our present study where rotational features are ignored.

The significant terms in $H_{\mathrm{J-T}}$ are thus the ones that depend on E_g- or F_{2g}-matrix elements for each nonbonding valence electron. These matrix elements for the electron ε, say, can be most conveniently expressed in terms of the components of this electron's effective orbital angular momentum, $\mathbf{l}'_\varepsilon$, by letting $\mathbf{T}_\varepsilon(2a)$ and $\mathbf{T}_\varepsilon(2b)$ be the two matrices that transform according to the E_g-representation and $\mathbf{T}_\varepsilon(5a)$, $\mathbf{T}_\varepsilon(5b)$, and $\mathbf{T}_\varepsilon(5c)$ be those that transform according to the F_{2g}-representation, where

$$\mathbf{T}_\varepsilon(jr) = \sum_{\mu\nu}(jr|\mu\nu)\mathbf{l}'_{\varepsilon\mu}\mathbf{l}'_{\varepsilon\nu},$$

with

$$(2a|++) = (2a|--) = \sqrt{3}/4;$$
$$(2b|+-) = (2b|-+) = -\tfrac{1}{4} \text{ and } (2b|00) = 1;$$
$$(5b|+0) = (5b|0+) = -(5b|-0) = -(5b|0-) = i/2;$$
$$(5c|+0) = (5c|0+) = (5c|-0) = (5c|0-) = -\tfrac{1}{2};$$

and

$$(5a|++) = -(5a|--) = i/2.$$

All other coefficients vanish. The specific coefficients given here are chosen to be consistent with the work of Moffitt and Thorson.[9] In particular, these authors, in discussing F_{2g}-states, define a vector matrix λ with components λ_a, λ_b, and λ_c, say:

$$\lambda_b = \begin{bmatrix} 0 & 0 & 0 \\ 0 & 0 & -i \\ 0 & i & 0 \end{bmatrix}, \lambda_c = \begin{bmatrix} 0 & 0 & i \\ 0 & 0 & 0 \\ -i & 0 & 0 \end{bmatrix}, \lambda_a = \begin{bmatrix} 0 & -i & 0 \\ i & 0 & 0 \\ 0 & 0 & 0 \end{bmatrix}$$

By identifying λ_a with $\mathbf{l}'_{\varepsilon 0}$ and $(\lambda_b \pm i\lambda_c)$ with $\mathbf{l}'_{\varepsilon\pm}$ in the above equations, one generates Moffitt and Thorson's matrices ε_1, ε_2, τ_1, τ_2, and τ_3 corresponding to $\mathbf{T}(2b)$, $\mathbf{T}(2a)$, $\mathbf{T}(5b)$, $\mathbf{T}(5c)$, and $\mathbf{T}(5a)$, respectively. The advantage gained from the present formulation is that it makes clear the relationship between the tensor quantities $\mathbf{T}_\varepsilon(jr)$ and the unit second-rank tensor quantities $\mathbf{u}_\varepsilon(2, q)$, mentioned earlier in this section. In terms of $\mathbf{T}^{(n)}(jr) = \sum_{\varepsilon=1}^{n} \mathbf{T}_\varepsilon(jr)$ we have,

$$\mathbf{T}^{(n)}(2a) = \alpha_2[\mathbf{U}^{(n)}(2, 2) + \mathbf{U}^{(n)}(2, -2)]/\sqrt{2},$$
$$\mathbf{T}^{(n)}(2b) = \alpha_2\mathbf{U}^{(n)}(2, 0),$$
$$\mathbf{T}^{(n)}(5a) = i\alpha_5[\mathbf{U}^{(n)}(2, 2) - \mathbf{U}^{(n)}(2, -2)]/\sqrt{2},$$
$$\mathbf{T}^{(n)}(5b) = -i\alpha_5[\mathbf{U}^{(n)}(2, 1) + \mathbf{U}^{(n)}(2, -1)]/\sqrt{2},$$

and

$$\mathbf{T}^{(n)}(5c) = \alpha_5[\mathbf{U}^{(n)}(2, 1) - \mathbf{U}^{(n)}(2, -1)]/\sqrt{2},$$

where

$$\alpha_2 = \sqrt{30}/2$$

and

$$\alpha_5 = \sqrt{10}.$$

In order to give an equation for $H_{\mathrm{J-T}}$, we define potential energies, $V(jr)$, that transform in the same way as the coordinates, q_{jr}:

$$V(2a) = l_2 q_{2a} + \kappa_2(q_{2a}q_{2b} + q_{2b}q_{2a}),$$
$$V(2b) = l_2 q_{2b} + \kappa_2(q_{2a}^2 - q_{2b}^2),$$
$$V(5a) = l_5 q_{5a},$$
$$V(5b) = l_5 q_{5b},$$

and

$$V(5c) = l_5 q_{5c}.$$

Then,

$$H_{\mathrm{J-T}} = \sum_j \sum_r V(jr) \sum_\varepsilon \mathbf{T}_\varepsilon(jr)$$

or finally, since $V(jr)$ is the same for each nonbonding valence electron,

$$H_{\mathrm{J-T}} = \sum_j \sum_r V(jr) \mathbf{T}^{(n)}(jr).$$

Here, the quantities l_2 and l_5 are the linear vibronic coupling constants for *each* f_{2g}-electron with respect to the e_g and f_{2g} vibrational modes, respectively, while κ_2 is the quadratic coupling constant for *each* f_{2g}-electron with respect to the e_g vibrational mode. These are the only three fundamental coupling constants for which the presently available spectra provide evidence.

The quantities l_2, l_5, and κ_2 may be regarded as determined by single-electron integrals that are, to a first approximation at least, constant for a series of closely homologous compounds such as hexafluoride molecules formed by elements in a particular transition series. The evaluation of these quantities from a study of the size of energy differences associated with, say, vibronic splittings for a certain electronic energy level of one compound in this type of homologous series therefore permits the prediction of vibronic splittings to be expected for the remaining electronic energy levels of that particular compound and of all other compounds in the series. This type of prediction depends, of course, on some previous assignment of the parameters, such as ζ_d and G, that determine the gross separations between electronic energy levels. Verification of the predicted vibronic effects for many electronic states in an extended series of homologous compounds would provide a crucial test of the correctness of any phenomenological description of vibronic properties for these molecules. At present we can only make such a test for the lowest energy electronic states in hexafluorides of the second and third transition series. The results of these tests do, however, give strong support to the validity of the approach to vibronic problems developed by Moffitt and by Longuet-Higgins with their respective co-workers.

C. Perturbation Theory

The energy levels and the properties of the eigenfunctions for the phenomenological Hamiltonian

$$H = H_E + H_V + H_{\mathrm{J-T}}$$

are to be evaluated in the remainder of this section by means of the perturbation formula developed by Moffitt and Thorson.[9] Because it is not immediately clear how their elegant procedure can be used to obtain wave functions as well as energy levels, we next give a derivation of their formula and several related, useful expressions. It turns out that this Moffitt–Thorson technique is useful quite generally for determining effects of perturbations in systems with evenly spaced energy levels. The simple harmonic oscillator and a Zeeman magnetic energy pattern are probably the two commonest examples of this type of system. The derivation of the Moffitt–Thorson formula seems, however, most conveniently given for an even more general type of problem.

Consider a Hamiltonian operator $\mathscr{H}$ that is the sum of an operator W, with well known energy levels and eigenfunctions, and some perturbation operator P:

$$\mathscr{H} = W + P.$$

If P can be analyzed into terms,

$$P = \sum_x P_x \text{ (where } x \text{ is a number),}$$

that obey the following commutation rules with respect to W,

$$[W, P_x] = xEP_x,$$

then perturbation formulas like that of Moffitt and Thorson can be used to obtain expressions for the energy levels and other properties of $\mathscr{H}$. The energy E is an arbitrary scaling constant. (If W represents a harmonic oscillator with mass m and force constant k, then it is convenient to take $E = \hbar\sqrt{k/m}$.)

As the first step in finding the energy levels of $\mathscr{H}$, we solve for $\mathscr{H}^{(T)}$, a transformed version of $\mathscr{H}$ that commutes with W and is defined as follows:

$$[W, \mathscr{H}^{(T)}] = 0$$

and

$$\mathscr{H}^{(T)} = S^{\dagger}\mathscr{H}S,$$

where $S = \exp\{iT\}$ and $T = T^{\dagger}$ so that $S^{\dagger} = \exp\{-iT\}$. These conditions almost completely define the Hermitian transformation operator T. In order to show this, we rely on the identity that for any operator, η, the following series expansion of

$$\eta^{(T)} = S^{\dagger}\eta S$$

is valid provided the series converges:

$$\eta^{(T)} = \eta + i[\eta, T] + i[i[\eta, T], T]/2! + \ldots + (i[)^t\eta(,T])^t/t! + \ldots$$

It is necessary to expand T and $\mathscr{H}^{(T)}$ according to the various orders of perturbation theory indicated by the positive integer p:

$$T = \sum_{p=0}^{\infty} T^{(p)}$$

and

$$\mathscr{H}^{(T)} = \sum_{p=0}^{\infty} \mathscr{H}^{(p)}.$$

For example, $\mathscr{H}^{(0)} = W$; and we can set $\mathscr{H}^{(1)} = P_0$ to obtain $[W, \mathscr{H}^{(T)}] = 0$ through first order in the perturbation. Each term $T^{(p)}$ is to be analyzed into quantities with well defined commutation properties with respect to W as follows:

$$T^{(p)} = \sum_x T_x^{(p)}$$

so that

$$[W, T_x^{(p)}] = xET_x^{(p)}.$$

Combining these equations, we find for $xE \neq 0$ the following expressions through second order for T and third order for $\mathscr{H}^{(T)}$:

$$T_x^{(1)} = iP_x/xE,$$
$$T_x^{(2)} = i\{[P_x, P_0] + \sum_{y\neq 0} (x/y)[P_y, P_{(x-y)}]\}/2x^2E^2,$$
$$\mathscr{H}^{(2)} = \sum_x [P_x, P_{-x}]/2xE,$$
$$\mathscr{H}^{(3)} = \sum_x \{[[P_x, P_0], P_{-x}] + \sum_{y\neq 0} (2x/y)[[P_y, P_{(x-y)}], P_{-x}]\}\{6x^2E^2\}^{-1}.$$

To obtain higher-order solutions is just a matter of algebra. For all $p \neq 0$, $T_0^{(p)} = 0$, while $[T_0^{(0)}, P_x] = 0$ for all values of x.

If the energy levels of W are nondegenerate then the energy levels of $\mathscr{H}$ are obtained directly from this procedure. If, however, there are sets of states, all with the same eigenvalue for W, then these states may be connected to each other by nonzero matrix elements of $\mathscr{H}^{(T)}$. In this case, an additional transformation may be required within each of these manifolds of states degenerate with respect to W in order to diagonalize $\mathscr{H}^{(T)}$ entirely, for whatever order of perturbation theory is required. The important feature of this procedure is that it reduces the off-diagonal elements of $\mathscr{H}^{(T)}$ to an easily managed form provided that no states need be considered that are infinitely degenerate with respect to W. The above procedure is closely similar to a Van Vleck transformation,[79] but its special simplicity arises from the commutation rules obeyed by the various terms in the perturbing energy:

$$[W, P_x] = xEP_x$$

with

$$P = \sum_x P_x.$$

(It seems worth noting that the derivation of these formulas would apply equally well for a continuous set of values for x. In each case, the validity of the procedure seems to depend only on the convergence of the perturbation series which are obtained for T and for $\mathscr{H}^{(T)}$.)

In order to apply the preceding perturbation formulas to determine the properties of the Hamiltonian of interest here,

$$H = H_E + H_V + H_{\text{J-T}},$$

it would seem most natural to regard $\langle H_V \rangle$ as W and $H_E + H_V - \langle H_V \rangle + H_{\mathrm{J-T}}$ as P. Here $\langle H_V \rangle$ is an average vibrational part of the energy, i.e., $\langle H_V \rangle$ attributes the same force constants and size to every electronic energy level of the molecule. Thus, $[H_E, \langle H_V \rangle] = 0$; and the analysis of P into terms with the desired commutation properties could be made without any unusual complications, since H_E would then simply be part of P_0. Certainly, this is the method that we should follow if the energy levels for H_E were about as far apart as those for $\langle H_V \rangle$. This method would, however, mean solving the electronic and the vibronic parts of the problem simultaneously. Therefore, it does not seem to be the most desirable way of approaching a vibronic analysis for the lowest electronic energy levels of metal hexafluoride molecules, because in these molecules the lowest energy levels are rather well isolated from the other energy levels. (One possible exception to this last statement would be the case of the still unknown hexafluoride of palladium, for which, as noted above, the predicted location of the first electronically excited energy level is only some 800 cm^{-1} above the ground state.)

Because of the relatively large energies separating those electronic states that are lowest with respect to H_E from the remaining electronic states for all the molecules analyzed here, we can use conventional second-order perturbation theory[79] to define an operator $h^{(n)}_{\mathrm{J-T}}$ (gr J') that gives the effect of $H_{\mathrm{J-T}}$ within the $(2J' + 1)$ states of lowest energy for a molecule with $n\, f_{2g}$-electrons. In order to make this definition concise, we write $H_{\mathrm{J-T}}$ as follows:

$$H_{\mathrm{J-T}} = \sum_q (-1)^q \mathbf{U}^{(n)}(2, q) \sum_j [L_j(2, -q) + K_j(2, -q)],$$

where $\mathbf{U}^{(n)}(2, q)$ is one of the unit tensor operators of rank 2 defined earlier and where we have set

$$\begin{aligned}
L_2(2, \pm 2) &= \alpha_2 l_2 q_{2a}/\sqrt{2},\\
L_5(2, \pm 2) &= \mp i\alpha_5 l_5 q_{5a}/\sqrt{2},\\
L_5(2, \pm 1) &= \alpha_5 l_5 \{\pm q_{5c} + i q_{5b}\}/\sqrt{2},\\
L_2(2, 0) &= \alpha_2 l_2 q_{2b},\\
K_2(2, \pm 2) &= \alpha_2 \kappa_2 (q_{2a} q_{2b} + q_{2b} q_{2a})/\sqrt{2},\\
K_2(2, 0) &= \alpha_2 \kappa_2 (q_{2a}^2 - q_{2b}^2),\\
L_2(2, \pm 1) &= L_5(2, 0) = K_2(2, \pm 1) = K_5(2, q) = 0.
\end{aligned}$$

Then we have the following expression for $h_{\text{J-T}}^{(n)}(\text{gr } J')$:

$$h_{\text{J-T}}^{(n)}(\text{gr } J') = U^{(n)}(2)_{gr}\sum_{Q}(-1)^{Q}\mathbf{U}_{J'}(2, Q)\sum_{j}\{L_j(2, -Q) + K_j(2, -Q)\} + \sum_{K}[E_K{}^{(n)}(\text{gr } J')]^{-1}\sum_{Q}(-1)^{Q}\mathbf{U}_{J'}(K, Q)\sum_{j}R_j(K, -Q),$$

where $U^{(n)}(2)_{gr}$, as defined above, is equal to $(\text{gr } J'\|\mathbf{U}^{(n)}(2)\|\text{gr } J')$ and we have introduced here the unit tensor operators of rank K,

$$\mathbf{U}_{J'}(K, Q).$$

Only the operator with Q equal to K is given explicitly here. The remaining ones can be generated by means of the commutation relationship:

$$[\mathbf{J}'_{-}, \mathbf{U}_{J'}(K, Q)] = \{(K + Q)(K - Q + 1)\}^{\frac{1}{2}}\mathbf{U}_{J'}(K, Q - 1)$$

for any of these unit tensor operators:

$$\mathbf{U}_{J'}(4, 4) = (\mathbf{J}'_{+})^4[7\,!(2J' - 4)\,!/4\,!(2J' + 5)\,!]^{\frac{1}{2}},$$
$$\mathbf{U}_{J'}(2, 2) = (\mathbf{J}'_{+})^2[3\,!(2J' - 2)\,!/(2J' + 3)\,!]^{\frac{1}{2}},$$
$$\mathbf{U}_{J'}(0, 0) = [2J' + 1]^{\frac{1}{2}}(\mathbf{J}'_{+}\mathbf{J}'_{-} + \mathbf{J}'_{-}\mathbf{J}'_{+} + 2\mathbf{J}'_0\mathbf{J}'_0)/[(2J' + 2)(2J' + 1)J'].$$

The matrix elements of $\mathbf{J}'_0$ and $\mathbf{J}'_{\pm}$ are simply the following ones:

$$(J'\bar{M}'|\mathbf{J}'_0|J'M') = M'\delta(\bar{M}', M')$$

and

$$(J'\bar{M}'|\mathbf{J}'_{\pm}|J'M') = [(J' \mp M')(J' \pm M' + 1)]^{\frac{1}{2}}\,\delta(\bar{M}', M' \pm 1).$$

We define effective energy denominators $E_K{}^{(n)}(\alpha J')$ in terms of Wigner 6-j coefficients as follows:

$$[E_K{}^{(n)}(\alpha J')]^{-1} = (2K + 1)^{\frac{1}{2}}(-1)^{K+2J'}\sum_{\bar{\alpha},\bar{J}'}\begin{Bmatrix}2 & 2 & K\\ J' & J' & \bar{J}'\end{Bmatrix}(\alpha J'\|\mathbf{U}^{(n)}(2)\|\bar{\alpha}\bar{J}')(\bar{\alpha}J'\|\mathbf{U}^{(n)}(2)\|\alpha J')/[E_{\bar{\alpha}} - E_{\alpha}].$$

The values of the required 6-j coefficients are given in Table XVII along with the values of $E_K{}^{(n)}(\text{gr } J')$ appropriate for $x = \zeta/(G + \zeta) = 0.3$ and for $x = 0.5862$, the two cases of interest here.

$$R_j(K, Q) = (-1)^{Q}(2K + 1)^{\frac{1}{2}}\sum_{q,\bar{q}}\begin{pmatrix}2 & 2 & K\\ q & \bar{q} & -Q\end{pmatrix}L_j(2, q)L_j(2, \bar{q}).$$

TABLE XVII

A. Values of those 6-j coefficients that are useful for vibronic problems of hexafluoride molecules and that do not vanish by definition

$$\begin{Bmatrix} 2 & 2 & 0 \\ 0 & 0 & 2 \end{Bmatrix} = 0.4472136,$$

$$\begin{Bmatrix} 2 & 2 & 0 \\ 1 & 1 & 1 \end{Bmatrix} = 0.2581989 = -\begin{Bmatrix} 2 & 2 & 0 \\ 1 & 1 & 2 \end{Bmatrix},$$

$$\begin{Bmatrix} 2 & 2 & 2 \\ 1 & 1 & 1 \end{Bmatrix} = 0.1527525 = \begin{Bmatrix} 2 & 2 & 2 \\ 1 & 1 & 2 \end{Bmatrix},$$

$$\begin{Bmatrix} 2 & 2 & 0 \\ 2 & 2 & 0 \end{Bmatrix} = 0.2000000 = -\begin{Bmatrix} 2 & 2 & 0 \\ 2 & 2 & 1 \end{Bmatrix} = \begin{Bmatrix} 2 & 2 & 0 \\ 2 & 2 & 2 \end{Bmatrix}$$

$$= \begin{Bmatrix} 2 & 2 & 2 \\ 2 & 2 & 0 \end{Bmatrix} = \begin{Bmatrix} 2 & 2 & 4 \\ 2 & 2 & 0 \end{Bmatrix},$$

$$\begin{Bmatrix} 2 & 2 & 2 \\ 2 & 2 & 1 \end{Bmatrix} = -0.1000000, \qquad \begin{Bmatrix} 2 & 2 & 2 \\ 2 & 2 & 2 \end{Bmatrix} = -0.0428571,$$

$$\begin{Bmatrix} 2 & 2 & 4 \\ 2 & 2 & 1 \end{Bmatrix} = 0.1333333, \qquad \begin{Bmatrix} 2 & 2 & 4 \\ 2 & 2 & 2 \end{Bmatrix} = 0.0571429,$$

$$\begin{Bmatrix} 2 & 2 & 0 \\ \frac{1}{2} & \frac{1}{2} & \frac{3}{2} \end{Bmatrix} = 0.3162278 = -\begin{Bmatrix} 2 & 2 & 0 \\ \frac{1}{2} & \frac{1}{2} & \frac{5}{2} \end{Bmatrix},$$

$$\begin{Bmatrix} 2 & 2 & 0 \\ \frac{3}{2} & \frac{3}{2} & \frac{1}{2} \end{Bmatrix} = 0.2236068 = -\begin{Bmatrix} 2 & 2 & 0 \\ \frac{3}{2} & \frac{3}{2} & \frac{3}{2} \end{Bmatrix} = \begin{Bmatrix} 2 & 2 & 0 \\ \frac{3}{2} & \frac{3}{2} & \frac{5}{2} \end{Bmatrix},$$

$$\begin{Bmatrix} 2 & 2 & 2 \\ \frac{3}{2} & \frac{3}{2} & \frac{1}{2} \end{Bmatrix} = 0.1870829, \qquad \begin{Bmatrix} 2 & 2 & 2 \\ \frac{3}{2} & \frac{3}{2} & \frac{3}{2} \end{Bmatrix} = 0,$$

$$\begin{Bmatrix} 2 & 2 & 2 \\ \frac{3}{2} & \frac{3}{2} & \frac{5}{2} \end{Bmatrix} = -0.1336306,$$

$$\begin{Bmatrix} 2 & 2 & 0 \\ \frac{5}{2} & \frac{5}{2} & \frac{1}{2} \end{Bmatrix} = -0.1825742 = -\begin{Bmatrix} 2 & 2 & 0 \\ \frac{5}{2} & \frac{5}{2} & \frac{3}{2} \end{Bmatrix} = \begin{Bmatrix} 2 & 2 & 0 \\ \frac{5}{2} & \frac{5}{2} & \frac{5}{2} \end{Bmatrix},$$

$$\begin{Bmatrix} 2 & 2 & 2 \\ \frac{5}{2} & \frac{5}{2} & \frac{1}{2} \end{Bmatrix} = -0.1632993, \qquad \begin{Bmatrix} 2 & 2 & 2 \\ \frac{5}{2} & \frac{5}{2} & \frac{3}{2} \end{Bmatrix} = 0.0583212 = \begin{Bmatrix} 2 & 2 & 2 \\ \frac{5}{2} & \frac{5}{2} & \frac{5}{2} \end{Bmatrix},$$

$$\begin{Bmatrix} 2 & 2 & 4 \\ \frac{5}{2} & \frac{5}{2} & \frac{1}{2} \end{Bmatrix} = -0.1054093, \qquad \begin{Bmatrix} 2 & 2 & 4 \\ \frac{5}{2} & \frac{5}{2} & \frac{3}{2} \end{Bmatrix} = -0.1204678,$$

$$\begin{Bmatrix} 2 & 2 & 4 \\ \frac{5}{2} & \frac{5}{2} & \frac{5}{2} \end{Bmatrix} = -0.0677631.$$

TABLE XVII (*continued*)

B. $E_K^{(n)}(\text{gr } J')$ in cm^{-1} for $x = 0.3$ and $G + \zeta = 4270\ \text{cm}^{-1}$

n	J'	$K = 0$	$K = 2$	$K = 4$
1	$\frac{3}{2}$	12,890	6890	
2	2	6852	4958	−10,748
3	$\frac{3}{2}$	1,340,701	−1,478,532	
4	0	12,794		
5	$\frac{1}{2}$	9114		

C. $E_K^{(n)}(\text{gr } J')$ in cm^{-1} for $x = 0.5862$ and $G + \zeta = 5800\ \text{cm}^{-1}$

n	J'	$K = 0$	$K = 2$	$K = 4$
1	$\frac{3}{2}$	34,212	18,287	
2	2	15,074	12,333	76,540
3	$\frac{3}{2}$	95,774	−130,506	
4	0	28,242		
5	$\frac{1}{2}$	24,191		

For example,

$$R_5(0, 0) = (\alpha_5 l_5)^2(q_{5a}^2 + q_{5b}^2 + q_{5c}^2)/\sqrt{5}.$$

Next we must translate the nuclear displacement coordinates, q_{jr}, appearing in $h_{\text{J-T}}^{(n)}$ (gr J'), into quantities that have more desirable commutation rules with respect to $\langle H_V \rangle$. As a preliminary to this translation, we consider the corresponding problem for a one-dimensional harmonic oscillator. All of the arguments used in this treatment of the one-dimensional oscillator have obvious extensions to any number of dimensions and, in particular, to the two- and three-dimensional cases represented by the e_g and the f_{2g} modes of a hexafluoride molecule. To make our discussion simple, we regard these modes now one dimension at a time.

Consider the Hamiltonian h_0 for a simple harmonic oscillator with a mass m and a force constant k:

$$h_0 = p^2/2m + kq^2/2,$$

in terms of a displacement coordinate q and its conjugate momentum p, such that $[q, p] = i\hbar$. Setting ν equal to $\hbar\sqrt{k/m}$ yields $E_n = (n + \frac{1}{2})\nu$ as the allowed energy levels of h_0. What are called raising and lowering operators, x_1 and x_{-1} respectively, are defined as follows in terms of the constant $a = (mk)^{\frac{1}{4}}$:

$$x_{\pm 1} = (aq \mp ip/a)(2\hbar)^{-\frac{1}{2}}.$$

These definitions lead to the following equations:

$$h_0 = (x_1 x_{-1} + x_{-1} x_1)\nu/2$$

and

$$[x_{-1}, x_1] = 1,$$

so that

$$[h_0, x_{\pm 1}] = \pm\, \nu x_{\pm 1}.$$

The inverse relationships are simply

$$q = (2\hbar)^{\frac{1}{2}}(x_1 + x_{-1})/2a$$

and

$$p = (2\hbar)^{\frac{1}{2}} ia(x_1 - x_{-1})/2$$

so that any operator P, say, that can be expressed in terms of positive, integral powers of p and/or q can be analyzed by inspection into a form appropriate to the commutator perturbation procedure:

$$P = P_0 + \sum_{k=1,2\ldots} (P_k + P_{-k}),$$

where P_0 contains all terms with equal powers of x_1 and x_{-1} (for example, $x_1^2 x_{-1}^2$) while P_k contains terms with k more powers of x_1 than powers of x_{-1} (for example, ${x_1}^{k+1} x_{-1}$) and, of course, P_{-k} contains all those terms with k more powers of x_{-1} than of x_1 (for example, $x_1^2 x_{-1}^{k+2}$). Thus, $[h_0, P_x] = x\nu P_x$, as required.

The terms in $h^{(n)}_{\text{J-T}}(\text{gr } J')$ with which we are concerned here depend on either the first or the second power of a coordinate q_{jr} where for $j = 2$, $r = a$ or b, and for $j = 5$, $r = a$, b, or c. Thus, $h^{(n)}_{\text{J-T}}(\text{gr } J')$ can be analyzed by inspection in terms of the quantities

$$x_{jr,\pm} = (a_j q_{jr} \mp ip_{jr}/a_j)(2\hbar)^{-\frac{1}{2}},$$

where $a_j = (m_j k_j)^{\frac{1}{4}}$. Immediately, we can express each of the quantities $X_j(K, Q)$, where X stands for either L, K, or R, in the form $X_j(K, Q) = \sum_s X_{j,s}(K, Q)$ so that

$$[\langle H_V \rangle, X_{j,s}(K, Q)] = s\nu_j X_{j,s}(K, Q).$$

For example,

$$L_2(2, \pm 2) = (\alpha_2 l_2 \hbar^{\frac{1}{2}}/2a_2)(x_{2a,1} + x_{2a,-1}).$$

Now, after rather extensive algebraic manipulation, the perturbed Hamiltonian, $\mathscr{H} = \langle H_V \rangle + h^{(n)}_{\text{J-T}}(\text{gr } J')$, takes the following form through second order:

$$\mathscr{H}^{(0)} = (n_1 + \tfrac{1}{2})\nu_1 + (n_2 + 1)\nu_2 + (n_5 + \tfrac{3}{2})\nu_5$$

$$\begin{aligned}\mathscr{H}^{(1)} = {} & (n_2 + 1)\nu_2[\beta_2^{(n)}\mathbf{U}(0, 0) + \gamma_2^{(n)}\mathbf{U}(4, 1a)] \\ & + (n_5 + \tfrac{3}{2})\nu_5[\beta_5^{(n)}\mathbf{U}(0, 0) + \gamma_5^{(n)}\mathbf{U}(4, 1a)] \\ & + \nu_2\{\phi^{(n)}[M_{2,+1}\mathbf{U}(2, 2-) + M_{2,-1}\mathbf{U}(2, 2+)] \\ & + \eta^{(n)}[M_{2,+1}\mathbf{U}(4, 2-) + M_{2,-1}\mathbf{U}(4, 2+)]\}\end{aligned}$$

$$\begin{aligned}\mathscr{H}^{(2)} = {} & -\lambda_2\theta_2^{(n)}\{\mathbf{U}(2, 2+)\mathbf{U}(2, 2-) + \mathbf{U}(2, 2-)\mathbf{U}(2, 2+) \\ & - 2M_{2,0}[\mathbf{U}(2, 2+), \mathbf{U}(2, 2-)]\} - \lambda_5\theta_5^{(n)}\{[\mathbf{U}(5\,0)]^2 \\ & - \mathbf{U}(5+)\mathbf{U}(5-) - \mathbf{U}(5-)\mathbf{U}(5+) + M_{5,0}[\mathbf{U}(5+), \\ & \mathbf{U}(5-)] + M_{5,-1}[\mathbf{U}(5\,0), \mathbf{U}(5-)]/\sqrt{2} + M_{5,1}[\mathbf{U}(5\,0), \\ & \mathbf{U}(5+)]/\sqrt{2}\}.\end{aligned}$$

Here we have used the following definitions for $j = 2$ or 5:

$$\begin{aligned}\beta_j^{(n)} &= -\alpha_j^2\lambda_j/(\sqrt{5}E_0^{(n)}(\text{gr } J')), \\ \gamma_2^{(n)} &= -\alpha_2^2\lambda_2\sqrt{30}/(10E_4^{(n)}(\text{gr } J')), \\ \gamma_5^{(n)} &= -\alpha_5^2\lambda_5\sqrt{30}/(15E_4^{(n)}(\text{gr } J')), \\ \phi^{(n)} &= \sqrt{2}\alpha_2\rho_2 U^{(n)}(2)_{gr} - 2\alpha_2^2\lambda_2/(\sqrt{7}E_2^{(n)}(\text{gr } J')), \\ \eta^{(n)} &= -\alpha_2^2\lambda_2\sqrt{21}/(7E_4^{(n)}(\text{gr } J')), \\ \theta_j^{(n)} &= [\alpha_j U^{(n)}(2)_{gr}]^2/2, \\ \lambda_j &= l_j^2/k_j \text{ and } \rho_2 = \kappa_2/k_2;\end{aligned}$$

$$\begin{aligned}\mathbf{U}(0, 0) = \mathbf{U}_{J'}(0, 0), \quad \mathbf{U}(4, 1a) = {} & [\sqrt{5}\mathbf{U}_{J'}(4, 4) \\ & + \sqrt{14}\mathbf{U}_{J'}(4, 0) + \sqrt{5}\mathbf{U}_{J'}(4, -4)]/\sqrt{24},\end{aligned}$$

$$\mathbf{U}(2, 2\pm) = [\pm i\mathbf{U}_{J'}(2, 2) + \sqrt{2}\mathbf{U}_{J'}(2, 0) \pm i\mathbf{U}_{J'}(2, -2)]/2,$$

$$\mathbf{U}(4, 2\pm) = [\sqrt{7}\mathbf{U}_{J'}(4, 4) \pm i\sqrt{12}\mathbf{U}_{J'}(4, 2) - \sqrt{10}\mathbf{U}_{J'}(4, 0) \pm i\sqrt{12}\mathbf{U}_{J'}(4, -2) + \sqrt{7}\mathbf{U}_{J'}(4, -4)]/4\sqrt{3},$$

$$\mathbf{U}(5\,0) = -i[\mathbf{U}_{J'}(2, 2) - \mathbf{U}_{J'}(2, -2)]/\sqrt{2}$$

and

$$\mathbf{U}(5\pm) = \mp i\mathbf{U}_{J'}(2, \pm 1).$$

(We have omitted here some of the terms in $\mathscr{H}^{(1)}$ that depend on l_5^2 and all terms in $\mathscr{H}^{(2)}$ that depend on l_2^4 and l_5^4. There are no experimental data about these terms; thus we say no more about them.) In the preceding equations, we have used the vibrational operators, $M_{2,0}$, $M_{2,\pm1}$, $M_{5,0}$, and $M_{5,\pm1}$, defined as follows:

$$M_{2,0} = [-i/2](x_{2a,1}x_{2b,-1} - x_{2b,1}x_{2a,-1}),$$

$$M_{2,\pm1} = \tfrac{1}{2}[x_{2a,1}x_{2a,-1} - x_{2b,1}x_{2b,-1} \pm i(x_{2a,1}x_{2b,-1} + x_{2b,1}x_{2a,-1})],$$

$$M_{5,0} = -i(x_{5b,1}x_{5c,-1} - x_{5c,1}x_{5b,-1}),$$

and

$$M_{5,\pm1} = \pm (x_{5a,1}x_{5b,-1} - x_{5b,1}x_{5a,-1}) + i(x_{5a,1}x_{5c,-1} - x_{5c,1}x_{5a,-1}).$$

It is useful to think of these vibrational operators as angular momenta because they satisfy the following commutation rules for either $j = 2$ or $j = 5$:

$$[M_{j,0}, M_{j,\pm1}] = \pm M_{j,\pm1}$$

and

$$[M_{j,1}, M_{j,-1}] = 2M_{j,0}.$$

Furthermore, it is easy to verify that

$$M_{2,0}^2 + (M_{2,1}M_{2,-1} + M_{2,-1}M_{2,1})/2 = n_2(n_2 + 2)/4,$$

where

$$n_2 = (x_{2a,1}x_{2a,-1} + x_{2b,1}x_{2b,-1}).$$

Thus the matrix elements of $M_{2,0}$ and $M_{2,\pm1}$ take the following form in the representation for which n_2 and $M_2 = (q_{2a}p_{2b} - q_{2b}p_{2a})/\hbar$ are diagonal:

$$(\bar{n}_2\bar{m}|M_{2,0}|n_2m) = \tfrac{1}{2}m\delta(\bar{n}_2, n_2)\delta(\bar{m}, m)$$

while

$$(\bar{n}_2\bar{m}|M_{2,\pm1}|n_2m) = \tfrac{1}{2}[(n_2 \mp m)(n_2 \pm m + 2)]^{\frac{1}{2}} \times \delta(\bar{n}_2, n_2)\delta(\bar{m}, m \pm 1).$$

The situation is slightly more complicated for a three-dimensional oscillator because, for example,

$$\mathbf{M}_5^2 = M_{5,0}^2 + (M_{5,-1}M_{5,1} + M_{5,1}M_{5,-1})/2 = n_5(n_5 + 1) - D = M_5(M_5 + 1),$$

where

$$n_5 = x_{5a,1}x_{5a,-1} + x_{5b,1}x_{5b,-1} + x_{5c,1}x_{5c,-1}$$

and

$$D = [(x_{5a,1})^2 + (x_{5b,1})^2 + (x_{5c,1})^2][(x_{5a,-1})^2 + (x_{5b,-1})^2 + (x_{5c,-1})^2].$$

Thus, there are in general several possible values for M_5 ranging from n_5 down, by steps of two units, to either 1 or 0 depending on whether n_5 is odd or even. It is easily verified that $\mathbf{M}_5^2$, $M_{5,0}$, and $M_{5,\pm1}$ all commute with n_5 and that $\mathbf{M}_5^2$ commutes with $M_{5,0}$ and $M_{5,\pm1}$. Thus, the matrix elements of these operators are given most simply in a representation for which n_5, M_5, and $M_{5,0}$ are all diagonal. These matrix elements are

$$(\bar{n}_5\bar{M}_5\bar{m}_0|M_{5,0}|n_5M_5m_0) = m_0\delta(\bar{n}_5, n_5)\delta(\bar{M}_5, M_5)\delta(\bar{m}_0, m_0)$$

and

$$(\bar{n}_5\bar{M}_5\bar{m}_0|M_{5,\pm1}|n_5M_5m_0) = [(M_5 \mp m_0)(M_5 \pm m_0 + 1)]^{\frac{1}{2}} \times \delta(\bar{n}_5, n_5)\delta(\bar{M}_5, M_5)\delta(\bar{m}_0, m_0 \pm 1).$$

We have now given complete directions for writing down the matrix elements of $\mathscr{H}^{(0)} + \mathscr{H}^{(1)} + \mathscr{H}^{(2)}$ in a form in which $\mathscr{H}^{(0)}$ is diagonal but in which $\mathscr{H}^{(1)}$ and $\mathscr{H}^{(2)}$ may have nonzero matrix elements connecting states with the same value of $\mathscr{H}^{(0)}$. Before we complete the diagonalization of $\mathscr{H}^{(0)} + \mathscr{H}^{(1)} + \mathscr{H}^{(2)}$ for the cases of interest here, $J' = 1, \frac{3}{2}$, and 2, we must give an equation for the matrix elements of a vibrational coordinate in this partially diagonalized basis.

We use the truncated formula

$$q^{(T)} = q + i[q, T^{(1)}] - [[q, T^{(1)}], T^{(1)}]/2$$

with $T^{(1)} = \sum_j \sum_r \sum_\mu L_{jr,\mu}/\mu v_j$, to obtain the following expressions:

$$q_{jr}^{(T)} = q_{jr} + q_{jr}^{(E)} + q_{jr}^{(P)},$$

where

$$q_{jr}^{(E)} = -d_j \sum_Q (-1)^Q \mathbf{U}_{J'}(2, Q)[L_{jr}(2, -Q)/q_{jr}],$$

and

$$q_{jr}^{(P)} = d_j U^{(n)}(2)_{gr} \sum_{j',r',Q,Q'} (-1)^{Q+Q'}[L_{jr}(2, -Q)/2v_{j'}q_{jr}]$$
$$\times [\mathbf{U}_{J'}(2, Q), \mathbf{U}_{J'}(2, Q')][L_{j'r',+1}(2, -Q') - L_{j'r',-1}(2, -Q')],$$

with

$$d_j = (\hbar/2)^{\frac{1}{2}} U^{(n)}(2)_{gr}/v_j a_j.$$

Two features of this equation for $q_{jr}^{(T)}$ deserve comment. The first is that the term $q_{jr}^{(E)}$ has nonzero matrix elements between states with the same value of n_j and with different values of M'. Thus, the presence of $q^{(E)}$ in $q^{(T)}$, the transformed nuclear displacement coordinate, makes clear that it is impossible to speak of separate nuclear motions and electronic motions when a molecule exhibits vibronic coupling. Dramatic evidence of the importance of the term $q^{(E)}$ is seen in the infrared absorption spectrum of OsF_6. There, transitions between the nearly degenerate components of vibronic multiplets combine with a transition of the infrared-active bending vibration to give unusual absorption bands. Detailed assignments of these combination bands are given in Section IV, but we do not try to assess their intensities quantitatively.

The second remarkable feature in the equation for $q_{jr}^{(T)}$ is the term $q_{jr}^{(P)}$. This term leads to a variation of transition intensities within a vibronic multiplet. The intensities of the longer wavelength components seem generally to increase while those of the shorter wavelength components decrease as compared to their values in the absence of vibronic coupling. The corresponding variation of intensities has been observed and is discussed in Section IV. Again we offer no quantitative comparisons on this point between theory and experiment. Our primary interest now is in calculating the positions of the vibronic energy levels themselves.

D. Vibronic Energy Levels for Metal Hexafluoride Molecules

The complexity of the energy levels associated with vibronic coupling increases as the effective, electronic angular momentum, J', increases. For $J' = 0$, as in the electronic ground state of PtF_6, the situation is quite simple and has already been described by Child.[50] In this case, there is no splitting of energy levels and no spectral anomalies can be seen. This does not mean, however, that there is no vibronic coupling. There is indeed important coupling between each $J' = 0$ state for PtF_6 and at least one of the $J' = 2$ quintets, as one may see from the entries in Table XVI. This coupling affects the two non-totally symmetric motions of PtF_6 that are Raman-active. It lowers their frequencies in comparison to the values that they would have in the absence of vibronic coupling. Assuming the values $\lambda_2 = 336\ \text{cm}^{-1}$ and $\lambda_5 = 141\ \text{cm}^{-1}$ derived below for ReF_6 apply also to PtF_6, we calculate that the lowering of the frequency for the f_{2g} mode in PtF_6 is 5.5 cm^{-1}, while that for the e_g mode of PtF_6 is 25.1 cm^{-1}. Our value for the f_{2g} mode is in close agreement with that of 4 cm^{-1} estimated by Child, but our value for the e_g mode is an order of magnitude greater than Child's estimate of 2 cm^{-1}.[50] An investigation of the change in these frequencies for PtF_6 on going from its lower $J' = 0$ state to its higher $J' = 0$ state would provide data about the correctness of these predicted frequency reductions for the electronic ground state of this molecule, but, no such comparison can be made at present.

For $J' = \frac{1}{2}$, the energy levels in the presence of vibronic coupling are still quite simple. Just as in the case of $J' = 0$, splittings do not arise, but the frequencies of the vibrational modes to which the Jahn–Teller theorem applies are reduced. In particular, writing ν_2 and ν_5 for the values to be expected in the absence of vibronic coupling for frequencies of the e_g and the f_{2g} modes, respectively, we have the following expression for vibronic energy levels when $J' = \frac{1}{2}$:

$$E^{(n,\frac{1}{2})}(n_2, n_5) = (n_2 + 1)\nu_2(1 + \beta_2^{(n)}/\sqrt{2}) + (n_5 + \tfrac{3}{2})\nu_5(1 + \beta_2^{(n)}/\sqrt{2}).$$

The lowest electronic state of an extremely unlikely molecule, AuF_6, would be a $J' = \frac{1}{2}$ doublet. The values of $\beta_j^{(5)}$ that would apply to this molecule are predicted to be as follows:

$$\beta_2^{(5)} = -0.0469 \quad \text{and} \quad \beta_5^{(5)} = -0.0261.$$

These are the numbers that result from the assumption that the values of λ_2 and λ_5 appropriate to AuF_6 would be the same as those assigned below for ReF_6. (Assuming the constancy of λ_j is not in general the same as assuming the constancy of l_j, because $\lambda_j/\bar{\lambda}_j = (l_j\bar{\nu}_j/\bar{l}_j\nu_j)^2$.)

For $J' = 1$, the vibronic energy levels begin to be more complicated. The energy levels involving excitation of only the f_{2g} nuclear motions, i.e., those energy levels for which $n_2 = 0$, form the sort of regular pattern that is characteristic of a scalar coupling of two angular momentum operators, $\mathbf{J}'\cdot\mathbf{M}_5$, in the present notation. As noted by Moffitt and Thorson,[9] this simple pattern is not affected by any further coupling that is linear in the e_g nuclear displacement coordinates. The pattern of these energy levels is, however, disturbed by low-symmetry terms in the energy that depend quadratically on the e_g coordinates.

Since such quadratic coupling terms do seem to be important for metal hexafluoride molecules, the energy levels for $J' = 1$ in these molecules must be calculated from the following matrix operator equation:

$$E = E^{(n,1)}(n_2, n_5)\begin{bmatrix} 1 & 0 & 0 \\ 0 & 1 & 0 \\ 0 & 0 & 1 \end{bmatrix} + B_2^{(n,1)}\{M_{2,1}\begin{bmatrix} 1 & 0 & -i\sqrt{3} \\ 0 & -2 & 0 \\ -i\sqrt{3} & 0 & 1 \end{bmatrix}$$

$$+ M_{2,-1}\begin{bmatrix} 1 & 0 & i\sqrt{3} \\ 0 & -2 & 0 \\ i\sqrt{3} & 0 & 1 \end{bmatrix}\} + A_5^{(n,1)}\{2M_{5,0}\begin{bmatrix} 1 & 0 & 0 \\ 0 & 0 & 0 \\ 0 & 0 & -1 \end{bmatrix}$$

$$+ M_{5,1}\begin{bmatrix} 0 & 0 & 0 \\ \sqrt{2} & 0 & 0 \\ 0 & \sqrt{2} & 0 \end{bmatrix} + M_{5,-1}\begin{bmatrix} 0 & \sqrt{2} & 0 \\ 0 & 0 & \sqrt{2} \\ 0 & 0 & 0 \end{bmatrix}\}$$

where the vibrational angular momenta $M_{j,\mu}$ are as defined above, while

$$E^{(n,1)}(n_2, n_5) = (n_2 + 1)\nu_2(1 + \beta_2^{(n)}/\sqrt{3}) + (n_5 + \tfrac{3}{2})\nu_5(1 + \beta_5^{(n)}/\sqrt{3}) - 2\lambda_2\theta_2^{(n)}/15 - \lambda_5\theta_5^{(n)}/5,$$

$$A_5^{(n,1)} = \lambda_5\theta_5^{(n)}/20 \quad \text{and} \quad B_2^{(n,1)} = \nu_2\phi^{(n)}/2\sqrt{15}\,.$$

Several of the lowest energy levels that result for $J' = 1$ are shown in Fig. 6. In this figure the scale is chosen to correspond to the lower F_{2g} electronic triplet of RuF_6, for which (assuming

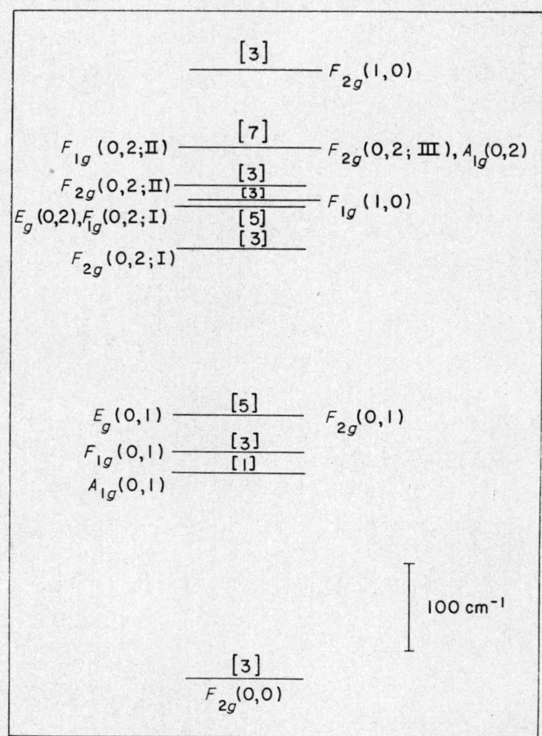

Fig. 6. The lowest vibronic energy levels for $J' = 1$, drawn approximately to scale for the lower F_{2g} electronic state of RuF_6. (The degeneracy of each energy level is indicated by the number in square brackets.)

that λ_2 and λ_5 have the same values for RuF_6 as those assigned below for TcF_6),

$$E^{(2,1)}(n_2, n_5) = (n_2 + 1)\ 624\ \text{cm}^{-1} + (n_5 + \tfrac{3}{2})\ 283\ \text{cm}^{-1} - 39.4\ \text{cm}^{-1} - 47.0\ \text{cm}^{-1},$$

$$A_5^{(2,1)} = 11.76\ \text{cm}^{-1} \quad \text{and} \quad B_2^{(2,1)} = -37.89\ \text{cm}^{-1}.$$

Because the vibronic coupling actually observed for the metal hexafluoride molecules may be described as mild, useful designations of the vibronic energy levels in these molecules may be given as follows:

$$\Gamma(n_2, n_5; \alpha),$$

where Γ stands for the representation of $O_h^*(V)$ according to which the state(s) of this energy level transform. Here the numbers n_2 and n_5 indicate the energy level for nuclear motion with which this vibronic energy level is associated in the Moffitt–Thorson perturbation formula. Finally, the label α is only used to distinguish between different energy levels for which the other three labels, Γ, n_2, and n_5, are duplicated. For example, the lowest energy level in Fig. 6 is $F_{2g}(0, 0)$, the label α being dropped because no other energy level for $J' = 1$ has the same label.

For $J' = \frac{3}{2}$, the vibronic energy levels are very involved; nevertheless, these energy levels can still be specified, for the cases of interest here, by rather simple algebraic formulas closely analogous to those given by Moffitt and Thorson[9] for the $G_{\frac{3}{2}g}$ and $G_{\frac{3}{2}u}$ electronic states. The formulas given here reduce to those of Moffitt and Thorson when $B_2^{(n,\frac{3}{2})}$ is set equal to zero. In order to give concisely the algebraic expression for these vibronic energy levels, we need the following definitions: r is an integer between zero and n_2, inclusive; s is an integer between zero and $n_5/2$, inclusive; $M_5 = n_5 - 2s$; for $M_5 = 0$, $V_5 = \frac{1}{2}$; and for $M_5 \geq 1$, $V_5 = M_5 \pm \frac{1}{2}$.

For each set of vibrational quantum numbers, n_2 and n_5, there are $(2V_5 + 1)$ states at each of the following energies:

$$\begin{aligned} E_\sigma^{(n,\frac{3}{2})}(n_2, n_5; r, M_5, V_5) &= E^{(n,\frac{3}{2})}(n_2, n_5) - A_2^{(n,\frac{3}{2})} \\ &+ \sigma\{[A_2^{(n,\frac{3}{2})}(n_2 + 1 - 2r)]^2 + [B_2^{(n,\frac{3}{2})}]^2 r(n_2 - r + 1)\}^{\frac{1}{2}} \\ &+ [3 - 4V_5(V_5 + 1) + 4M_5(M_5 + 1)]A_5^{(n,\frac{3}{2})}/4, \end{aligned}$$

where σ is either $+1$ or -1. For $r = 0$, both choices for σ lead to the same positive value for the root: $A_2^{(n,\frac{3}{2})}(n_2 + 1)$.

$$A_2^{(n,\frac{3}{2})} = \lambda_2 \theta_2^{(n)}/10,$$

$$B_2^{(n,\frac{3}{2})} = \nu_2 \phi^{(n)}/\sqrt{10},$$

and

$$A_5^{(n,\frac{3}{2})} = \lambda_5 \theta_5^{(n)}/10;$$

while

$$E^{(n,\frac{3}{2})}(n_2, n_5) = (n_2 + 1)\nu_2(1 + \beta_2^{(n)}/2) + (n_5 + \tfrac{3}{2})\nu_5(1 + \beta_5^{(n)}/2) - A_2^{(n,\frac{3}{2})} - (\tfrac{3}{2})A_5^{(n,\frac{3}{2})}.$$

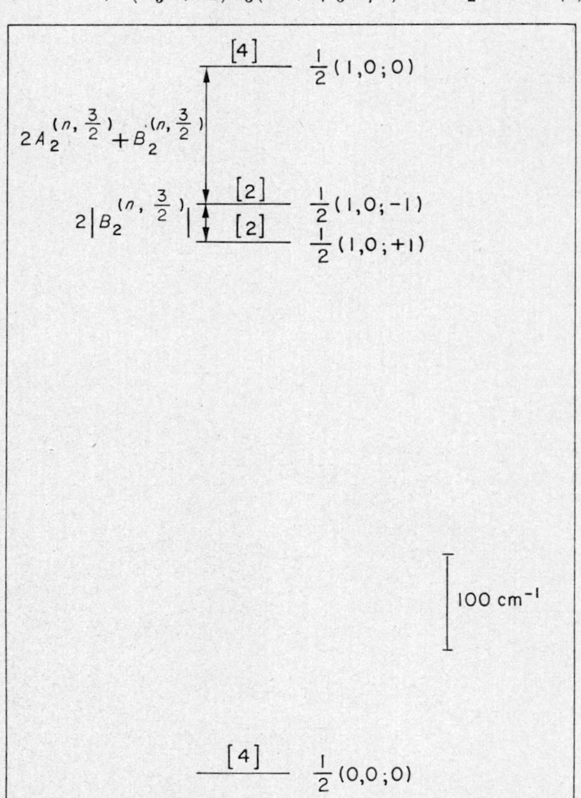

Fig. 7. The lowest vibronic energy levels involving the e_g vibrational mode for $J' = \frac{3}{2}$, drawn approximately to scale for ReF_6. (The degeneracy of each energy level is indicated by the number in square brackets.)

As is discussed in Section V, this closed expression for the vibronic energy levels of a molecule with an effective electronic angular momentum of $\frac{3}{2}$ allows the evaluation of the corresponding vibronic partition function, leading to estimates of the effects of vibronic coupling on the thermodynamic properties of ReF_6, IrF_6, TcF_6, and RhF_6.

Our principal concern in this Section is, however, the location of the lowest energy levels in these molecules. Figures 7 and 8 show these energy levels for independent excitations of the e_g and of the f_{2g} modes, respectively. For a molecule with $J' = \frac{3}{2}$ and in

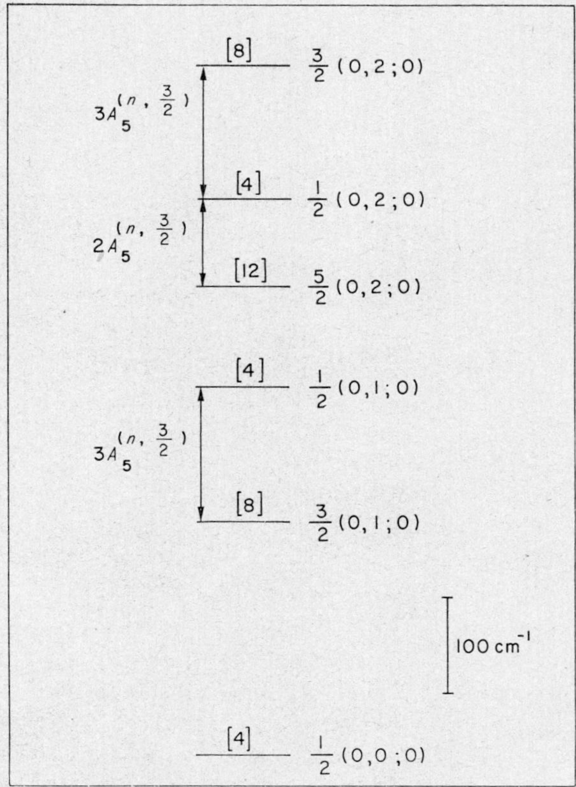

Fig. 8. The lowest vibronic energy levels involving the f_{2g} vibrational mode for $J' = \frac{3}{2}$, drawn approximately to scale for ReF_6. (The degeneracy of each energy level is indicated by the number in square brackets.)

which the e_g and the f_{2g} modes are simultaneously excited, the proper energy-level diagram can be obtained by simply taking the sum of these pictured energy levels. This great simplification does not happen in general for vibronic problems, as we noted above for $J' = 1$ and as we see below for $J' = 2$.

In the special case of $J' = \frac{3}{2}$, it is most convenient to label the vibronic energy levels with the symbols

$$V_5(n_2, n_5;\ \sigma r).$$

For $r \neq 0$, each of these energy levels is $(2V_5 + 1)$-fold degenerate, while for $r = 0$, $V_5(n_2, n_5;\ +0)$ and $V_5(n_2, n_5;\ -0)$ both refer to the same $2(2V_5 + 1)$-degenerate energy level, which is simply called $V_5(n_2, n_5;\ 0)$.

Figures 7 and 8 are drawn approximately to scale for ReF_6 by taking

$$\begin{aligned} \nu_2(1 + \beta_2^{(1)}/2) &= 671\ \text{cm}^{-1}, \\ \nu_5(1 + \beta_5^{(1)}/2) &= 295\ \text{cm}^{-1}, \\ A_2^{(1,\frac{3}{2})} &= 84\ \text{cm}^{-1}, \\ B_2^{(1,\frac{3}{2})} &= -21\ \text{cm}^{-1}, \\ A_5^{(1,\frac{3}{2})} &= 47\ \text{cm}^{-1}. \end{aligned}$$

These figures make it apparent that the energies $A_j^{(n,\frac{3}{2})}$ and $B_2^{(n,\frac{3}{2})}$ are natural choices for the spectral parameters to describe vibronic splittings for the $J' = \frac{3}{2}$ electronic states. In discussing the observed spectra of ReF_6 and TcF_6, we find it desirable to use simpler symbols Δ_2, δ_2, and Δ_5 for $A_2^{(1,\frac{3}{2})}$, $|B_2^{(1,\frac{3}{2})}|$, and $A_5^{(1,\frac{3}{2})}$, respectively. In Section IV, we assign $\Delta_2 = 84\ \text{cm}^{-1}$, $\delta_2 = 21\ \text{cm}^{-1}$, and $\Delta_5 = 47\ \text{cm}^{-1}$ for ReF_6; and for TcF_6, $\Delta_2 = 88\ \text{cm}^{-1}$, $\delta_2 = 42\ \text{cm}^{-1}$, and $\Delta_5 = 42\ \text{cm}^{-1}$. The difference between these values for δ_2 is well outside the experimental uncertainty of these assignments, but it is hard to assess the significance of the small changes for Δ_2 and Δ_5. Nevertheless, we find it desirable to use two slightly different sets of vibronic parameters: one for hexafluoride molecules formed by elements in the second transition series and the other for those formed by elements in the third transition series. Each set of vibronic parameters is based on the assumed constancy of λ_2 and λ_5 throughout such a homologous series of compounds. We use $\lambda_j(4df_{2g})$ to refer to the second transition series and $\lambda_j(5df_{2g})$ to refer to the third transition series. In each case, the foregoing definitions make it clear that

$$\lambda_2 = 4\Delta_2$$

and

$$\lambda_5 = 3\Delta_5.$$

Thus, we have

$$\lambda_2(4df_{2g}) = 352\ \text{cm}^{-1} \quad \text{and} \quad \lambda_5(4df_{2g}) = 126\ \text{cm}^{-1};$$

while

$$\lambda_2(5df_{2g}) = 336\ \text{cm}^{-1} \quad \text{and} \quad \lambda_5(5df_{2g}) = 141\ \text{cm}^{-1}.$$

The corresponding values of $\beta_j^{(n)}$, $\gamma_j^{(n)}$, $\phi^{(n)}$, $\eta^{(n)}$ and $\theta_j^{(n)}$ are summarized for each transition series in the following tables.

As an example of the use of these tables, we now estimate the value of δ_2 for TcF_6 and ReF_6 on the basis of the assumption that κ_2 is effectively zero in each case. The values of $\beta_2^{(1)}$ given in Tables XVIII and XIX imply that the observed frequencies of

TABLE XVIII. Vibronic Parameters for the Lowest Electronic Multiplets of Hexafluoride Molecules from the Second Transition Series: $x = \zeta/(G + \zeta) = 0.3$, $\zeta + G = 4270\ \text{cm}^{-1}$, $\lambda_2(4df_{2g}) = 352\ \text{cm}^{-1}$, and $\lambda_5(4df_{2g}) = 126\ \text{cm}^{-1}$

n	1	2	3	4	5
J'	$\frac{3}{2}$	2	$\frac{3}{2}$	0	$\frac{1}{2}$
$\beta_2^{(n)}$	−0.09159	−0.17231	−0.00088	−0.09228	−0.12954
$\beta_5^{(n)}$	−0.04372	−0.08224	−0.00042	−0.04404	−0.06183
$\gamma_2^{(n)}$		+0.06421			
$\gamma_5^{(n)}$		+0.04281			
$\phi^{(n)}$	−0.28964[a]	−0.40251[a]	+0.00135[a]		
$\eta^{(n)}$		+0.16080			
$\theta_2^{(n)}$	2.500	1.961	2.3×10^{-5}		
$\theta_5^{(n)}$	3.333	2.613	3.2×10^{-5}		

[a] Value based on the assumption that κ_2 is effectively zero.

639 cm^{-1} and 671 cm^{-1} for the e_g mode in TcF_6 and ReF_6, respectively, are some 4.6% and some 1.6% lower, respectively, than these frequencies would be in the absence of vibronic coupling.

TABLE XIX. Vibronic Parameters for the Lowest Electronic Multiplets of Hexafluoride Molecules from the Third Transition Series: $x = 0.5862$, $(G + \zeta) = 5800\ \text{cm}^{-1}$, $\lambda_2(5df_{2g}) = 336\ \text{cm}^{-1}$, and $\lambda_5(5df_{2g}) = 141\ \text{cm}^{-1}$

n J'	1 $\frac{3}{2}$	2 2	3 $\frac{3}{2}$	4 0	5 $\frac{1}{2}$
$\beta_2^{(n)}$	−0.0331	−0.0752	−0.01184	−0.0401	−0.0469
$\beta_5^{(n)}$	−0.0184	−0.0418	−0.00658	−0.0223	−0.0261
$\gamma_2^{(n)}$		−0.0181			
$\gamma_5^{(n)}$		−0.0067			
$\phi^{(n)}$	−0.1048[a]	−0.1554[a]	−0.01468[a]		
$\eta^{(n)}$		−0.0217			
$\theta_2^{(n)}$	2.500	1.092	0.02583		
$\theta_5^{(n)}$	3.333	1.453	0.03447		

[a] Value based on the assumption that κ_2 is effectively zero.

Thus, we estimate that ν_2 for TcF_6 would be 669.7 cm^{-1} and that ν_2 for ReF_6 would be 682.3 cm^{-1}.

The corresponding values of $B_2^{(1,\frac{3}{2})}$ are then −61.3 cm^{-1} for TcF_6 and −22.6 cm^{-1} for ReF_6. Thus, we calculate values of 61 cm^{-1} and 23 cm^{-1} for δ_2 to compare with the observed values of 42 cm^{-1} and 21 cm^{-1}, respectively. This comparison seems to support the view that there is no need to include an additional adjustable parameter, κ_2, to account for what are known as pseudo Jahn–Teller terms in the molecular energy. These terms that depend quadratically on the e_g nuclear displacement coordinates are a direct consequence, in second-order perturbation theory, of the true Jahn–Teller terms that are linear in the e_g nuclear displacement coordinates.

The sign of the quantity $B_2^{(1,\frac{3}{2})}$ comes out unambiguously negative according to the preceding reasoning. Unfortunately, however, it seems quite unlikely that this prediction could be checked experimentally. On the other hand, the signs of $A_2^{(1,\frac{3}{2})}$ and $A_5^{(1,\frac{3}{2})}$ are given just as unambiguously as positive.

Very striking experimental verification of the positive sign of $A_2^{(1,\frac{3}{2})}$ is seen in the spectra of ReF_6 and of TcF_6. Each spectral transition that involves the e_g mode for these molecules is split into primarily two bands spaced about equally above and below the frequency at which this transition would be expected to occur by comparison with the more normal homologous compounds in each series. Always it is the lower frequency of these two bands that is the broader component in this spectral effect. This observation proves that the sign of $A_2^{(1,\frac{3}{2})}$ is positive for these molecules, as one can see from Fig. 7, which has been drawn for a positive value of $A_2^{(1,\frac{3}{2})}$. In this figure the $\frac{1}{2}(1,0;+1)$ and $\frac{1}{2}(1,0;-1)$ levels both lie below the $\frac{1}{2}(1,0;0)$ level. For $A_2^{(1,\frac{3}{2})}$ negative this order would be reversed. We return to the details of these assignments for TcF_6 and for ReF_6 in Section IV.

We are led to conclude that the vibronic features of TcF_6 and ReF_6 can be rationalized on the basis of two sets of not widely different energy parameters: λ_2 near 345 cm^{-1} and λ_5 near 135 cm^{-1}. The question remains whether the spectra of the remaining members of the series can be understood on the basis of these parameters. The cases of IrF_6 and RhF_6 can be considered without further ado because each of these molecules has a $J' = \frac{3}{2}$ electronic quartet at lowest energy. For RhF_6, we calculate the following set of spectral parameters:

$$\Delta_2 = 0.0008 \text{ cm}^{-1}, \quad \delta_2 = 0.025 \text{ cm}^{-1}, \quad \Delta_5 = 0.0004 \text{ cm}^{-1};$$

while for IrF_6, we find

$$\Delta_2 = 0.87 \text{ cm}^{-1}, \quad \delta_2 = 3.02 \text{ cm}^{-1}, \quad \Delta_5 = 0.49 \text{ cm}^{-1}.$$

In both cases the calculated spectral parameters are far smaller than the presently available vibrational spectra for these molecules could reveal. Thus it is reasonable that RhF_6 and IrF_6 appear quite "normal" vibrationally despite their electronically degenerate ground multiplets with $J' = \frac{3}{2}$.

For $J' = 2$ the vibronic energy levels are far more involved than any that we have considered up to now. The matrix operator equation that gives these energy levels is as follows:

$$E = E^{(n,2)}(n_2, n_5) \begin{bmatrix} 1 & 0 & 0 & 0 & 0 \\ 0 & 1 & 0 & 0 & 0 \\ 0 & 0 & 1 & 0 & 0 \\ 0 & 0 & 0 & 1 & 0 \\ 0 & 0 & 0 & 0 & 1 \end{bmatrix}$$

$$+ \bar{E}^{(n,2)}(n_2, n_5) \begin{bmatrix} 1 & 0 & 0 & 0 & 5 \\ 0 & -4 & 0 & 0 & 0 \\ 0 & 0 & 6 & 0 & 0 \\ 0 & 0 & 0 & -4 & 0 \\ 5 & 0 & 0 & 0 & 1 \end{bmatrix}$$

$$+ A_2{}^{(n,2)} M_{2,0}\sqrt{2} \begin{bmatrix} 0 & 0 & -i & 0 & 0 \\ 0 & 0 & 0 & 0 & 0 \\ i & 0 & 0 & 0 & i \\ 0 & 0 & 0 & 0 & 0 \\ 0 & 0 & -i & 0 & 0 \end{bmatrix}$$

$$+ B_2{}^{(n,2)}\{M_{2,1} \begin{bmatrix} 2 & 0 & -i\sqrt{2} & 0 & 0 \\ 0 & -1 & 0 & -i\sqrt{3} & 0 \\ -i\sqrt{2} & 0 & -2 & 0 & -i\sqrt{2} \\ 0 & -i\sqrt{3} & 0 & -1 & 0 \\ 0 & 0 & -i\sqrt{2} & 0 & 2 \end{bmatrix}$$

$$+ M_{2,-1} \begin{bmatrix} 2 & 0 & i\sqrt{2} & 0 & 0 \\ 0 & -1 & 0 & i\sqrt{3} & 0 \\ i\sqrt{2} & 0 & -2 & 0 & i\sqrt{2} \\ 0 & i\sqrt{3} & 0 & -1 & 0 \\ 0 & 0 & i\sqrt{2} & 0 & 2 \end{bmatrix} \}$$

$$+ \bar{B}_2{}^{(n,2)}\{M_{2,1} \begin{bmatrix} -1 & 0 & -3i\sqrt{2} & 0 & 7 \\ 0 & 4 & 0 & 4i\sqrt{3} & 0 \\ -3i\sqrt{2} & 0 & -6 & 0 & -3i\sqrt{2} \\ 0 & 4i\sqrt{3} & 0 & 4 & 0 \\ 7 & 0 & -3i\sqrt{2} & 0 & -1 \end{bmatrix}$$

$$+ M_{2,-1} \begin{bmatrix} -1 & 0 & 3i\sqrt{2} & 0 & 7 \\ 0 & 4 & 0 & -4i\sqrt{3} & 0 \\ 3i\sqrt{2} & 0 & -6 & 0 & 3i\sqrt{2} \\ 0 & -4i\sqrt{3} & 0 & 4 & 0 \\ 7 & 0 & 3i\sqrt{2} & 0 & -1 \end{bmatrix} \}$$

$$
+ A_5^{(n,2)}\{2M_{5,0}\begin{bmatrix} 6 & 0 & 0 & 0 & 0 \\ 0 & -5 & 0 & 0 & 0 \\ 0 & 0 & 0 & 0 & 0 \\ 0 & 0 & 0 & 5 & 0 \\ 0 & 0 & 0 & 0 & -6 \end{bmatrix}
$$

$$
+ M_{5,-1}\begin{bmatrix} 0 & 0 & 0 & 8 & 0 \\ -2 & 0 & 0 & 0 & 8 \\ 0 & 3\sqrt{6} & 0 & 0 & 0 \\ 0 & 0 & 3\sqrt{6} & 0 & 0 \\ 0 & 0 & 0 & -2 & 0 \end{bmatrix}
$$

$$
+ M_{5,1}\begin{bmatrix} 0 & -2 & 0 & 0 & 0 \\ 0 & 0 & 3\sqrt{6} & 0 & 0 \\ 0 & 0 & 0 & 3\sqrt{6} & 0 \\ 8 & 0 & 0 & 0 & -2 \\ 0 & 8 & 0 & 0 & 0 \end{bmatrix}\}
$$

where

$$
\begin{aligned}
E^{(n,2)}(n_2, n_5) &= (n_2 + 1)\nu_2(1 + \beta_2^{(n)}/\sqrt{5}) \\
&\quad + (n_5 + \tfrac{3}{2})\nu_5(1 + \beta_5^{(n)}/\sqrt{5}) - 2\lambda_2\theta_2^{(n)}/25 \\
&\quad - 3\lambda_5\theta_5^{(n)}/25, \\
\bar{E}^{(n,2)}(n_2, n_5) &= (n_2 + 1)\nu_2\gamma_2^{(n)}/6\sqrt{10} + (n_5 + \tfrac{3}{2})\nu_5\gamma_5^{(n)}/6\sqrt{10} \\
&\quad - \lambda_2\theta_2^{(n)}/175 + \lambda_5\theta_5^{(n)}/175 \\
A_2^{(n,2)} &= 4\lambda_2\theta_2^{(n)}/35 \\
B_2^{(n,2)} &= \nu_2\phi^{(n)}/2\sqrt{35} \\
\bar{B}_2^{(n,2)} &= \nu_2\eta^{(n)}/12\sqrt{7} \\
A_5^{(n,2)} &= \lambda_5\theta_5^{(n)}/140
\end{aligned}
$$

In working with the experimental spectra in Section IV, it seems useful to define the following spectral parameters:

$$
\bar{\Delta}_2 = A_2^{(2,2)}, \quad \bar{\Delta}_5 = 10A_5^{(2,2)},
$$

$$
\bar{\delta}_2 = 2|B_2^{(2,2)} - 4\bar{B}_2^{(2,2)}|, \quad \text{and} \quad \bar{\bar{\delta}}_2 = 4|B_2^{(2,2)} + 3\bar{B}_2^{(2,2)}|.
$$

For OsF_6, these definitions lead to

$$\bar{\Delta}_2 = 42\ \text{cm}^{-1}, \qquad \bar{\Delta}_5 = 14.6\ \text{cm}^{-1},$$
$$\bar{\delta}_2 = 14\ \text{cm}^{-1}, \qquad \bar{\bar{\delta}}_2 = 42\ \text{cm}^{-1},$$

based on the entries in Table XIX. For RuF_6, on the other hand, $\bar{\Delta}_2 = 79\ \text{cm}^{-1}$, $\bar{\Delta}_5 = 23.5\ \text{cm}^{-1}$, $\bar{\delta}_2 = 76\ \text{cm}^{-1}$, and $\bar{\bar{\delta}}_2 = 52\ \text{cm}^{-1}$, based on the entries in Table XVIII.

Figure 9 shows the lowest vibronic energy levels for $J' = 2$, drawn approximately to scale for OsF_6. To calculate the positions of the energy levels labelled $F_{1g}(0, 1; \text{I})$, $F_{1g}(0, 1; \text{II})$, $F_{2g}(0, 1; \text{I})$ and $F_{2g}(0, 1; \text{II})$ in Fig. 9, one must find the roots of the following

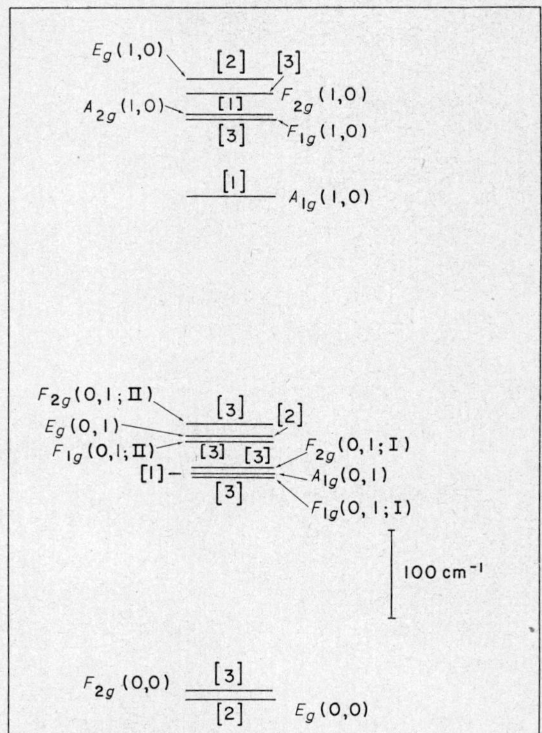

Fig. 9. The lowest vibronic energy levels for $J' = 2$, drawn approximately to scale for OsF_6. (The degeneracy of each energy level is indicated by the number in square brackets.)

secular determinents:

$$\begin{vmatrix} -E(F_{1g}) + E(0, 1) + 6\bar{E}(0, 1), & -\sqrt{72}A_5 \\ -\sqrt{72}A_5, & -E(F_{1g}) + E(0, 1) - 4\bar{E}(0, 1) \end{vmatrix} = 0$$

and

$$\begin{vmatrix} -E(F_{2g}) + E(0, 1) + 6\bar{E}(0, 1), & -\sqrt{216}A_5 \\ -\sqrt{216}A_5, & -E(F_{2g}) + E(0, 1) - 4\bar{E}(0, 1) \end{vmatrix} = 0.$$

These theoretical results are compared with the experimental spectra in the next section.

IV. VIBRATIONAL SPECTRA OF ABNORMAL HEXAFLUORIDE MOLECULES

The infrared absorption spectra of four gaseous metal hexafluorides are abnormal. These are the hexafluorides of Tc, Ru, Re, and Os. In this section we analyze their experimental spectra in terms of the theoretical energy levels first calculated by Moffitt and Thorson[9] and reviewed in Section III. Certain abnormalities observed in the spectra of these molecules have been previously referred to as "dynamic Jahn–Teller effects". Now we make detailed spectral assignments for these anomalous features and substantially confirm the correspondence between the experimental observations and the theoretical predictions.

According to ligand-field theory, the lowest electronic energy level for each of these four molecules is orbitally degenerate for an octahedrally symmetric nuclear configuration. Thus, the theorem of Jahn and Teller applies to these molecules and states that there will be forces tending to distort their highly symmetrical configurations. Instead of being large and producing a static distortion, however, these Jahn–Teller forces are rather small and lead to an inextricable coupling of the electronic and the nuclear motions. As a result of this coupling, spectral abnormalities appear for the e_g and f_{2g} vibrational modes of these molecules, in correspondence with the symmetry arguments presented by Jahn and Teller. If the Jahn–Teller forces were large and a static distortion were produced, different abnormalities would be present in the spectra. For example, if a static distortion of the nuclear framework from O_h to D_{4h} symmetry resulted, then five infrared-active fundamentals would be observed instead of two. But this does not occur. For the four abnormal hexafluoride molecules, just two infrared-active fundamentals are observed, as in the case of the eleven normal hexafluoride molecules.

The approximate magnitude of the energy coupling together electronic and nuclear motions in ReF_6 (and IrF_6) has been estimated in a note by Child,[50] but up until now only an oral account* has been given of the detailed connection between observed spectral features for these hexafluorides and the type of vibronic energy levels first studied by Moffitt and his co-workers. In the remaining pages of this section, we describe the procedures that we used to derive the magnitude of the vibronic coupling from the experimental spectra and our detailed identification of vibronic features in these spectra.

A. General Features

The magnitude of the vibronic coupling was derived in two ways. In one procedure, the (2 + 3) band profiles of the normal and abnormal hexafluoride molecules were compared and their differences related to vibronic effects. In the other procedure, the systematic variation of observed frequencies with the number of nonbonding valence electrons for the normal hexafluoride molecules was used to estimate those values of ν_2 and ν_5 to be expected for the abnormal molecules in the absence of vibronic coupling. The reliability of this latter procedure was significantly increased by the critical review of the vibrational spectra of the normal hexafluoride molecules (Section II).

(1) *The* (1 + 3) *and* (2 + 3) *Band Profiles*

Two bands having almost identical intensities and profiles are characteristic features of the infrared spectra of all the normal hexafluoride molecules. These bands correspond to the binary

* "Evidence for Vibronic Coupling in Metal Hexafluoride Molecules", presented at the St. Louis Meeting of the American Physical Society, March 26, 1963, Symposium of the Division of Chemical Physics, entitled "The Jahn–Teller Effect: Electronic Degeneracy and Vibronic Properties of Molecules".

transitions (1 + 3) and (2 + 3), and occur in the spectral region between 1140 and 1710 cm^{-1}, separated in frequency by amounts ranging from 27 to 132 cm^{-1}. They are usually the most intense binary bands, except for a few cases where the (2 + 6) band is of somewhat greater intensity. For each molecule, the band at longer wavelength corresponds to the (2 + 3) transition and the band at shorter wavelength to the (1 + 3) transition.

The spectra of the abnormal hexafluoride molecules differ from those of the normal hexafluorides with respect to the appearance

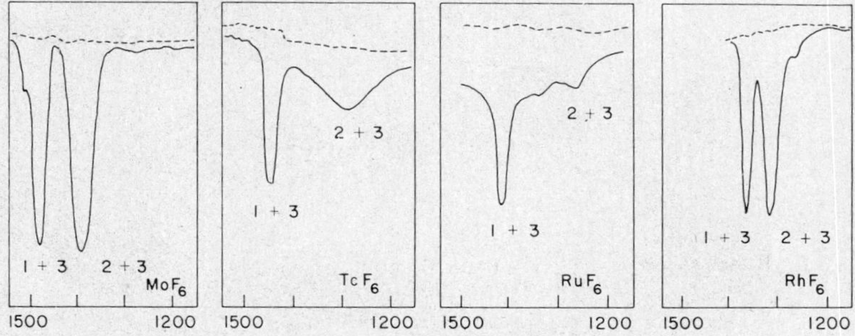

Fig. 10. Band profiles for (1 + 3) and (2 + 3) for the 4*d* transition series hexafluoride molecules (Weinstock, Claassen, and Chernick[22]).

of these two bands. While the profile of the (1 + 3) band is similar to that for the normal molecules, the appearance of the (2 + 3) band is markedly altered in each case. This contrast between the band profiles of the (1 + 3) and the (2 + 3) transitions has been used as a qualitative criterion to distinguish abnormal from normal hexafluoride molecules. The anomaly of the (2 + 3) profile was first pointed out by Weinstock and Claassen[19] for the spectra of OsF_6 and ReF_6. Subsequently, when the spectra of TcF_6[21] and RuF_6[22] were obtained, similar anomalies were observed to complete the correlation of this anomaly with the existence of an orbitally degenerate ground state for the molecule.

This striking difference between the infrared spectra of normal and vibronic hexafluoride molecules can be seen in Figs. 10 and 11, where the band profiles for the (1 + 3) and (2 + 3) transitions of the 4*d* and 5*d* transition series hexafluorides are shown. For the five normal molecules the two bands are seen to be characteristically of nearly identical shape and intensity. For ReF_6, the (2 + 3) band is noticeably broader than the (1 + 3) band but still of about the same maximum absorbance. For OsF_6 and TcF_6, the broadening of the (2 + 3) band profile is more pronounced and its peak absorbance much less than that of the (1 + 3) band. For RuF_6, the (2 + 3) band is so strongly affected that its presence was not definitely established in the experimental paper. When the infrared spectrum of RuF_6 is considered in detail later

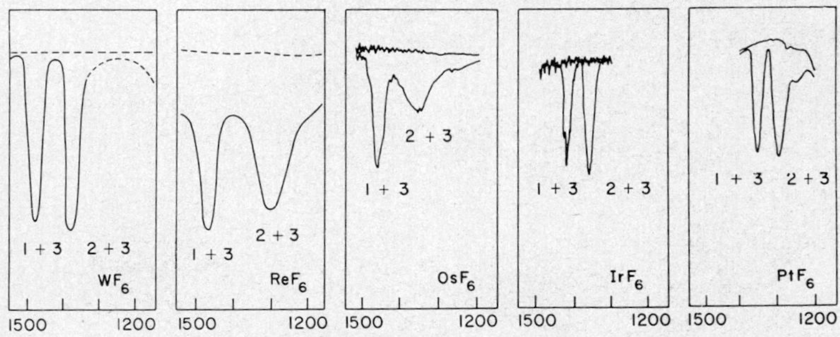

Fig. 11. Band profiles for (1 + 3) and (2 + 3) for the 5*d* transition series hexafluoride molecules (Weinstock, Claassen, and Chernick[22]).

in this section, a method for resolving the (2 + 3) band of RuF_6 from the other spectral absorptions in this region is given. The (2 + 3) band is then identified as a very weak shoulder on the high wavelength side of the (1 + 3) band (Fig. 24).

A more quantitative demonstration of the abnormal (2 + 3) band profiles can be made by comparing the (1 + 3) and (2 + 3) band widths at half absorbance. This comparison is made in Table XX, where the values of the (1 + 3) and (2 + 3) band half widths reported by Weinstock *et al.*[22] for the 4*d* and 5*d* transition series hexafluoride molecules are given in the first two columns. The half width of the (1 + 3) band is seen to be about the same for both the normal and vibronic molecules, with the exception of ReF_6. The larger (1 + 3) band width for ReF_6 can be discounted in this discussion; it will be shown to be the result of an accidental degeneracy that occurs because of the vibronic coupling. For the (2 + 3) transitions, the band half widths are

TABLE XX. Band Widths at Half Absorbance

Compound	Width for (1 + 3), cm^{-1}	Width for (2 + 3), cm^{-1}	Vibronic splitting, cm^{-1}
MoF_6	20	31	0
TcF_6	23	113[a]	84
RuF_6	21	–	–
RhF_6	19	27	0
Average	21	29	
WF_6	17	24	0
ReF_6	36[a]	68[a]	43
OsF_6	23	84[a]	59
IrF_6	18	26	0
PtF_6	20	26	0
Average	20	25	

[a] Not used for the averages.

fairly constant for the normal molecules, but are much larger for the three vibronic molecules, in agreement with the qualitative observations.

The data in Table XX can also be used to estimate a vibronic splitting for the e_g mode of each molecule. These estimates are given in the last column of the table, and were derived by us by subtracting the average value for the (2 + 3) band half width of the normal molecules in each series from the (2 + 3) band half width of each abnormal molecule in that series. Weinstock *et al.*[22] also estimated vibronic splittings from these data and obtained similar results for OsF_6 and TcF_6, but a significantly lower value for ReF_6. In their procedure, the (1 + 3) band half widths were used to normalize the data, and the lower value for ReF_6 resulted because of the abnormally high value of the (1 + 3) band half width for ReF_6, on which we have already commented. Except for this difficulty for ReF_6, the two procedures are essentially equivalent.

The values given for the vibronic splittings in Table XX will be applied in the interpretation of the spectra of the vibronic molecules in the following pages of this section. For ReF_6 and TcF_6, the excess half width of the $(2 + 3)$ band is set equal to $2\delta_2$, twice the value of the quadratic splitting parameter for each molecule. For OsF_6, the excess half width of the $(2 + 3)$ band cannot be used directly to derive a quadratic splitting parameter. However, later in this section, we show the excess band width to be consistent with splittings arising from both linear and quadratic coupling.

(2) *The Frequency Systematics*

The variation of each of the six fundamental vibration frequencies of the metal hexafluoride molecules as a function of the

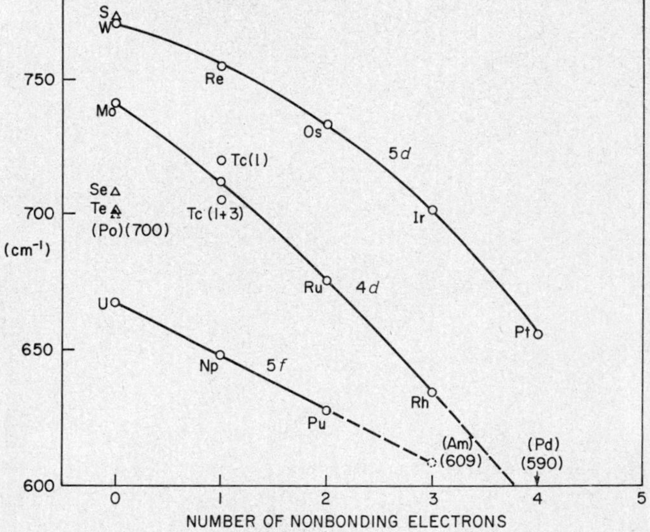

Fig. 12. The variation of the Raman-active fundamental vibration frequency, $\nu_1(a_{1g})$, with the number of nonbonding valence electrons for the hexafluoride molecules.

number of nonbonding valence electrons in these molecules is shown in Figs. 12–17. The frequencies plotted for the eight normal hexafluoride molecules were derived in Section II. The frequencies plotted for the four abnormal molecules are derived later in this

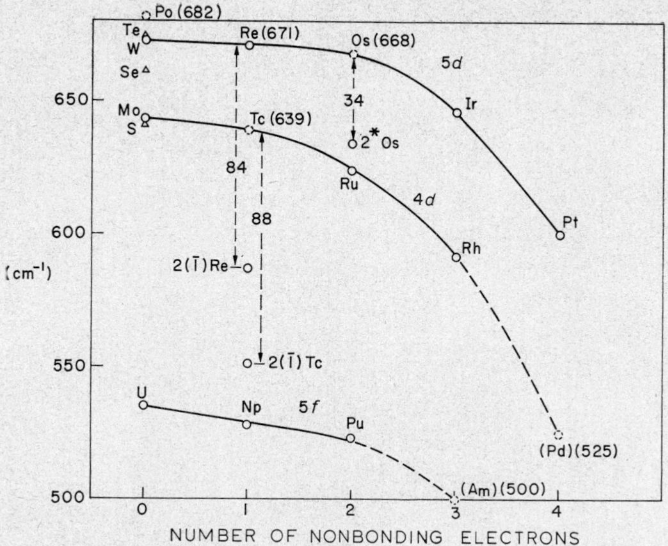

Fig. 13. The variation of the Raman-active fundamental vibration frequency, $\nu_2(e_g)$, with the number of nonbonding valence electrons for the hexafluoride molecules. The frequency displacements from the solid curves for ReF_6, TcF_6, and OsF_6 arise from linear vibronic coupling in these molecules.

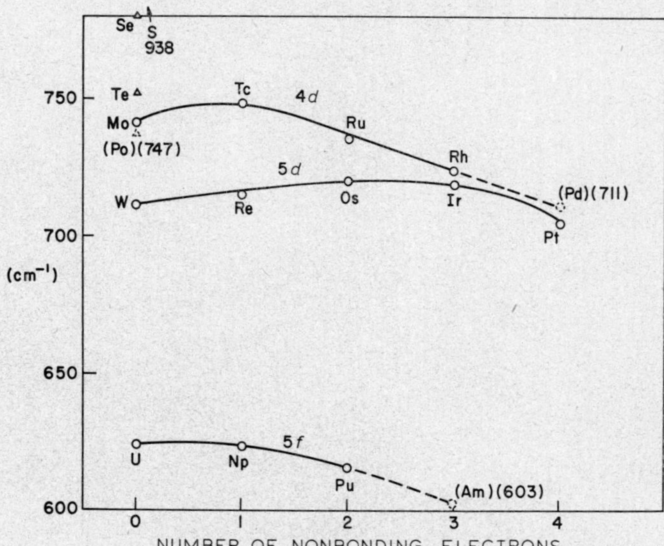

Fig. 14. The variation of the infrared-active fundamental vibration frequency, $\nu_3(f_{1u})$, with the number of nonbonding valence electrons for the hexafluoride molecules.

section. For these molecules, the assignments for ν_2 and ν_5 require special attention because of vibronic effects. The fundamental vibration frequencies for the nonmetal hexafluoride molecules are also indicated in the figures for the sake of completeness.

For ν_1, ν_3, and ν_4, the frequencies vary continuously within each of the three transition series (4*d*, 5*d*, and 5*f*) and are connected

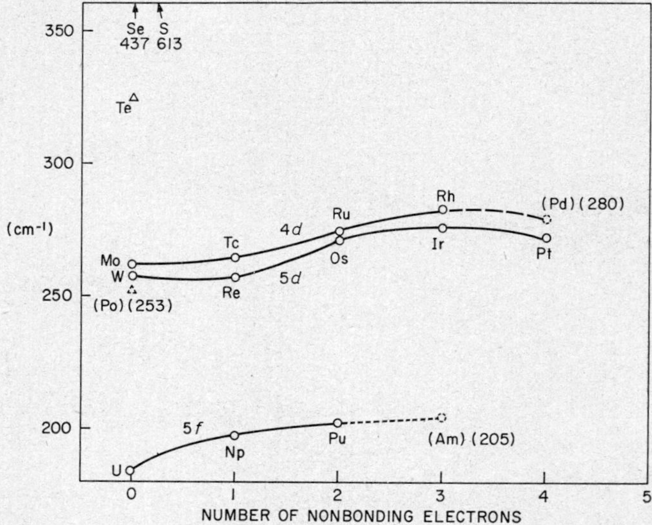

Fig. 15. The variation of the infrared-active fundamental vibration frequency, $\nu_4(f_{1u})$, with the number of nonbonding valence electrons for the hexafluoride molecules.

together by continuous curves. Two different values of ν_1 for TcF_6 are plotted, neither of which agrees with the systematics. One value is obtained from the liquid Raman spectra and the other derived from the binary band (1 + 3).[21] The curve that is drawn passes through the average of these two values. No explanation for the discrepancy, other than experimental uncertainty (± 7 cm^{-1}), is offered.

The data for the spectrally inactive frequency, ν_6, are incomplete. As was discussed in Section II, the frequency values for ν_6 are derived from the combination bands (2 + 6), (2 − 6), and (5 + 6), which are sometimes difficult to assign. Since ν_2 and ν_5 are both involved in vibronic coupling, this difficulty is compounded for the Jahn–Teller molecules. Some confidence can be placed in the value of ν_6 plotted for ReF_6, but assignments of ν_6 from the experimental data are not presently possible for RuF_6, OsF_6, and TcF_6. From the data that are plotted, the variation of the

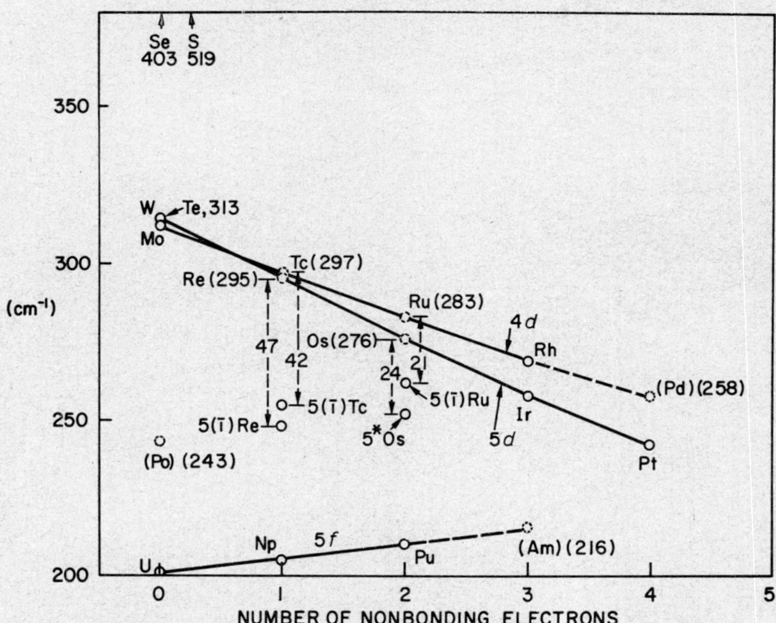

Fig. 16. The variation of the Raman-active fundamental vibration frequency, $\nu_5(f_{2g})$, with the number of nonbonding valence electrons for the hexafluoride molecules. The frequency displacements from the solid curves for ReF_6, TcF_6, OsF_6, and RuF_6 arise from linear vibronic coupling in these molecules.

frequency ν_6 with the number of nonbonding electrons appears regular (this is in accordance with symmetry expectations) and we have estimated values of ν_6 for these three molecules. The estimated value of ν_6 for OsF_6 is probably reliable to ± 3 cm^{-1}, because it is the only missing value in the 5*d* series. The values for TcF_6 and RuF_6 are interpolated on the assumption that the 4*d* frequency values generally parallel the 5*d* frequency values. This appears to be a reasonable assumption because of the correspondence between the frequency values for the two isoelectronic pairs, MoF_6 and WF_6, and RhF_6 and IrF_6. However, the estimated values of ν_6 for TcF_6 and RuF_6 may well be uncertain to ± 5 cm^{-1}.

The variation of vibrational frequency with the number of nonbonding electrons shows very marked deviations for the e_g (ν_2)

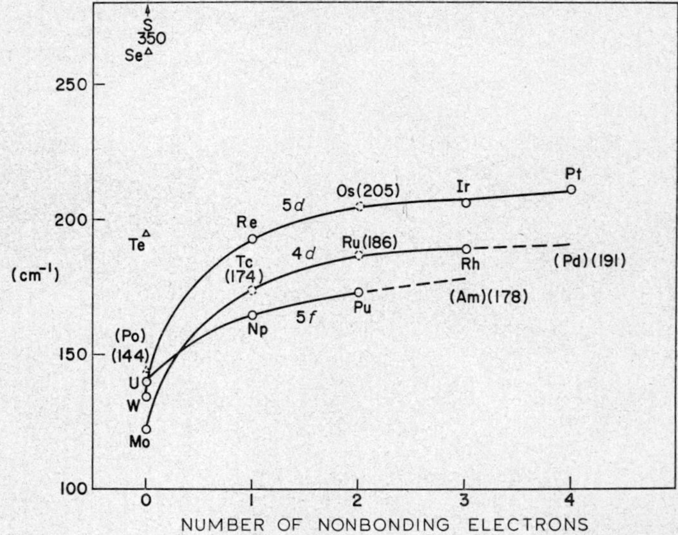

Fig. 17. The variation of the inactive fundamental vibration frequency, $\nu_6(f_{2u})$, with the number of nonbonding valence electrons for the hexafluoride molecules.

and f_{2g} (ν_5) vibrational modes. As was the case for the $(2 + 3)$ band profiles, the deviations occur only for the four vibronic molecules and never for the normal hexafluoride molecules. This correspondence between frequency abnormality for ν_2 and ν_5 and orbital degeneracy in the electronic ground state of these molecules provides further striking qualitative evidence of vibronic coupling in these molecules.

In Fig. 13, the variation of ν_2 with the number of nonbonding electrons is plotted. Irregularities are noted only in the 4*d* and 5*d* series, but not in the actinide (5*f*) series, where none of the three molecules has an orbitally degenerate electronic ground state. For ReF_6, TcF_6, and OsF_6, the value of ν_2, derived from the frequency of the apparent $(2 + 3)$ transition, is plotted and is seen to be 84, 88, and 34 cm^{-1} lower than the value (ν_2^0) that would be expected from the systematics. It is shown later in this section that this apparent shift of frequency for ReF_6 and TcF_6 is the result of a linear Jahn–Teller effect and it is used to determine the value of the splitting parameter that we call Δ_2. For OsF_6, the apparent lowering of the ν_2 frequency involves both linear and quadratic vibronic coupling and cannot be used to derive spectral splitting parameters directly.

For RuF_6, an experimental value for ν_2 had not previously been obtained. The value that is plotted was derived by us from spectra of RuF_6 by resolving the $(2 + 3)$ band of Weinstock *et al.*[22] from several interfering spectral features. The frequency of ν_2 thus obtained for RuF_6 corresponds to the unshifted value (ν_2^0) that is suggested by the systematics. In this respect, RuF_6 differs from OsF_6. The electronic ground state which influences the observed features for OsF_6 is a $J' = 2$ quintet, symmetries E_g and F_{2g}. In RuF_6, the corresponding quintet is split to a greater extent than in OsF_6 (Section III). Only the doublet component (E_g) undergoes linear splitting involving the e_g vibrational mode; the triplet component (F_{2g}) does not. Therefore, the observation of an unshifted peak for $(2 + 3)$ is consistent with a F_{2g} electronic ground state for RuF_6.

In order to complete the correlation between the theoretical energy-level scheme given by Moffitt and Thorson[9] and the observed spectra of ReF_6 and TcF_6, two components, $2(1)$ and $2(\bar{1})$, have to be identified for the ν_2 transitions at energies, respectively, above and below the systematic prediction, ν_2^0. Although the $[2(\bar{1}) + 3]$ absorption band is easily recognized in the infrared, the $[2(1) + 3]$ band seemed at first to be missing. For ReF_6, this puzzle was explained by finding that the $2(1)$ frequency is accidentally coincident with ν_1, so that $2(1)$ transitions are usually masked by the more intense ν_1 transitions. We have obtained a confirmation of this explanation by identifying in the published spectra a band corresponding to the $[2(1) - 6]$ transition. This band can not be masked by a corresponding ν_1 transition, $(1 - 6)$, because this latter transition is not infrared-active. For TcF_6, a re-examination of the infrared spectrum reveals a very weak shoulder on the high-frequency side of the (1 + 3) band that was not previously reported and that corresponds to the [2(1) + 3] transition, to complete the picture. For OsF_6, the vibronic levels involving the e_g transition are more numerous than for ReF_6 and TcF_6 and, consequently, the various e_g components are not resolved in presently available spectra.

The values of ν_2^0 derived from these systematics are probably good to $\pm 3\ \text{cm}^{-1}$. Where the frequency is obtained by interpolation, the point is designated by a broken circle and the estimated frequency indicated in parentheses. For ReF_6 and TcF_6, the estimated values of ν_2^0 also agree with the value of ν_2^0 taken as the average of the $2(\bar{1})$ and 2(1) values.

In Fig. 16, the variation of ν_5 with the number of nonbonding electrons is plotted. Again, there is a regular variation of frequency for the actinide (5*f*) series and anomalous behavior in the 4*d* and 5*d* series. The apparent value of ν_5 observed for the vibronic molecules is taken either directly from Raman shifts or derived from the value of the (3 + 5) transition. For ReF_6, TcF_6, and RuF_6, values are assigned as $5(\bar{1}) = \nu_5^0 - \Delta_5$, and can be used to derive Δ_5, the spectral splitting parameter, for these molecules. For OsF_6, the pattern of vibronic transitions is again very complex and only the general correspondence between theory and experiment can be demonstrated. The vibronic transitions involving the f_{2g} vibration are generally very complicated and a much less complete analysis is possible. The assignments of ν_5^0 are again reliable to $\pm 3\ \text{cm}^{-1}$.

(3) *Other Hexafluoride Molecules*

In view of the general regularity found for the variation of vibrational frequency within each group of hexafluoride molecules, a reasonable estimate of the vibrational frequencies for still unknown hexafluoride molecules can be made by extrapolation of the values plotted in Figs. 12–17. This has been done for PoF_6, PdF_6, and AmF_6, and the estimated values for each vibrational frequency of these molecules is indicated beside the symbol of the central element (both in parentheses) in these figures. These estimated frequencies are also summarized in Table XXVI of Section V.

Preliminary evidence for the existence of CrF_6 has recently been reported by Glemser.[18] The compound could not be studied in the vapor phase because of rapid decomposition at room temperature. However, he observed three infrared absorptions with the solid at the boiling point of nitrogen. An intense absorption at 790 cm^{-1} was assigned as ν_3. Two weaker absorptions at 1440 and 1510 cm^{-1} were assigned as $(2 + 3)$ and $(1 + 3)$. From these latter bands, the values of 650 and 720 cm^{-1} are derived for ν_2 and ν_1. Only very approximate values of the vibrational frequencies for CrF_6 can be estimated by extrapolation of the frequencies of MoF_6 and WF_6. However, Glemser's values for ν_1 and ν_3 agree with the trend in the corresponding frequencies of MoF_6 and WF_6, while his value for ν_2 appears to be 30 cm^{-1} too high on the same basis. Glemser's preliminary values for CrF_6 are given in Table XXVI (Section V) as well as estimates of ν_4, ν_5, and ν_6 for CrF_6. We have based these latter estimates on the fact that the values of ν_4, ν_5, and ν_6 for MoF_6 are very close to the corresponding values for WF_6. If this trend continues for CrF_6 the margin of error in our guesses should not be very great.

The least expected hexafluoride molecule to be discovered is XeF_6.[17] Like the other binary xenon fluorides, this compound is readily prepared by heating xenon and fluorine together under the proper conditions. XeF_6 is relatively stable to dissociation, but is highly reactive. It is the least volatile of the hexafluoride molecules; its vapor pressure is only 22 mm at 23°C.

Vibrational spectra have been obtained with XeF_6 in the vapor and solid states and with solutions of XeF_6. The spectra obtained for XeF_6 are different in many respects from those of all the other hexafluoride molecules and a controversy presently exists about the proper way in which to describe the structure of this compound. Because of the unsettled status of the structure of XeF_6, we shall not discuss its vibrational spectra in this article. For further information about XeF_6, the reader is referred to Claassen's and Smith's discussions of the case for a symmetrical or unsymmetrical structure.[84]

B. Detailed Spectral Assignments

As previously discussed, the vibrational spectra of ReF_6, OsF_6, TcF_6, and RuF_6 show anomalies for the e_g and f_{2g} vibrational transitions which are caused by vibronic coupling. Consequently, many of the spectral bands involving these vibrational modes are so diffuse and reduced in intensity that they are difficult to observe. To add to this intrinsic difficulty, the compounds involved are extremely reactive and their spectra are complicated by spurious absorptions that arise from compounds formed by reaction between the hexafluorides and the window materials of the spectral cells. Most of the spectra which we shall discuss next were taken with a full appreciation of the possibility of these spurious absorptions being present and proper precautions were taken either to exclude them or to identify them when they were present. However, the spectra obtained were not analyzed in terms of a theory of vibronic coupling, so that many of the subtle vibronic features have not been assigned. In view of the general reliability of these spectra we have undertaken to re-examine them in an effort to find additional spectral features that arise from vibronic coupling. We were greatly aided in this effort by the authors of the three most recent experimental papers on these molecules. They have made their original spectral traces entirely available to us. Their kind generosity in doing this has contributed immeasurably to the completeness and reliability of the re-analyses which we have given their data. We are greatly indebted to and want to thank H. H. Claassen, who was an author in all three cases, and his collaborators, J. G. Malm, H. Selig, and C. L. Chernick, for their cooperation with us. (One of the present authors (B.W.) was also a collaborator in two of these experimental studies.)

(1) *Rhenium Hexafluoride*

Rhenium hexafluoride is the first abnormal hexafluoride molecule which we shall consider. Vibronic coupling occurs in $ReF_6(5d^1)$ because its lowest electronic states form a quartet, symmetry $G_{\frac{3}{2}g}$, with mainly orbital character for its degeneracy.

ReF_6 is readily prepared by direct fluorination of rhenium metal. However, special care is required to prevent contamination by ReF_7 and $ReOF_5$. The vapor of ReF_6 appears colorless, but the liquid and the solid are pale yellow. Aside from the question of its purity, the study of the vibrational spectra of ReF_6 does not offer unusual experimental difficulty, and both infrared and

Raman spectra have been obtained with ReF_6.[55,81] These spectra differ in a subtle way from those of normal hexafluoride molecules because of vibronic coupling, and some of the qualitative aspects of these differences have been pointed out.[19,81]

In the following pages, we critically review the published vibrational spectra of ReF_6 and give detailed assignments of the observed features that are associated with vibronic transitions. A fairly complete evaluation of the nature and magnitude of vibronic coupling for this molecule is obtained.

Gaunt[55] first reported vibrational spectra for ReF_6. He measured the infrared spectrum of ReF_6 vapor from 400–5000 cm^{-1} and the Raman spectrum of liquid ReF_6. Gaunt concluded that ReF_6 has a regular octahedral structure because its infrared spectrum was similar to that of other hexafluoride molecules he had studied and because he observed three Raman shifts, in accordance with the requirements of O_h symmetry. Although Gaunt stated that the band shapes and relative intensities of the Raman shifts for ReF_6 are the same as those of the other hexafluoride molecules, this is not the case. The Raman spectrum of liquid ReF_6 was obtained by Gaunt without regard to polarization. It showed ν_1 to be strong and sharp and ν_2 and ν_5 to be much weaker and diffuse. These results should be compared with the Raman spectra of MoF_6 and WF_6, reported by Burke, Smith, and Nielsen,[41] which were also measured without regard to polarization. In the spectra of Burke *et al.*, ν_1, ν_2, and ν_5 are all strong; ν_1 and ν_2 are of almost equal intensity with the ν_2 band somewhat broader; ν_5 is less intense and broader than ν_1 and ν_2. The Raman spectrum of ReF_6 is thus quite different from that of MoF_6 and of WF_6. Gaunt had based his conclusion on the similarity between his Raman spectrum of ReF_6 and depolarization studies reported by Tanner and Duncan[51] for MoF_6 and WF_6 (see Section II). In O_h symmetry, ν_2 and ν_5 are expected to be much weaker than ν_1 in depolarization measurements and the similarity between Gaunt's ReF_6 spectrum and those of Tanner and Duncan was only coincidental. Gaunt also reported that there is no absorption for ReF_6 in the range 2000–5000 cm^{-1}, but an electronic transition beginning just below 5000 cm^{-1} has been reported.[15]

Weinstock and Claassen[19] were the first to note that Gaunt's ReF_6 spectra are abnormal. In their study of the vibrational spectra of OsF_6,[20] they had observed several anomalies associated with ν_2 that they attributed to a dynamical Jahn–Teller effect. For ReF_6, they suggested that the diffuse character of ν_2 in the Raman spectrum and the anomalous band shape of $(2 + 3)$ in the infrared spectrum had a similar explanation. They noted no corresponding unusual features involving ν_5 in Gaunt's spectra, although, as they point out, the Jahn–Teller theorem applies to both the e_g and f_{2g} vibrational modes.

About a year later, Child[50] correctly estimated the magnitude of the vibronic coupling for each of these modes. These estimates were based on his observation that the frequencies reported for ν_2 and ν_5 in ReF_6 were about 50 cm^{-1} lower than expected by comparison with the corresponding frequencies for WF_6 and PtF_6. Child did not offer any more detailed analysis for the observed spectral features of ReF_6.

Claassen, Malm, and Selig[81] have also reported vibrational spectra for ReF_6. They extended the infrared measurements to longer wavelengths, covering the range 6–43 μ. They obtained the Raman shifts with ReF_6 vapor, rather than with the liquid as Gaunt had done. The numerical values of the three Raman shifts reported in the two studies are in close agreement. However, the Raman bands obtained for ν_2 and ν_5 with the vapor are more diffuse than those observed with the liquid. Claassen, Malm, and Selig point out the unusual diffuseness of ν_2, which they describe as barely discernible above the noise, but do not discuss it further.

In their measurements, Claassen *et al.* observed ν_4 directly at 257 cm^{-1}. Gaunt had derived a value of 393 cm^{-1} for ν_4 from a band at 993 cm^{-1} that he assigned as $(2 + 4)$. This band is absent from the spectra of Claassen *et al.* and was therefore caused by an impurity. Claassen *et al.*[81] also mention that other impurity bands of both ReF_7 and $ReOF_5$ are present in Gaunt's spectra. The question of the purity of Gaunt's sample of ReF_6 had been raised earlier by Malm and Selig.[80] They pointed out that the vapor pressure reported by Gaunt for his sample of ReF_6 suggests the presence of a considerable amount of ReF_7 in his sample. (Gaunt characterized his sample as having a vapor pressure of 200 mm at its melting point; the vapor pressure of pure ReF_6 at its melting point is 423 mm.)

Because of the impure character of Gaunt's sample, we have not used his ReF_6 spectra for the assignment of vibronic transitions. On the other hand, the spectra of Claassen, Malm, and Selig,[81] are well suited to our purpose. These authors were careful to prevent contamination of their samples by impurities, particularly by volatile compounds such as ReF_7 and $ReOF_5$, containing Re—F bonds. Therefore, a number of subtle features

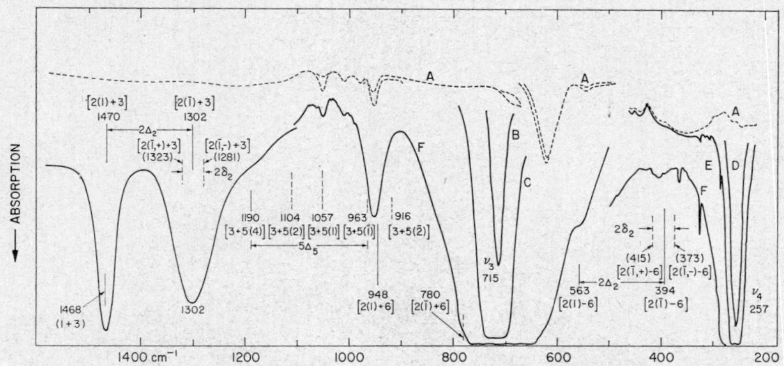

Fig. 18. Infrared spectrum of ReF_6 gas (Claassen, Malm, and Selig[81]). Sample pressures: A (background), B(1 mm), C(3 mm), D(9 mm), E(100 mm), F(630 mm). Frequencies for vibronic transitions that are assigned in this article and that are discussed in the text are indicated by vertical lines.

that appear in their spectra can be assigned as vibronic transitions with a reasonable degree of confidence that they are not spurious features.

The results of Gaunt, (G),[55] and of Claassen *et al.*, (CMS),[81] are summarized in Table XXI. The additional bands that we have assigned from our analysis of the spectra of Claassen *et al.* are given in the column labeled, "This Paper". The infrared spectrum of ReF_6 published by Claassen *et al.*[81] is shown in Fig. 18; the assignments given in Table XXI are indicated by vertical lines in this figure.

The assignments for ν_1, ν_3, and ν_4 given by Claassen *et al.*[81] for ReF_6 remain unchanged in the present analysis. The interpretation of the bands involving ν_2 and ν_5 differs from that of Gaunt and that of Claassen *et al.* because of the vibronic effects. Since ν_6

is derived from binary sum or difference bands involving ν_2 and ν_5, the assignments for ν_6 are also changed.

The e_g vibrational mode

Figure 7 of Section III shows that three transitions are possible from the ground state of ν_2 ($n_2 = 0$) to the first excited state ($n_2 = 1$) for ReF_6 ($J' = \frac{3}{2}$). The energies of these transitions are:

$$2(1) = \nu_2^0 + \Delta_2$$
$$2(\bar{1}, +) = \nu_2^0 - \Delta_2 + \delta_2$$
$$2(\bar{1}, -) = \nu_2^0 - \Delta_2 - \delta_2$$

where ν_2^0 is the frequency derived from the systematics, Δ_2 arises from the linear vibronic coupling, and δ_2 from quadratic coupling. The abbreviations Δ_2, δ_2, and Δ_5 used in this section stand for the vibronic splitting parameters $A_2^{(1,\frac{3}{2})}$, $|B_2^{(1,\frac{3}{2})}|$, and $A_5^{(1,\frac{3}{2})}$, which were defined in Section III.

When the splitting of $\pm\delta_2$ is unresolved, the designation

$$2(\bar{1}) = \nu_2^0 - \Delta_2$$

is used.

The value of ν_2^0 is taken to be 671 cm^{-1} for ReF_6 by interpolation (Fig. 13). The broad peak at 1302 cm^{-1} is reassigned here as $[2(\bar{1}) + 3]$, instead of $(2 + 3)$. Then $2(\bar{1}) = 587$ cm^{-1} and $\Delta_2 = 84$ cm^{-1}. In these spectra, the quadratic splitting is unresolved. However, on the assumption that the excess width of this band, given in Table XX, is equal to $2\delta_2$, we estimate δ_2 to be 21 cm^{-1}. The value of $2(\bar{1}, -)$ is then 566 cm^{-1} and $2(\bar{1}, +)$, 608 cm^{-1}.

In Table XX, the half-width of $(1 + 3)$ for ReF_6 was seen to be 16 cm^{-1} broader than the average half-width of $(1 + 3)$ for the other 5*d* hexafluoride molecules. We now explain this as the result of an accidental degeneracy of $(1 + 3)$ with $[2(1) + 3]$. Using $\Delta_2 = 84$ cm^{-1}, $2(1)$ is 755 cm^{-1}, so that $[2(1) + 3]$ is 1470 cm^{-1}. The calculated value of $(1 + 3)$ is also 1470 cm^{-1}, compared to the value of 1468 cm^{-1} for the observed band. The observed broadness of $(1 + 3)$ for ReF_6 thus qualitatively confirms the assignments for ν_2^0 and Δ_2 for ReF_6. A further verification of these parameters and of the accidental degeneracy of $(1 + 3)$ and

[2(1) + 3] is obtained when the (2 — 6) transitions are considered later in this section.

In Fig. 19 we have reproduced the spectrophotometer scan of Claassen *et al.* for the Raman spectrum of ReF_6 vapor. The frequencies of ν_2^0, 2(1), 2($\bar{1}$, +), and 2($\bar{1}$, —) derived from the

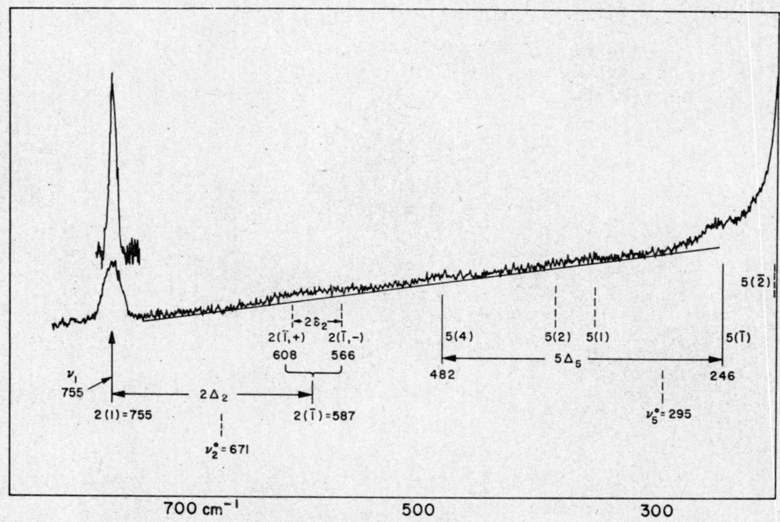

Fig. 19. Raman spectrum of ReF_6 gas at about 3 atmospheres pressure and 70°C (Claassen, Malm, and Selig[81]). Frequencies for vibronic transitions that are assigned in this article and that are discussed in the text are indicated by vertical lines.

systematics and the infrared data are indicated by vertical lines in this figure. The accidental degeneracy of 2(1) with ν_1 is evident. In order to display the diffuseness of 2($\bar{1}$), we have drawn a straight line below the background scattering and parallel to it. Raman scattering above this background extends from 530 to 630 cm^{-1}, and is now identified as a composite of the 2($\bar{1}$, —) and 2($\bar{1}$, +) transitions, whose assigned frequencies are 566 and 608 cm^{-1}. These Raman features are thus consistent with our assignments of the vibronic parameters. Claassen *et al.* chose 596 $\pm$ 5 cm^{-1} as the peak of this broad Raman band, but we suggest 587 cm^{-1} as a better assignment to agree with the infrared data.

The f_{2g} vibrational mode

The vibrational transitions for ReF_6 ($J' = \frac{3}{2}$) involving ν_5 are complicated by the fact that near room temperature less than one-fourth of the molecules are in the lowest vibrational state of ν_5 ($n_5 = 0$), and transitions from both the ground level and the first excited levels of ν_5 are important. The allowed vibrational transitions from the ground state to the first vibrational state and from the first vibrational state to the second can be obtained from Fig. 8. Their energies are summarized by the following relations:

$$5(\bar{2}) = \nu_5^0 - 2\Delta_5$$
$$5(\bar{1}) = \nu_5^0 - \Delta_5$$
$$5(1) = \nu_5^0 + \Delta_5$$
$$5(2) = \nu_5^0 + 2\Delta_5$$
$$5(4) = \nu_5^0 + 4\Delta_5$$

where ν_5^0 is the value predicted for ν_5 from the systematics and Δ_5 is the splitting parameter arising from linear vibronic coupling. Quadratic coupling for ν_5 is not considered here, because we cannot assign specific features in the present spectra to quadratic splittings.

The Raman shift at 246 cm^{-1}, assigned by Gaunt and by Claassen *et al.* as ν_5, is assigned here as $5(\bar{1})$. This is the most prominent feature for ν_5. Similarly, the most intense infrared band involving ν_5 (952 cm^{-1}), assigned in the two experimental papers as (3 + 5), is assigned here as $[3 + 5(\bar{1})]$. The value of ν_5^0 derived from the systematics, Fig. 16, is 295 cm^{-1}. Based on our Raman assignment, $\Delta_5 = 295 - 246 = 49$ cm^{-1}. Based on our infrared assignment, $\Delta_5 = 295 + 715 - 952 = 58$ cm^{-1}. This discrepancy of 9 cm^{-1} is probably outside the uncertainty of the measurements, and a method for deciding which value is more nearly correct is presented below.

In Fig. 19 the reference line drawn below and parallel to the background scattering curve was extended to the ν_5 region. In addition to the Raman shift at 246 cm^{-1}, a weak excursion above this background, centered at 482 cm^{-1}, can be identified. We have assigned this band as the vibronic transition 5(4). The value of Δ_5 derived from this band is 47 cm^{-1}. We have also re-analyzed infrared spectral traces of Claassen *et al.* (Fig. 18) in the region

900–1300 cm^{-1} by subtracting the background from the observed absorption curve. This procedure reveals a shoulder on the 1302 cm^{-1} band at 1190 cm^{-1} that we have assigned as $[3 + 5(4)]$. The value of Δ_5 derived from this band is 45 cm^{-1}.

These values of Δ_5 are in agreement with the value of 49 cm^{-1} derived from the Raman band $5(\bar{1})$. We have assigned Δ_5 the value of 47 cm^{-1} to give the best agreement with all of the spectral data. Table XXI gives the five calculated transitions for ν_5 based on $\nu_5^0 = 295$ cm^{-1} and $\Delta_5 = 47$ cm^{-1}. The calculated values for 5(2) and 5(1) are shown as dashed vertical lines in Fig. 19 and the values for $[3 + 5(2)]$ and $[3 + 5(1)]$ are similarly indicated in Fig. 18. Although there is a suggestion of vibrational bands in the experimental spectra near these frequencies, the evidence is too tenuous to make definite assignments.

We have disregarded the value of $\Delta_5 = 58$ cm^{-1} derived from the 952 cm^{-1} band, because this band is probably a composite. This possibility was revealed when the background was subtracted from the observed absorption band. The other band that is close to the $[3 + 5(\bar{1})]$ transition in energy is the $[2(1) + 6]$ transition.

The inactive fundamental, ν_6

The evaluation of ν_6 for ReF_6 is complicated by vibronic effects. Claassen *et al.* assigned a very weak broad band at 403 cm^{-1} as $(2 - 6)$ and, using their Raman assignment of $\nu_2 = 596$ cm^{-1}, obtained a value of 193 cm^{-1} for ν_6. This should be a reasonable value for ν_6 because the ν_2 vibronic transition involved in each case is $2(\bar{1})$. Coincidentally, the value we derive for ν_6 is also 193 cm^{-1}, although our frequency assignments for the $[2(\bar{1}) - 6]$ band and $2(\bar{1})$ differ from those of Claassen *et al.* Gaunt observed a peak at 413 cm^{-1}, and obtained a low value of 170 cm^{-1} for ν_6 by assigning this band as $(5 + 6)$. This band in Gaunt's spectrum is relatively sharp rather than broad and may be caused by an impurity.

From the previous analysis of the ν_2 vibronic transitions of ReF_6, 2(1) was assigned the same frequency as ν_1. This coincidence explained the abnormal band width of $(1 + 3)$, and also implied that 2(1) would be masked by ν_1 in the Raman effect. However, for the infrared band $[2(1) - 6]$, a 2(1) transition can be observed without interference from ν_1, because the corresponding band involving ν_1, $(1 - 6)$, is not infrared-active. Gaunt reported a band at 549 cm^{-1}, which is probably $[2(1) - 6]$, although he assigns it as $(1 + 4 - 2)$. Claassen *et al.* do not make any frequency assignments in this spectral region, but there is a prominent shoulder on the high wavelength side of ν_3 in their spectra at 563 cm^{-1}, that we assign as $[2(1) - 6]$. The observation of this band, then, experimentally verifies the occurrence of the 2(1) vibronic transition and its accidental degeneracy with ν_1.

We have also re-examined two spectral traces of Claassen *et al.* in the region of the $[2(\bar{1}) - 6]$ band to look for evidence of the δ_2 splitting. When the background is subtracted from this band, it then appears twinned and its apparent midpoint shifts from 403 cm^{-1} to 394 cm^{-1}. Only a poor estimate can be made of the width of this band at half-absorbance, but the value obtained of 42 cm^{-1} compares favorably with the value of 43 cm^{-1} derived for $2\delta_2$ from the data on the $[2(\bar{1}) + 3]$ band. This agreement may be fortuitous, and, in any case, no correction for the undetermined "normal" band width for $(2 - 6)$ has been made.

Using the assignments $[2(1) - 6] = 563$ cm^{-1} and $[2(\bar{1}) - 6] = 394$ cm^{-1}, we derive the value of 193 cm^{-1} for ν_6.

We have previously mentioned the lack of agreement between the calculated value for $[3 + 5(\bar{1})]$ of 963 cm^{-1} and the observed band at 952 cm^{-1}. We can now calculate the frequency for the $[2(1) + 6]$ transition and obtain 948 cm^{-1}. The closeness of the calculated frequency values for the $[3 + 5(\bar{1})]$ and $[2(1) + 6]$ bands suggests that the band at 952 cm^{-1} is probably a composite of these two transitions, and explains the structure of the band that was mentioned earlier.

The calculated value of the $[2(\bar{1}) + 6]$ transition is 780 cm^{-1} and is apparently too close to ν_3 (715 cm^{-1}) to be observed with certainty.

Other bands

There remain vibronic bands involving ν_5, for which little if any evidence can be seen in the spectra. Of the complex of bands arising from $(5 + 6)$, $[5(\bar{1}) + 6]$ should be the most intense. A very weak absorption is observed at 441 cm^{-1}, the calculated value for this transition, but a certain assignment cannot be made.

TABLE XXI. ReF_6

Mode	Assigned value	(G)[55]	(CMS)[81]	This paper	Intensity
1	755	(753)	755		
1, l	753	753			
2(1)	(755)				
2°	(671)				
$2(\bar{1})$	587	(600)[a]	596[a]	587	
$2(\bar{1})$, l	600	600[a]			
3	715	716	715		
4	257	(393)	257		
$5(\bar{2})$	(201)				
$5(\bar{1})$	248	(246)[a]	246[a]		
5°	(295)				
5(1)	(342)				
5(2)	(389)				
5(4)	483			482	
$5(\bar{1})$, l	246	246[a]			
6	(193)				
Δ_2	84				
δ_2	21				
Δ_5	47				
1 + 3	1470	1471	1468	a d	m
1 + 4	(1012)				n o
1 − 4	(498)				n o
2(1) + 3	1470			a d	a d
$2(\bar{1})$ + 3	1302	1306[a]	1302[a]		m
$2(\bar{1})$ + 4	(844)				n o
2(1) + 6	948	956[a]	952[a]	a d	m w
$2(\bar{1})$ + 6	(780)				n o
3 + $5(\bar{2})$	(916)				n o
3 + $5(\bar{1})$	963	956[a]	952[a]	a d	m w
3 + 5(1)	(1057)				v v w
3 + 5(2)	1104				v v w
3 + 5(4)	1198			1190	v w
3 − $5(\bar{1})$	(467)				n o
$5(\bar{1})$ + 6	(441)				n o
2(1) − 6	562	549[a]		563	v w
$2(\bar{1})$ − 6	394	413[a]	403[a]	394[a]	v w

[a] These assignments are discussed in the text.

For the (4 + 5) group, an absorption is not observed at the value of 500 cm^{-1} calculated for the most intense component, $[4 + 5(\bar{1})]$. For the transition $[3 - 5(\bar{1})]$, a very weak shoulder can be seen in the region of 467 cm^{-1}, the calculated value for this transition.

Summary

A new interpretation of the vibrational spectra of ReF_6 has been made. These spectra provide the most complete evidence that is currently available for the type of vibronic coupling discussed in Section III. With the aid of the frequency systematics, assignments of the linear splitting parameters, Δ_2 and Δ_5, have been made that are reliable to about 2 cm^{-1}. A value of the quadratic splitting parameter, δ_2, has been estimated from the (2 + 3) band half-width data that is probably accurate to about 3 cm^{-1}.

A theoretical estimate of δ_2 for ReF_6 was obtained in Section III. This calculation was based on the values $\Delta_2 = 84$ cm^{-1} and $\zeta_d = 3400$ cm^{-1}. The value of δ_2 obtained in this way is about 23 cm^{-1}, in satisfying agreement with the value of $\delta_2 = 21$ cm^{-1} derived here.

This analysis confirms the theories of vibronic coupling of Moffitt and collaborators and of Longuet-Higgins and collaborators. Further experimental studies at higher dispersion will be necessary to identify many of the spectral assignments more explicitly. The interpretations of the spectra of TcF_6, OsF_6, and RuF_6 which follow will be seen to rely heavily on the assignments that have been made here for ReF_6.

(2) *Technetium Hexafluoride*

Technetium hexafluoride ($4d^1$) is the second transition series analogue of ReF_6 ($5d^1$). In Section III, electronic levels of TcF_6 are assigned the same symmetries as those of ReF_6 and their energies estimated from preliminary electronic spectra. The problem of vibronic coupling for TcF_6 ($J' = \frac{3}{2}$) is similar to that for ReF_6 ($J' = \frac{3}{2}$).

Technetium is a synthetic element and consequently the hexafluoride has only recently been prepared (1961).[60] The preparation of TcF_6 offers no particular difficulty and is accomplished by heating technetium metal in a fluorine atmosphere.

Vibrational spectra of TcF_6 have been reported by Claassen, Selig, and Malm,[21] who measured the infrared spectrum of the vapor from 6–43 μ and the Raman spectrum of liquid TcF_6. Only ν_1 was observed in the Raman effect. Self absorption by orange-colored liquid TcF_6 contributes to the failure to observe Raman bands for the e_g and f_{2g} vibrational modes. The Raman spectrum of gaseous TcF_6 was not studied because of the potential biological hazard provided by the radioactivity of technetium.

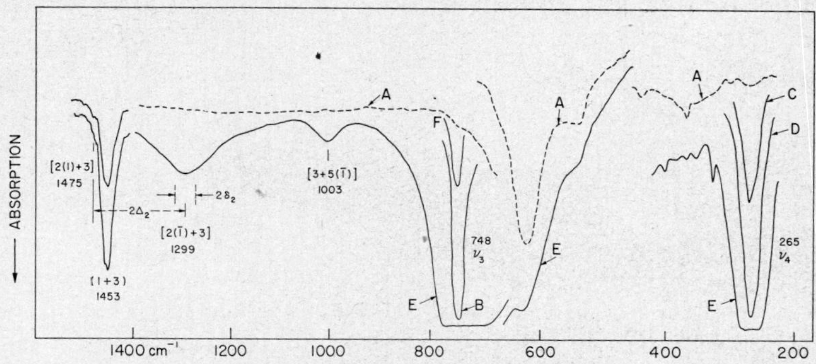

Fig. 20. Infrared spectrum of TcF_6 gas (Claassen, Selig, and Malm[21]). Sample pressures: A(background), B(1 mm), C(3.7 mm), D(13 mm), E(166 mm), F(<1 mm). Frequencies for vibronic transitions that are assigned in this article and that are discussed in the text are indicated by vertical lines.

Claassen *et al.* suggest that TcF_6 does not have a lower symmetry than O_h because only two fundamentals are seen in the infrared. However, they point out the striking difference between the infrared spectra of MoF_6 and TcF_6 in respect to the combination bands involving ν_2 or ν_5. For TcF_6, the combination bands of ν_2 or ν_5 are either very broad or missing. Additionally, the failure to detect ν_2 and ν_5 in the Raman spectrum is cited. They attribute these anomalies involving ν_2 and ν_5 to a dynamic Jahn–Teller effect, in agreement with the interpretation given for analogous effects observed for the e_g vibration in ReF_6 and OsF_6.[19]

Qualitatively, the vibronic effects are more pronounced for TcF_6 than for ReF_6. Consequently, fewer band assignments can be made from the spectra and a less extended analysis is possible. The data are summarized in Table XXII, together with the new assignments made here. These new assignments are indicated as vertical lines in Fig. 20, which is a composite of several infrared spectral scans obtained by Claassen *et al.*[21]

The a_{1g} and f_{1u} vibrational modes

The infrared-active fundamentals, ν_3 and ν_4, are seen directly and require no further comment. The value of 720 cm^{-1} is obtained for ν_1 with liquid TcF_6. The frequency derived for ν_1 from the binary band $(1 + 3)$ is 705 cm^{-1}. Claassen *et al.* explain this difference as a liquid–gas shift and assign 705 cm^{-1} to ν_1 for the gas. However, 15 cm^{-1} is considerably higher than the usual value of ± 2 cm^{-1} observed for liquid–gas shifts and may instead be experimental uncertainty. We have assigned $\nu_1 = 712$ cm^{-1} from the systematics, Fig. 12; coincidentally, this value is the average of 705 and 720 cm^{-1}.

The e_g vibrational mode

The e_g vibronic levels for TcF_6 are qualitatively the same as for ReF_6. The center of the broad band at 1299 cm^{-1} is assigned as $[2(\bar{1}) + 3]$, and $2(\bar{1}) = \nu_2^0 - \Delta_2$ is evaluated as 551 cm^{-1}. From the systematics, ν_2^0 is taken as 639 cm^{-1} and Δ_2 is then 88 cm^{-1}. For $2(1) = \nu_2^0 + \Delta_2$, we obtain the value of 727 cm^{-1}, from which $[2(1) + 3]$ is calculated to be 1475 cm^{-1}. A re-examination of the spectra of TcF_6 reveals a shoulder on the $(1 + 3)$ band centered at 1475 cm^{-1} in confirmation of the splitting derived for ν_2. The position of this band is indicated in Fig. 20. The width of the $(1 + 3)$ band was seen in Table XX to be normal and is unaffected by the closeness of the $[2(1) + 3]$ band because of the weak intensity of the latter band. This is in contrast to the behavior found for ReF_6, where the accidental degeneracy of the $(1 + 3)$ and $[2(1) + 3]$ bands resulted in a 16 cm^{-1} broadening of the $(1 + 3)$ band width.

In Table XX, the quadratic vibronic splitting $(2\delta_2)$ for TcF_6 was derived as 84 cm^{-1}. From the value of the parameter $\delta_2 =$ 42 cm^{-1}, we obtain for $2(\bar{1}, +)$, 593 cm^{-1}, and for $2(\bar{1}, -)$, 509 cm^{-1}.

There is no other definite evidence for e_g transitions in the published spectra. The calculated value of $[2(\bar{1}) + 4]$ is 816 cm^{-1}, which could only appear as a shoulder on the high-wave-number side of ν_3 (748 cm^{-1}). The calculated value of $[2(\bar{1}) + 6]$ is 725 cm^{-1}. An asymmetry toward lower wave numbers is suggested by

TABLE XXII. TcF_6

Mode	Assigned value	(CSM)[21]	This paper	Intensity
1	(712)[a]	(705)		
1, 1	720	720		
2(1)	(727)			
2°	(639)			
$2(\bar{1})$	(551)	(551)[a]		
3	748	748		
4	265	265		
$5(\bar{1})$	(255)	(255)[a]		
5°	(297)			
6	(174)			
Δ_2	88			
δ_2	42			
Δ_5	42			
1 + 3	1460	1453		m
2(1) + 3	1475		1475	v w
$2(\bar{1}) + 3$	1299	1299[a]		m
$3 + 5(\bar{1})$	1003	1003[a]		m w
$2(\bar{1}) + 4$	(816)			n o
$2(\bar{1}) + 6$	(725)			n o
$2(\bar{1}) - 6$	(377)			n o

[a] These assignments are discussed in the text.

comparison of the lowest pressure trace for ν_3 (F in Fig. 20) and the 1 mm pressure trace. While this suggests the possibility of a contribution from $[2(\bar{1}) + 6]$, a definite assignment can not be made. The calculated value for $[2(\bar{1}) - 6]$ is 377 cm^{-1}. Although there is absorption in this region of the spectrum, a definite assignment for the expected broad band is again not possible.

The f_{2g} and f_{2u} vibrational modes

The assignments for ν_5 are also much less complete than those made for ReF_6. The band at 1003 cm^{-1} is assigned as $[3 + 5(\bar{1})]$ by analogy with ReF_6. From this the value of $5(\bar{1}) = \nu_5^0 - \Delta_5$ is calculated to be 255 cm^{-1}. From the systematics, ν_5^0 is 297 cm^{-1}

and Δ_5 is then 42 cm^{-1}. There is no other definite evidence for bands involving the f_{2g} vibration in the spectrum.

Claassen *et al.*[21] do not derive a value for ν_6, because of the failure to observe any binary bands involving ν_6. From the systematics we have estimated the value of 174 cm^{-1} for ν_6.

Summary

The vibrational spectra of TcF_6 provide definite qualitative evidence for vibronic coupling. The $(2 + 3)$ band profile is abnormally broad and the apparent frequency values of the e_g and f_{2g} transitions are shifted from the values of ν_2^0 and ν_5^0 expected from the systematics. Although only limited spectral evidence is available, values of Δ_2, δ_2, and Δ_5 for TcF_6 can be estimated from the spectra by analogy with ReF_6. The assignment for Δ_2 is confirmed by the appearance of a very weak band in the infrared spectrum corresponding to the transition $[2(1) + 3]$.

In Section III we have used the observed value of $\Delta_2 = 88$ cm^{-1} to predict a corresponding value of δ_2. For TcF_6 this predicted value for δ_2 is 61 cm^{-1}. The agreement of this value with the value of $\delta_2 = 42$ cm^{-1} derived here is not good, but is reasonable in light of the preliminary state of our understanding of the electronic structure of TcF_6.

(3) *Osmium Hexafluoride*

Osmium hexafluoride ($5d^2$) is the third abnormal hexafluoride molecule that we shall consider. Vibronic coupling occurs for OsF_6 because the lowest electronic levels of this molecule (E_g and F_{2g}) are orbitally degenerate. These levels are of nearly the same energy and are treated as a $J' = 2$ multiplet.

OsF_6 is readily prepared by the fluorination of osmium metal, but it is less stable and more reactive than ReF_6. In the condensed phases, OsF_6 is yellow in color; in the vapor, the yellow color can be seen only through a long absorbing path. This compound was originally identified as OsF_8 in 1913.[82] It enjoyed the distinction of being the only known octafluoride for over forty years, until, in 1958, its formula was shown instead to be OsF_6.[83]

The infrared spectrum of OsF_6 from 6 to 50 μ has been reported by Weinstock, Claassen, and Malm.[20] These measurements were difficult to obtain because of reaction between OsF_6 and the window materials. Weinstock *et al.* also obtained the Raman spectrum of liquid OsF_6, using the 5461 A green Hg line for excitation, in order to minimize photochemical decomposition.

These authors assigned octahedral symmetry, point group O_h, to OsF_6, because the main features of its vibrational spectra are similar to those of other hexafluoride molecules. However, they noted important differences between OsF_6 and the normal hexafluoride molecules for the e_g vibrational mode, ν_2. For example, they did not detect a Raman shift for ν_2 with OsF_6, under conditions favorable for its observation, although Raman shifts were obtained for both ν_1 and ν_5. By contrast, with normal hexafluoride molecules ν_2 is readily observed, and it is ν_5 that may offer difficulty, because the Raman band for ν_5, being of low frequency, is obscured by the high background scattering near the exciting line. Weinstock *et al.* also called attention to the fact that binary sum bands involving ν_2 are less intense or markedly broader for OsF_6 than for normal hexafluoride molecules and drew particular attention to the abnormal band profile of $(2 + 3)$ which was discussed earlier in this section. These anomalies in regard to ν_2 in the OsF_6 spectra and similar anomalies that appeared in Gaunt's ReF_6 spectra[55] were ascribed to a dynamic Jahn–Teller effect by Weinstock and Claassen.[19] The corresponding anomalies which are possible for the f_{2g} vibration, ν_5, were not detected.

Theoretical vibronic energy levels for OsF_6 ($J' = 2$) are shown in Fig. 9 (Section III). A comparison with Figs. 7 and 8, where the corresponding vibronic levels for ReF_6 ($J' = \frac{3}{2}$) are given, makes it evident that many more vibronic transitions are possible for OsF_6 than for ReF_6. The resulting complexity of the spectra makes it difficult to deduce reliable values for the vibronic parameters of OsF_6 directly from the observed spectral bands as was done for ReF_6. Instead, we have evaluated the vibronic parameters of OsF_6 from theory (Section III) on the assumption that the relevant one-electron integrals have the same values for OsF_6 as for ReF_6. The results of these calculations are $\bar{\Delta}_2 = 42\ \text{cm}^{-1}$, $\bar{\delta}_2 = 14.0\ \text{cm}^{-1}$, $\bar{\bar{\delta}}_2 = 42\ \text{cm}^{-1}$, and $\bar{\Delta}_5 = 14.6\ \text{cm}^{-1}$. These symbols are defined in Section III: $\bar{\Delta}_2 = A_2^{(2,2)}$; $\bar{\delta}_2 = 2|B_2^{(2,2)} - 4\bar{B}_2^{(2,2)}|$; $\bar{\bar{\delta}}_2 = 4|B_2^{(2,2)} + 3\bar{B}_2^{(2,2)}|$; and $\bar{\Delta}_5 = 10A_5^{(2,2)}$.

In order to compare theory and experiment, we have used the

above vibronic parameters and the frequencies $\nu_2^0 = 668$ and $\nu_5^0 = 276\ \text{cm}^{-1}$ obtained from the systematics to derive theoretical values for the vibronic transitions. These frequencies were compared with the experimental data of Weinstock, Claassen, and Malm and assignments made for the reported absorption bands.

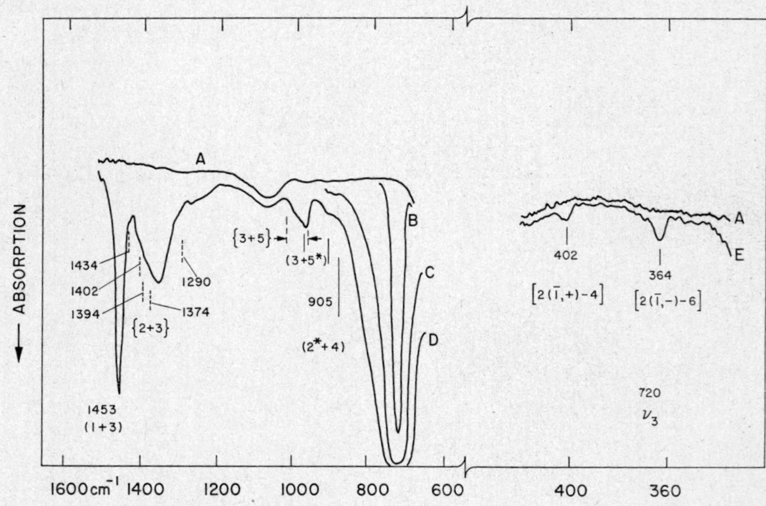

Fig. 21. Infrared spectrum of OsF_6 vapor (Weinstock, Claassen, and Malm[20]). Sample pressures: A(background), B(5 mm), D(90 mm), E(300 mm), F(600 mm). Frequencies for vibronic transitions that are assigned in this article and that are discussed in the text are indicated by vertical lines.

From these assignments, experimental values $\bar{\Delta}_2 = 46\ \text{cm}^{-1}$ and $\bar{\bar{\delta}}_2 = 52\ \text{cm}^{-1}$ were deduced. Experimental values for $\bar{\delta}_2$ and $\bar{\Delta}_5$ cannot be deduced from the presently available spectra. The data of Weinstock *et al.* are summarized in Table XXIV. Our assignments for the fundamentals, binary transitions, and vibronic parameters are given in the second column of this table. For these assignments, we have used the experimental values for $\bar{\Delta}_2$ and $\bar{\bar{\delta}}_2$ and the theoretical values for $\bar{\delta}_2$ and $\bar{\Delta}_5$. These assignments are compared with the observed spectra in Fig. 21. Details of our re-analysis of the spectra of Weinstock *et al.* and of our assignments are given next.

The fundamental vibration frequencies ν_1, ν_3, and ν_4

The values assigned by Weinstock *et al.*[20] for the directly observed fundamentals, $\nu_1 = 733$ and $\nu_3 = 720\ \text{cm}^{-1}$, remain unchanged in the present analysis. However, a re-examination of the grating spectrometer scans that were obtained for ν_4 has resulted in the discovery of a number of unexpected new features.

One of these spectral scans of ν_4 for OsF_6 is shown in Fig. 22, where it is compared with a similar scan of ν_4 for PtF_6.[20] These spectra were obtained at the same time by Weinstock *et al.* using R. A. Oetjen's far infrared spectrometer and with assistance from

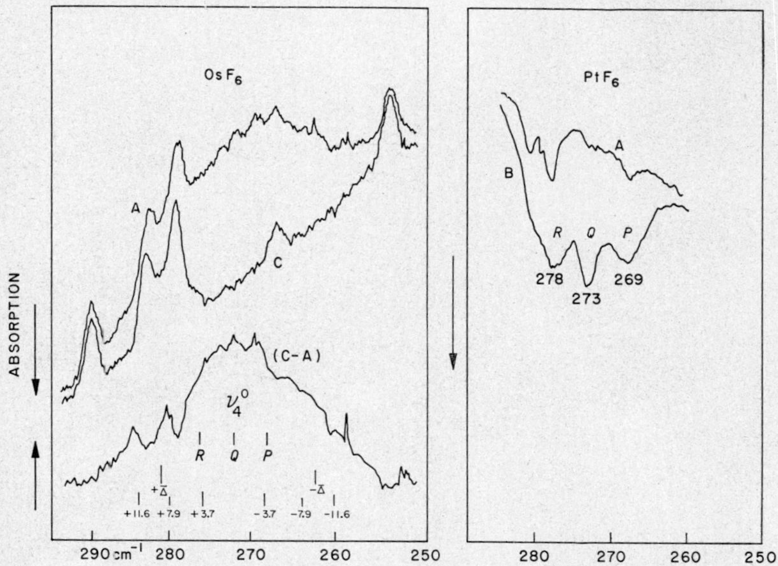

Fig. 22. The infrared-active fundamental, ν_4, for OsF_6 and PtF_6 (Weinstock, Claassen, and Malm[20]). A is the background scan for both molecules; for C, the sample pressure of OsF_6 was 13 mm and that of PtF_6, 20 mm. For OsF_6, curve (C − A) was obtained by subtracting the background scan (A) from the sample scan (C); the sense of absorption is inverted for curve C to facilitate its display in the figure. The small vertical lines beneath (C − A) in the figures are frequencies that are assigned in this article and that are discussed in the text for the P, Q, and R bands for OsF_6 and vibronic transitions that accompany the ν_4 transition. The vibronic transitions labeled $\pm\bar{\Delta}$ originate from the lowest levels (E_g and F_{2g}); those labeled with a frequency shift originate from thermally populated $n_5 = 1$ levels.

M. Vance and R. Brown. It is evident from the figure that the two molecules give quite different spectral profiles for ν_4. The band shape for PtF_6 is symmetrical with well defined P, Q, and R branches, as might be expected for a spherical-top molecule. However, the ν_4 band for OsF_6 is diffuse and the individual rotational branches cannot be easily detected. This unexpected difference between the ν_4 band profiles of OsF_6 and PtF_6 led us to explore the possibility that for OsF_6 the ν_4 transition may be complicated vibronicly.

The vibronic levels that pertain to this discussion and their relative energies are given in Table XXIII. The method used to

TABLE XXIII. The Energies of Vibronic Levels for OsF_6 with $n_2 = 0$ and $n_5 = 0$ or 1

Levels, $\Gamma(n_2, n_5; \alpha)$	Energies, cm^{-1} [$E(J' = 2) + \nu_2^0 + 1.5\nu_5^0$ taken as zero]
$F_{2g}(0, 1; \text{II})$	$\nu_5^0 - 23.6$
$E_g(0, 1)$	$\nu_5^0 - 35.6$
$F_{1g}(0, 1; \text{II})$	$\nu_5^0 - 41.2$
$F_{2g}(0, 1; \text{I})$	$\nu_5^0 - 71.6$
$A_{1g}(0, 1)$	$\nu_5^0 - 79.5$
$F_{1g}(0, 1; \text{I})$	$\nu_5^0 - 83.2$
$F_{2g}(0, 0)$	$- 50.3$
$E_g(0, 0)$	$- 59.5$

calculate these energies was given in Section III. Each level is identified by its symmetry (Γ), and by vibrational quantum numbers, given in parentheses. For the lower pair of levels, the e_g quantum number, n_2, is zero and the f_{2g} quantum number, n_5, is zero. For the upper group of six states, $n_2 = 0$ and $n_5 = 1$. In this group, the roman numerals I and II simply differentiate between levels of the same symmetry. These six upper levels are also considered here because some of them are significantly populated at room temperature, e.g., the $F_{1g}(0, 1; \text{I})$ level, which is 252 cm^{-1} above the $E_g(0, 0)$ ground level ($\nu_5^0 = 276$ cm^{-1}).

We consider first the ν_4 transitions that start from the $E_g(0, 0)$ and $F_{2g}(0, 0)$ vibronic levels. The four possible transitions are shown in Fig. 23. Two transitions occur without change in vibronic state at the unshifted frequency ν_4^0. The symbols $(E_g \rightarrow E_g + \nu_4^0)$ and $(F_{2g} \rightarrow F_{2g} + \nu_4^0)$ represent these transitions. The other two transitions take place with simultaneous change of the vibronic state of the molecule. One, $(E_g \rightarrow F_{2g} + \nu_4^0)$, occurs at the frequency $(\nu_4^0 + \bar{\Delta})$, where $\bar{\Delta} = (0.5\bar{\Delta}_2 - 0.8\bar{\Delta}_5) =$

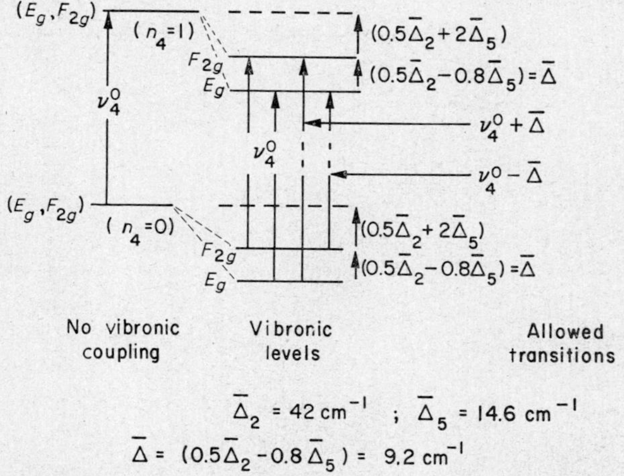

Fig. 23. Vibronic transitions originating from the two lowest levels of OsF_6 (E_g and F_{2g}) that accompany the ν_4 transition. The values for the vibronic parameters are derived from theory.

9.2 cm^{-1}; the other, $(F_{2g} \rightarrow E_g + \nu_4^0)$, occurs at the frequency $(\nu_4^0 - \bar{\Delta})$.

The other ν_4 transitions that we consider start from the thermally populated $(n_5 = 1)$ states. For this discussion, we limit ourselves to the three lowest states, $F_{1g}(0, 1; \mathrm{I})$, $A_{1g}(0, 1)$, and $F_{2g}(0, 1; \mathrm{I})$, because they have the highest thermal population. Again, the ν_4 transitions that take place without change in vibronic state occur at the unshifted frequency ν_4^0. There are six ν_4 transitions that take place with a simultaneous change in vibronic state and that occur close to the frequency of the ν_4^0 transition. These are: $[F_{1g}(0, 1; \mathrm{I}) \rightarrow A_{1g}(0, 1) + \nu_4^0]$, which occurs at $\nu_4^0 + 3.7$ cm^{-1}; $[F_{1g}(0, 1; \mathrm{I}) \rightarrow F_{2g}(0, 1; \mathrm{I}) + \nu_4^0]$ at $\nu_4^0 + 11.6$ cm^{-1}; $[A_{1g}(0, 1) \rightarrow F_{2g}(0, 1; \mathrm{I}) + \nu_4^0]$ at $\nu_4^0 + 7.9$ cm^{-1};

$[A_{1g}(0, 1) \rightarrow F_{1g}(0, 1; \mathrm{I}) + \nu_4^0]$ at $\nu_4^0 - 3.7\ \mathrm{cm}^{-1}$; $[F_{2g}(0, 1; \mathrm{I}) \rightarrow A_{1g}(0, 1) + \nu_4^0]$ at $\nu_4^0 - 7.9\ \mathrm{cm}^{-1}$; and $[F_{2g}(0, 1; \mathrm{I}) \rightarrow F_{1g}(0, 1; \mathrm{I}) + \nu_4^0]$ at $\nu_4^0 - 11.6\ \mathrm{cm}^{-1}$.

The comparison of these vibronic transitions with the observed spectral scans for OsF_6 is made in the following way. The absorption for the background scan, A, is first subtracted from the absorption scan, C, obtained with 13 mm OsF_6 pressure in the absorption cell. The resulting difference, (C — A), is plotted below the observed scans, A and C, in Fig. 22, with the sense of absorption for curve (C — A) inverted in the figure for convenience. The spectral assignments in this region are indicated in the figure by vertical lines. The value obtained for ν_4^0 is now 272 cm^{-1}, compared with the value of 268 cm^{-1} assigned by Weinstock *et al.* The peaks of the P and R branches for OsF_6 are indicated, shifted $\pm 4\ \mathrm{cm}^{-1}$ from ν_4^0 on the basis of the displacements observed with PtF_6 for these branches. The two ν_4 vibronic transitions starting from $E_g(0, 0)$ and $F_{2g}(0, 0)$ are indicated by the lines labeled $+\bar{\Delta}$ and $-\bar{\Delta}$. The $(\nu_4^0 + \bar{\Delta})$ transition is clearly resolved, and the frequency calculated for the $(\nu_4^0 - \bar{\Delta})$ transition is at the center of a shoulder appearing on the high-wavelength side of the ν_4^0 band. The six ν_4 transitions starting from the $n_5 = 1$ states are labeled by the frequencies calculated for their displacements from ν_4^0. The transitions at $(\nu_4^0 \pm 3.7)$ fall between the peaks of the P, Q, and R branches of ν_4^0. This helps to explain why the P, Q, and R branches of ν_4^0 are not as clearly resolved for OsF_6 as they are for PtF_6. The $(\nu_4^0 + 11.6)$ transition is now resolved. The $(\nu_4^0 + 7.9)$ transition is combined with the $(\nu_4^0 + \bar{\Delta})$ band, which is at $(\nu_4^0 + 9.2)$. Similarly, the $(\nu_4^0 - 7.9)$ transition is combined with the $(\nu_4^0 - \bar{\Delta})$ transition. The $(\nu_4^0 - 11.6)$ transition is centered at another shoulder on the high-wavelength side of the ν_4^0 band.

This rather complete agreement between the calculated values for the vibronic transitions accompanying the ν_4^0 transition and the resolved spectral scan is satisfying. Because of the complicated course of the background scan, it could be argued that the apparent spectral features that appear in curve (C — A) are fortuitous. However, Weinstock *et al.* obtained two other independent scans of the ν_4 region for OsF_6 under different conditions of sample pressure and scanning speed, and, when each of these scans is similarly analyzed, a resultant (C — A) curve is obtained that reproduces the same spectral features shown in Fig. 22. This reproducibility of the data is taken as good supporting evidence for the reliability of the interpretation that is given.

These transitions are the vibronic analogue of the more usual binary transitions observed in the spectra of normal molecules. For a truly harmonic oscillator, binary transitions are not allowed; they are enabled by anharmonicity. For OsF_6, the vibronic coupling of the f_{2g} mode together with the normal anharmonicity enables binary combinations of infrared-active fundamentals and vibronic transitions. Similar transitions should accompany the ν_3 fundamental, but are not observed in the present spectra probably because of poorer resolution in the ν_3 region of the spectrum.

The e_g vibrational mode

The vibronic energy levels for the e_g vibrational mode were calculated in Section III (Fig. 9). The ground state ($n_2 = 0$) is split into two levels and the first excited state ($n_2 = 1$) is split into five levels. There are ten transitions possible between these levels. We have assumed that the five transitions containing the representation E should be more intense than the other five transitions. The designations of the transitions containing E and their calculated frequencies are listed below in the column labeled "Theoretical value". The values used for the parameters are: $\nu_2^0 = 668$, $\bar{\Delta}_2 = 42$, $\bar{\delta}_2 = 14.0\ \text{cm}^{-1}$, and $\bar{\bar{\delta}}_2 = 42\ \text{cm}^{-1}$.

The lack of evidence for the e_g transitions in the Raman spectrum of OsF_6 is most likely a consequence of the large number of possible transitions. The spectral intensity becomes too weak

Assignment	"Theoretical value"	"Experimental value"
$2(\bar{1}, -) = (\nu_2^0 - \bar{\Delta}_2 - \bar{\bar{\delta}}_2) =$	584 cm^{-1}	570 cm^{-1}
$2(0, -) = (\nu_2^0 - \bar{\delta}_2) \quad =$	654	654
$2(\bar{1}, +) = (\nu_2^0 - \bar{\Delta}_2 + \bar{\bar{\delta}}_2) =$	668	674
$2(0, +) = (\nu_2^0 + \bar{\delta}_2) \quad =$	682	682
$2(1) = (\nu_2^0 + \bar{\Delta}_2) \quad =$	710	714

and diffuse to be observed because it is dissipated among many individual transitions and spread over a wide frequency range. It is very possible, however, that Raman shifts corresponding to some of these transitions will be observed with instruments of greater sensitivity.

Several bands are observed in the infrared spectrum that are binary transitions including the e_g vibrational mode. The most prominent of these is the $(2 + 3)$ band. The anomalous band profile of this transition and its significance in supplying the first experimental evidence of vibronic coupling has already been discussed. The maximum absorbance for this band is at 1352 cm^{-1}. Using $\nu_3 = 720\ cm^{-1}$, the frequency of the ν_2 component is 632 cm^{-1}. This value is near the average of the calculated frequencies for the $2(\bar{1}, -)$, $2(0, -)$, and $2(\bar{1}, +)$ transitions, 635 cm^{-1}. We designate this average as ν_2^*. The 1352 cm^{-1} band is then assigned as $(2^* + 3)$. The $(2^* + 3)$ band is asymmetrical with its center of gravity at a higher frequency than 1352 cm^{-1}, probably because of contributions from the other two e_g transitions, $2(0, +)$ and $2(1)$. The excess band width of the $(2^* + 3)$ transitions is 59 cm^{-1} (Table XX). By analogy with our previous interpretations for the excess band widths, this should correspond to the difference in frequency between the $2(\bar{1}, -)$ and $2(\bar{1}, +)$ transitions, 84 cm^{-1}. Although the agreement is not good, it is reasonable in view of our present lack of knowledge about the relative intensities of the vibronic transitions.

This complicated interpretation of the $(2 + 3)$ band of OsF_6 is to be compared with the much simpler explanation for the $(2 + 3)$ band of ReF_6. For ReF_6, the linear e_g splitting could be derived directly from the shift of band frequency and the quadratic e_g splitting directly from the excess band width. In contrast, for OsF_6, the shift in band frequency and the excess band width are both influenced by a combination of linear and quadratic splittings. This is typical of the greater difficulty that is met in the interpretation of the vibrational spectra of OsF_6 as compared with ReF_6, and that led us to compare calculated values of the vibronic transitions with the observed spectra rather than to try to derive vibronic parameters directly from the short-wavelength region of the spectrum.

The two next most intense binary bands involving the e_g

vibration are at 364 and 402 cm^{-1}. Weinstock *et al.* assigned the band at 364 cm^{-1} as (2 — 4) and the band at 402 cm^{-1} as (2 — 6). We now assign the 364 band as $[2(\bar{1}, -) - 6]$, theoretical value 379 cm^{-1}, and the 402 band as $[2(\bar{1}, +) - 4]$, theoretical value 396 cm^{-1}. This suggests that the $2(\bar{1}, \pm)$ transitions are the most intense e_g transitions, which is consistent with the interpretation of the $(2^* + 3)$ band. However, absorptions for the $[2(\bar{1}, +) - 6]$ transition, theoretical value 463 cm^{-1}, and the $[2(\bar{1}, -) - 4]$ transition, theoretical value 312 cm^{-1}, cannot be seen in the spectra of Weinstock *et al.* This apparent lack of intensity for these transitions is puzzling but may be illusory, since they occur in frequency regions that were unfavorable for observation.

Alternatively, the above assignments can be used to derive values for $\bar{\Delta}_2$ and $\bar{\bar{\delta}}_2$. For these evaluations, ν_2^0 is taken as 668 cm^{-1}, ν_4 as 272 cm^{-1}, and ν_6 as 205 cm^{-1}. From $[2(\bar{1}, -) - 6] = 364$, $(\bar{\Delta}_2 + \bar{\bar{\delta}}_2)$ is derived to be 99 cm^{-1}; and from $[2(\bar{1}, +) - 4] = 402$, $(\bar{\Delta}_2 - \bar{\bar{\delta}}_2)$ is derived to be -6 cm^{-1}. Then, $\bar{\Delta}_2$ is 46 cm^{-1} and $\bar{\bar{\delta}}_2$ is 52 cm^{-1}, compared with the theoretical value of 42 cm^{-1} for both of these parameters. Although the agreement between theory and experiment for these parameters is much poorer than is generally obtained for the hexafluoride molecules, it is reasonable in view of the limited experimental data available for OsF_6 and of the greater complexity of the theoretical analysis required for OsF_6.

The experimental values for $\bar{\Delta}_2$ and $\bar{\bar{\delta}}_2$ are reported in Table XXIV and are used for the comparisons with the observed spectral data. In the above listing of the e_g transitions, the frequencies given as "Experimental values" were calculated using these values instead of the corresponding theoretical ones.

Weinstock *et al.* also report a very weak band at 894 cm^{-1}, which they assign as (2 + 4). This is now assigned as $(2^* + 4)$, calculated value 907 cm^{-1}.

The f_{2g} vibrational mode

The energy levels for the ground state ($n_5 = 0$) and first excited state ($n_5 = 1$) of the f_{2g} vibrational mode for OsF_6 were given in Table XXIII. There are ten transitions between the ground state and the first excited state that contain the species F_{2g}. Additionally, a much larger number of transitions containing F_{2g} are possible from the first excited state of OsF_6, which is significantly populated at room temperature, to the second excited state.

Two spectral features involving f_{2g} transitions of OsF_6 have been reported. Weinstock *et al.* observed a diffuse Raman band with liquid OsF_6 centered at 252 cm^{-1} which they assigned as ν_5. This Raman shift was barely detectable in a 15 minute exposure, although the ν_1 band was distinctly seen. In a 45 minute exposure, the ν_1 shift was still sharp and clear, but the f_{2g} band was now completely obscured by the high background scattering near the exciting line. They also observed a broad asymmetrical band at 969 cm^{-1} in their infrared studies, which they assigned as $(3 + 5)$.

These bands can be attributed to vibronic transitions because they are significantly shifted from the frequencies of 276 and 996 cm^{-1} that would be predicted for them in the absence of vibronic splittings for ν_5. The difficulty of observing ν_5 in the Raman effect and the broad asymmetrical shape of the $(3 + 5)$ transition are also consistent with the large number of f_{2g} vibronic transitions predicted at the beginning of this discussion. It is not possible to derive a value for $\bar{\Delta}_5$ from these data, but the value, $\bar{\Delta}_5 = 14.6$ cm^{-1}, has been assigned from theory.

Subtracting $\nu_3 = 720$ from 969 cm^{-1} gives the value of 249 cm^{-1} for the center of the most intense component of the $(3 + 5)$ band. This agrees with the Raman value of 252 cm^{-1} obtained with liquid OsF_6. We assign both of these bands to the $F_{2g}(0, 0) \rightarrow A_{1g}(0, 1)$ transition (see Table XXIII). The theoretical value for this transition is 247 cm^{-1} and we refer to it in Table XXIV as ν_5^*. If we just consider the F_{2g} transitions from the $n_5 = 0$ states of OsF_6 to the $n_5 = 1$ states, there is one predicted transition, 243 cm^{-1}, of lower frequency than 247 cm^{-1} and there are eight transitions of higher frequency that extend out to 312 cm^{-1}. The observed asymmetry of the $(3 + 5)$ band agrees with this assignment. The band falls off rapidly in intensity on the low-frequency side of the maximum, but does not reach the base line for about 50 cm^{-1} on the high-frequency side.

The inactive fundamental ν_6

There are no bands other than $[2(\bar{1}, —) - 6]$ that we can assign to transitions involving ν_6. As mentioned above we have therefore assigned ν_6 for OsF_6 as 205 cm^{-1}, a value taken from the systematics. The strongest (2 + 6) transition, (2* + 6), would fall at 840 cm^{-1}, and the most intense (5 + 6) transition, (5* + 6), at 457 cm^{-1}.

Weinstock *et al.* report five other bands as weak and uncertain, so that they did not assign them, and we fail to assign them here also. Their frequencies are 325, 352, 544, 644, and 800 cm^{-1}.

Summary

Convincing qualitative evidence for vibronic coupling in OsF_6 is supplied by the anomalous (2 + 3) band profile, and by the

TABLE XXIV. OsF_6

Mode	Assigned value	(WMC)[2]	This paper	Intensity
1	(733)	(733)		
1, 1	733	733		
2*	(633)			
2°	(668)	(632)[a]		
3	720	720		
4	272	268	272	
5*	(252)			
5*, 1	252	252[a]		
5°	(276)	(252)[a]		
6	(205)	(230)		
$\bar{\Delta}_2$	46			
$\bar{\delta}_2$	(14.0)			
$\bar{\bar{\delta}}_2$	52			
$\bar{\Delta}_5$	(14.6)			
1 + 3	1453	1453		m
2* + 3	1353	1352[a]		m
2* + 4	905	894[a]		v w
$2(\bar{1}, +) - 4$	402	402[a]		v w
$2(\bar{1}, -) - 6$	365	364[a]		v w
3 + 5*	972	969[a]		w

[a] These assignments are discussed in the text.

differences between the apparent frequencies of ν_2 and ν_5 derived from the vibrational spectra and the corresponding frequencies expected from the systematics. The vibronic parameters for OsF_6 are difficult to derive directly from the spectral data, and have been calculated from theory (Section III) on the assumption that the relevant one-electron integrals have the same values for OsF_6 as for ReF_6. The expected vibronic transitions involving ν_2 and ν_5 have been derived using these parameters and values of ν_2^0 and ν_5^0 estimated from the systematics. The most prominent features in the vibrational spectra of OsF_6 reported by Weinstock *et al.*[20] have been assigned in terms of these expected transitions. From these assignments, we have then been able to derive "experimental" values for $\bar{\Delta}_2$ and $\bar{\bar{\delta}}_2$.

Rather unexpectedly, a new type of binary transition was discovered in the spectra. The absorption band profile for the infrared-active fundamental ν_4 is quite abnormal and shows side-bands that are interpreted as vibronic transitions accompanying the ν_4 absorption. This appears to be the vibronic analogue of the binary combination bands observed in the spectra of nonvibronic molecules.

(4) *Ruthenium Hexafluoride*

Ruthenium hexafluoride ($4d^2$) is the analogue of OsF_6 ($5d^2$) and is the final vibronic molecule to be considered in this article. Although the electronic levels of RuF_6 are closely analogous to those of OsF_6, our analysis of preliminary electronic spectra of RuF_6 (Section III) suggests that the energies separating E_g and F_{2g} electronic states in RuF_6 are much greater than those for the corresponding states of OsF_6. As a result, the vibronic spectra of these two molecules are expected to be quite different.

RuF_6 is difficult to prepare and was not isolated until 1961.[61] Vibrational spectra of RuF_6 are also difficult to obtain, because the compound is thermally unstable and very reactive. In these respects RuF_6 resembles PuF_6, RhF_6, and PtF_6. Like these compounds, its vapor is colored (reddish-brown) and photochemically unstable, so that Raman spectra have not been attempted. RuF_6 also has a relatively low vapor pressure, which increases the problems associated with its study.

Infrared spectra of RuF_6 vapor from 230–4000 cm^{-1} have been obtained by Weinstock, Claassen, and Chernick.[22] Two strong absorptions are observed and assigned as ν_3 and ν_4 by analogy with other hexafluorides. The absence of other intense bands in the infrared led the authors to conclude that RuF_6 has O_h symmetry in the vapor state.

However, the authors report that the infrared spectrum of RuF_6 shows only two distinct combination bands, or considerably fewer than that of any of the known hexafluorides. Only four of the six fundamentals could therefore be assigned from the data. An estimate of the frequency for ν_2 was also made by taking the average of the values of ν_2 for TcF_6 and RhF_6. The unusual character of the spectrum of RuF_6 was attributed by Weinstock *et al.* to a dynamic Jahn–Teller effect.

The e_g vibrational mode

Perhaps the most distinctive characteristic of the infrared spectrum of RuF_6 is the lack of intensity of the (2 + 3) band. As discussed earlier in this section, the juxtaposition and appearance of the (2 + 3) and (1 + 3) bands are hallmarks of the hexafluoride spectra. Weinstock *et al.*[22] state that their search for evidence of the (2 + 3) band was complicated by the appearance of impurity bands at 1250 and 1280 cm^{-1}, which are close to the frequency of 1308 cm^{-1} that they expected for the (2 + 3) band (based on $\nu_3 = 735$ cm^{-1} and their estimated value for ν_2 of 573 cm^{-1}). The bands at 1250 and 1280 cm^{-1} were attributed to impurities, because comparisons of spectra obtained with different RuF_6 samples revealed that the intensities of these bands varied independently of the (1 + 3) band intensity. Weinstock *et al.* also report continuous electronic absorption in the range 1000–3000 cm^{-1} with broad maxima at 1400 and 1950 cm^{-1}. They point out that the overlapping of electronic and vibrational spectra in this frequency range further complicates observation of the (2 + 3) band.

We have re-examined the infrared spectra of Weinstock *et al.*[22] for RuF_6 and have found evidence for the (2 + 3) transition. The (1 + 3) transition of RuF_6 has its maximum absorbance at 1410 cm^{-1}, and appears in the spectrum symmetrically placed on top of the broad 1400 cm^{-1} electronic transition (Fig. 24). Between the peak of the (1 + 3) band and the 1280 cm^{-1} impurity band, a shoulder can be seen on the low-frequency side of the (1 + 3) band in all the spectral scans of Weinstock *et al.* This shoulder follows the (1 + 3) band in intensity and can therefore be assigned to RuF_6 rather than to an impurity. This absorption

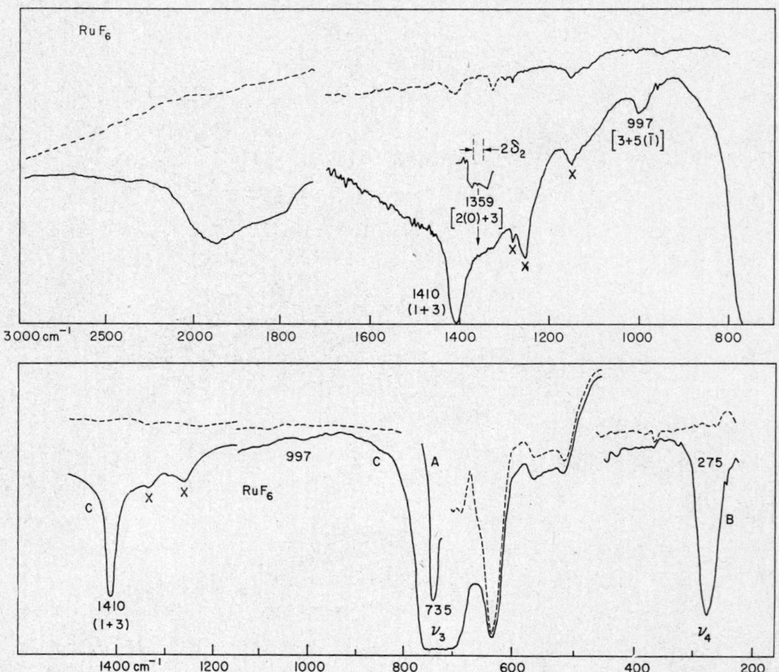

Fig. 24. Infrared spectrum of RuF_6 vapor (Weinstock, Claassen, and Chernick[22]). Upper: 62 cm path length, 77 mm sample pressure; lower: 10 cm path length, sample pressures—A(4 mm), B(23 mm), C(77 mm). The method that we used to resolve the band labeled [2(0) + 3] is described in the text.

was not reported in the experimental paper because of its low intensity and the uncertainty created by the profusion of other absorptions in that region of the spectrum. We now assign this shoulder to the (2 + 3) transition.

In order to make as quantitative an estimate as possible of the frequency and shape of the absorption corresponding to this shoulder, we have resolved the shoulder from the (1 + 3) band in the following way. If the (1 + 3) band is assumed to be symmetrical, the absorbance on the high-frequency side of the (1 + 3) band can be subtracted from that on the low-frequency side so that the resulting difference gives the approximate absorbance of the (2 + 3) band. The background absorbance must also be subtracted because of an absorption that appears at 1325 cm^{-1}. This procedure was followed for the three spectral scans obtained by Weinstock *et al.* with 62 cm absorbing path length and similar (2 + 3) band profiles were obtained in all cases. The appearance of the (2 + 3) band resolved in this way is shown in Fig. 24, where one of the infrared spectra obtained by Weinstock *et al.*[22] is reproduced.

There is a possibility that this resolved band is part of the broad electronic band centered at 1400 cm^{-1}. Coincidentally, however, this electronic transition appears to be symmetrical about the center of the (1 + 3) band. This coincidence may be inferred from the fact that the difference between the absorbance on the high-frequency side of the (1 + 3) band and that on the low-frequency side is constant for 20 cm^{-1}, that is, until the absorption associated with the (2 + 3) band begins.

The (2 + 3) band for RuF_6 is much weaker in intensity and quite different in appearance from the corresponding band for OsF_6. This difference is probably the result of different splittings for the five lowest electronic states in each of these molecules. For OsF_6, the splitting of the lowest $J' = 2$ state is only 9 cm^{-1} so that both the E_g and the F_{2g} electronic states contribute importantly to the observed vibronic transitions. For RuF_6, the splitting of the E_g and F_{2g} electronic states is much greater and only one of these is important in determining the lowest energy vibronic levels. We cannot decide whether the E_g or F_{2g} level is of lower energy from the available electronic spectra, but we can deduce that the F_{2g} level is the lower lying one from our analysis of the vibrational spectra.

The average frequency of the (2 + 3) transition for RuF_6 is assigned from the spectra as 1359 ± 2 cm^{-1}. Subtracting the value of $\nu_3 = 735$ cm^{-1} gives a value of 624 cm^{-1} for the ν_2 component. This is very close to the value for ν_2^0 of 628 cm^{-1} estimated from the systematics, and we now re-assign ν_2^0 for RuF_6 to be 624 cm^{-1}. The occurrence of the (2 + 3) band at a frequency corresponding to the unshifted value of ν_2^0 favors the

F_{2g} electronic states lying lower in energy than the E_g states. If the E_g electronic states were lower, the ν_2 vibronic transitions would occur at frequencies shifted from ν_2^0. However, for a F_{2g} electronic state ($J' = 1$), some ν_2 transitions occur at about the same frequencies as they would occur in the absence of vibronic coupling. These ν_2 transitions are unshifted because, for linear vibronic coupling, all the ν_2 levels are lowered in energy by a uniform amount. Therefore, although we cannot derive a value for $\bar{\Delta}_2$ from the spectra for RuF_6, we have obtained an experimental value of ν_2^0.

The (2 + 3) band for RuF_6 ($J' = 1$) is actually a composite of two vibronic transitions:

$$2(0, +) + 3 = \nu_2^0 + \bar{\delta}_2 + \nu_3$$

and

$$2(0, -) + 3 = \nu_2^0 - \bar{\delta}_2 + \nu_3$$

because the ν_2 ($n_2 = 1$) level is split into two levels by quadratic vibronic coupling. The width of the (2 + 3) band of RuF_6 at half absorbance is 58 cm^{-1}, and, when the value of 29 cm^{-1} is subtracted for the normal (2 + 3) band half-width, the value of 29 cm^{-1} is obtained for the excess band width. This excess band width is assigned as $2\bar{\delta}_2$. The value of $\bar{\delta}_2 = 14.5$ cm^{-1} obtained in this way is in poor agreement with the value of 76 cm^{-1} which was calculated in Section III. We cannot explain this discrepancy.

There is, however, a further complication of the (2 + 3) transition that arises from thermal population of the ν_5 ($n_5 = 1$) state. We shall not treat this matter in detail because it is not possible to derive any related features from the spectra. However, some qualitative comment is worthwhile. For $J' = 1$, there are three vibronic energy levels derived for $n_2 = 0$, $n_5 = 1$ and eight levels for $n_2 = 1$, $n_5 = 1$. Thus, many more ν_2 transitions can arise involving these levels than from those that involve $n_2 = 0$, $n_5 = 0$ and $n_2 = 1$, $n_5 = 0$. The ν_2 transitions involving the $n_5 = 1$ states may thus add significantly to the diffuse character of the (2 + 3) transition, and may account for the weak intensity of the (2 + 3) transition. If spectra could be obtained at lower temperatures, where the $n_5 = 1$ levels are not significantly populated, the (2 + 3) band might be more readily detected in the spectrum.

The f_{2g} vibrational mode

Study of the ν_5 transitions should also permit us to decide whether the E_g or the F_{2g} electronic states lie lower. For RuF_6, only one transition involving ν_5 is observed, a broad diffuse band at 997 cm^{-1} assigned as (3 + 5).[22] For a F_{2g} electronic ground state, the ν_5 transitions from the $n_5 = 0$ state to the $n_5 = 1$ state are $\nu_5^0 + \bar{\Delta}_5$, $\nu_5^0 - \bar{\Delta}_5$, and $\nu_5^0 - 2\bar{\Delta}_5$. Thermal population of $n_5 = 1$ states allows transitions to the $n_5 = 2$ states and further broadens the bands involving ν_5. The unshifted value, ν_5^0, derived from the systematics for RuF_6 is 283 cm^{-1}. If we assume that the (3 + 5) band is centered near $[3 + 5(\bar{1})]$, we derive for $5(\bar{1})$ the value 262 cm^{-1}, and for $\bar{\Delta}_5$ the value 21 cm^{-1}. In Section III, the theoretical value of $\bar{\Delta}_5$ was calculated to be 23 cm^{-1}, in good agreement with the value derived from the spectrum. For an E_g electronic ground state, a linear Jahn–Teller effect involving ν_5 is excluded by symmetry, resulting in a value for ν_5 close to ν_5^0 and relatively sharp spectral features for ν_5 transitions. Neither of these characteristics appears in the experimental spectra, so that the E_g electronic states are probably not the lower lying ones.

Other transitions

Further examination of the spectra of Weinstock *et al.* for RuF_6 shows a number of other very, very weak absorptions near values predicted for binary bands. However, in agreement with those authors, we believe that assignments for these bands from the present data would be extremely doubtful. It is our hope that the present analysis will serve as a stimulus to further work on this extremely difficult experimental problem, and that as a result additional data on RuF_6 will be forthcoming. The value of ν_6 for RuF_6 is estimated to be 186 cm^{-1} from the systematics. The experimental data and the assignments for RuF_6 are summarized in Table XXV.

Summary

The infrared spectrum of RuF_6 completes the picture for vibronic coupling in the metal hexafluoride molecules. Fewer transitions can be assigned in this spectrum than in that of any other molecule that we have considered. Of particular note is the lack of intensity of the (2 + 3) transition, which appears as a prominent band in the infrared spectra of all the other hexafluorides, including the other three vibronic molecules.

We have been able to deduce preliminary evidence for the (2 + 3) transition in the spectra of Weinstock *et al.*[22] From this we concluded that the lowest lying electronic states for RuF_6 ($4d^2$) are best described as $J' = 1$ (F_{2g}) rather than either as E_g or as $J' = 2$ (E_g and F_{2g}). The greater splitting of the E_g and F_{2g} electronic states in RuF_6 than in OsF_6 was inferred from preliminary electronic spectra for RuF_6 (Section III).

TABLE XXV. RuF_6

Mode	Assigned value	(WCC)[22]	This paper	Intensity
1	(675)	(675)		
2(0, −)	(613)			
2°	(624)	(573)[a]	(624)[a]	
2(0, +)	(635)			
3	735	735		
4	255	275		
$5(\bar{1})$	(278)	(262)[a]		
5°	(283)			
6	(186)			
$\bar{\Delta}_2$	(79)			
$\bar{\delta}_2$	(76)			
$\bar{\bar{\delta}}_2$	(52)			
$\bar{\Delta}_5$	21			
1 + 3	1410	1410		m
2(0) + 3	1359		1359[a]	v w
$3 + 5(\bar{1})$	993	997[a]		w

[a] These assignments are discussed in the text.

V. FORCE CONSTANTS AND THERMODYNAMIC PROPERTIES

In this last section we report force constants and vibronic modifications of the thermodynamic functions for metal hexafluoride molecules. The values given in this section are based on the spectral assignments discussed in Sections II and IV above.

A. Force Constants

In order to understand the meaning of vibrational frequencies in terms of the strength of chemical bonding and of nonbonded interactions, a normal mode analysis is necessary for the molecule of interest and its vibrational force constants must be deduced. For hexafluoride molecules there are only six observed vibrational frequencies: ν_j^{obs}, for $j = 1$ through 6. Most generally, for these molecules, however, there are seven independent force constants. We write these seven quantities as A, B, C, D, E, F, and G. In terms of the force constants defined by Claassen:[85]

f_d, the bond stretching constant;
f_α, the angle bending constant;

and the interaction constants,

f_{dd}, bond with a bond at right angles to it,
$f_{\alpha\alpha}$, angle with an angle adjacent and in the same plane,
$f_{d\alpha}$, angle with one of the bonds forming its sides,
δ, bond with an opposite bond,
γ, angle with a bond perpendicular to the plane but non-adjacent,
ρ, angle with a bond in its plane but not forming one of its sides,
κ, angle with an angle in a perpendicular plane, but with a bond in common,
ε, angle with an angle in a perpendicular plane, but with no bond in common;

the seven quantities that we use are given as follows:

$$\begin{aligned}
A &= f_d \\
B &= \delta \\
C &= 2f_{dd} \\
D &= 2(f_{d\alpha} - \rho) \\
E &= f_\alpha - \gamma \\
F &= 2(f_{\alpha\alpha} - \gamma) \\
G &= 2(\kappa - \varepsilon).
\end{aligned}$$

In order to give unambiguous values for these seven independent quantities, at least seven pieces of information are required; the observed vibrational frequencies for one isotopic type of molecule provide only six pieces of information. In the absence of isotopic data, Claassen suggested that it is reasonable to evaluate the force constants for a hexafluoride molecule based on the following assumption: the value of $|\kappa - \varepsilon| = |G/2|$ should be minimal with respect to variation of the quantity $f_{d\alpha} - \rho = D/2$. Claassen used a graphical method for implementing his assumption. We have solved for the force constants by using the equation $\partial G/\partial D = 0$ together with the six observed frequencies. This derivative condition reduces to the equation $2(f_{d\alpha} - \rho)/(f_{\alpha} - \gamma + 2\kappa - 2\varepsilon) = 4m/(M + 2m)$, where M and m are the masses of the metal atom and the fluorine atom, respectively.

The normal coordinates that result from this relationship between the force constants and masses are rather interesting: for the higher frequency f_{1u} mode, only the metal atom and the axial fluorine atoms move—the equatorial fluorine atoms do not move; while for the lower frequency f_{1u} mode, only the bond angles change—there is no change of metal–fluorine distances. This type of displacement coordinate was shown in Fig. 1 (Section II) for NpF_6, where $4m/(M + 2m) = 0.2763$.

In listing the force constants deduced in the preceding manner for hexafluoride molecules we have not used the traditional units of m dyne/A, thousandths of a dyne per Ångstrom displacement. Instead, we have used what seem the more natural units of cm^{-1}/cA^2, wave numbers per hundredths of an Ångstrom displacement squared.

$1\ cm^{-1}/cA^2 = 10^{15}$ hc m dyne/A $= 0.196276$ m dyne/A and 1 m dyne/A $= 5.09487\ cm^{-1}/cA^2$. The root-mean-square displacement for a vibrational energy level is especially easy to obtain from a knowledge of the cm^{-1}/cA^2 value of the force constant, k. If ν is the energy of this vibrational energy level in cm^{-1}, the root-mean-square displacement is $10^{-2}(\nu/k)^{\frac{1}{2}}$ Ångstrom.

The fundamental vibration frequencies used for the evaluation of the force constants given in Table XXVII are taken from the tables for the individual molecules given in Sections II and IV. These values are summarized in Table XXVI. In this table, the frequencies given for ν_2 and ν_5 for ReF_6, TcF_6, and OsF_6 are the

TABLE XXVI. Fundamental Vibration Frequencies of Hexafluoride Molecules, cm^{-1}

	$\nu_1(1)$ a_{1g}(R)	$\nu_2(2)$ e_g(R)	$\nu_3(3)$ f_{1u}(IR)	$\nu_4(3)$ f_{1u}(IR)	$\nu_5(3)$ f_{2g}(R)	$\nu_6(3)$ f_{2u}(not)
SF_6	770	640	939	614	522	(349)
SeF_6	708	(661)	780	437	(403)	(262)
TeF_6	701	674	752	325	313	(195)
(PoF_6)	[700]	[682]	[747]	[253]	[243]	[144]
MoF_6	741	643	741	262	(312)	(122)
TcF_6	(712)	(639)[a]	748	265	(297)[a]	[174]
RuF_6	(675)	(624)	735	275	(283)[a]	[186]
RhF_6	(634)	(592)	724	283	(269)	(189)
(PdF_6)	[590]	[525]	[711]	[280]	[258]	[191]
WF_6	(771)	(673)	711	258	(315)	(134)
ReF_6	755	(671)[a]	715	257	(295)[a]	(193)
OsF_6	(733)	(668)[a]	720	272	(276)[a]	[205]
IrF_6	(701)	(646)	719	276	(258)	(206)
PtF_6	(655)	(600)	705	273	(242)	(211)
UF_6	667	535	624	(184)	(201)	(140)
NpF_6	(648)	(528)	624	(198)	(205)	(165)
PuF_6	(628)	(523)	616	(203)	(211)	(173)
(AmF_6)	[609]	[500]	[603]	[205]	[216]	[178]
CrF_6	(720)	(650)	790	[266]	[309]	[110]

values of ν_2^0 and ν_5^0 taken from the systematics and they are marked with the superscript *a*. For RuF_6, ν_5^0 was also obtained from the systematics, but ν_2^0 was derived from the infrared spectrum. Frequencies that have not been observed directly, but are derived from combination bands, are placed in parentheses. Square brackets are used for frequencies that were estimated from systematics. The compounds PoF_6, PdF_6, and AmF_6 are parenthesized because the evidence for the existence of PoF_6 is rather tenuous and because PdF_6 and AmF_6 have not been prepared.

TABLE XXVII. Force Constants for Hexafluoride Molecules in cm^{-1}/cA^2, Based on the Assumption that $D/(E + G) = 4m/(M + 2m)$

	SF_6	SeF_6	MoF_6	TcF_6	RuF_6	RhF_6	TeF_6	WF_6
A	28.163	25.663	24.673	24.483	23.304	21.770	26.032	26.480
B	−1.125	0.666	1.673	0.860	0.331	−0.654	0.773	2.246
C	3.580	1.232	2.597	1.888	1.268	0.986	0.711	2.709
D	5.634	2.160	0.713	0.720	0.767	0.806	0.499	0.488
E	4.346	2.648	0.843	1.083	1.199	1.260	1.090	0.970
F	0.433	0.316	−0.555	−0.183	0.050	0.221	−0.317	−0.455
G	0.848	0.676	0.415	0.214	0.205	0.234	−0.002	0.454

	ReF_6	OsF_6	IrF_6	PtF_6	UF_6	NpF_6	PuF_6
A	26.353	26.185	25.181	23.033	19.412	19.029	18.451
B	1.802	1.190	0.209	−1.033	0.066	−0.314	−0.425
C	2.293	1.744	1.418	1.322	3.038	2.702	2.314
D	0.480	0.531	0.543	0.526	0.210	0.244	0.255
E	1.243	1.400	1.432	1.446	0.663	0.832	0.894
F	−0.006	0.307	0.476	0.605	0.083	0.229	0.255
G	0.173	0.194	0.213	0.168	0.100	0.050	0.035

B. Thermodynamic Properties

Calculations of the thermodynamic properties have been reported for a number of the hexafluoride molecules in the vapor state. These calculations will have to be modified for the molecules in which the fundamental vibrational frequencies used are different from those given in Table XXVI. The most significant changes result for MoF_6 and WF_6, for which a large change in the frequency ν_6 results from our re-analysis of the spectra (see Section II).

The thermodynamic properties of the molecules in the vapor state are usually determined experimentally from heat-capacity, heat-of-vaporization, and equation-of-state measurements. The overall correctness of the calculated values are then checked by comparing the calculated entropy for the ideal gas at one atmosphere pressure, S^0, with that derived from the experimental measurements at a number of temperatures. Excellent agreement has been obtained between the calculated and experimentally determined entropies in all cases that have been examined. The results for UF_6, NpF_6, WF_6, and MoF_6 were discussed in Section II. For the abnormal hexafluoride molecules, heat-capacity measurements have been completed for ReF_6,[54] TcF_6,[87] and OsF_6,[54] but the results are not presently available. Moreover, the calculation of the thermodynamic properties for these molecules, in which vibronic coupling is important, presents a rather unusual problem.

The vibrational and rotational partition function is usually evaluated with two simplifications: anharmonicity and the interaction between vibration and rotation are neglected. The partition function for vibration and rotation can then be factored into a product of terms, one for each of the normal vibrational modes and one for rotation. Each of these terms can be evaluated separately. The electronic degeneracy is included by multiplying the partition function by the degeneracy.

We shall confine ourselves in this article to the evaluation of the effect of vibronic coupling on the thermodynamic properties for the case $J' = \frac{3}{2}$. The influence of this vibronic coupling can be evaluated as follows.

Let $Q(\text{HYP})$ be the partition function of a hypothetical molecule with vanishingly small vibronic coupling. This hypothetical molecule is assumed to have the frequencies of vibration listed in Table XXVI for one of the molecules with a fourfold degenerate electronic multiplet lowest in energy. The function $Q(\text{HYP})$ can be constructed (as explained, for example, by Andrews[88]) by multiplying the usual product of translational, rotational and vibrational functions by a factor of 4 to account for the electronic degeneracy. The influence of vibronic coupling on the thermodynamic properties of the real molecule with vibronic coupling can be assessed in terms of the ratio r between the actual partition function for this molecule, $Q(\text{VIB})$, say, and the hypothetical one:

$$r = Q(\text{VIB})/Q(\text{HYP})$$

To a good approximation this ratio r depends only on the Jahn–Teller active vibrational modes and the electronic degeneracy factor—the other terms in the partition functions cancel out. For the hypothetical molecule, the important factor of the partition function is $Q_{e,2,5}(\mathrm{HYP})$:

$$Q_{e,2,5}(\mathrm{HYP}) = 4[1 - \exp(-2u_2)]^{-2}[1 - \exp(-2u_5)]^{-3}$$

where $u_2 = hc\nu_2^0/2kT$ and $u_5 = hc\nu_5^0/2kT$ and h, c, k, and T have their usual meanings. The convention of measuring the energy of the levels from that of the lowest level has been adopted.

For the vibronic molecule the corresponding factor $Q_{e,2,5}(\mathrm{VIB})$ is more complicated and depends on $X_2 = hc\Delta_2/2kT$, $x_2 = hc\delta_2/2kT$ and $X_5 = hc\Delta_5/2kT$ as well as on u_2 and u_5. The vibronic splitting parameters Δ_2, δ_2 and Δ_5 were defined and their assigned values are given in Section III for ReF_6, IrF_6, TcF_6, and RhF_6. The ratio r has the following form correct to second order in X_2 and x_2:

$$r = Q_{e,2,5}(\mathrm{VIB})/Q_{e,2,5}(\mathrm{HYP}) \simeq r_1 r_2 r_3 r_4,$$

where

$$r_1 = \tfrac{1}{2}\exp(4X_2 - 2u_2)[\exp(-2X_2 - 3X_5)],$$

$$r_2 = 2 + 4\exp(u_2 - X_2)[\sinh u_2]^2[\sinh(u_2 + X_2)]^{-1} + [2X_2{}^2 + x_2{}^2\exp(2u_2)][\sinh u_2]^{-2},$$

$$r_3 = \tfrac{1}{2}\exp X_5\,[\sinh u_5][\tanh u_5],$$

and

$$r_4 = \exp(-u_5)[\sinh(u_5 + X_5)]^{-2} + \exp u_5[\sinh(u_5 - X_5)]^{-2}.$$

The difference between the thermodynamic properties of the actual abnormal molecules ($J' = \frac{3}{2}$) and those of the hypothetical molecules can be obtained from r, viz:

$$-\Delta[G - E_0]/T = R\ln r$$

$$\Delta[H - E_0] = RT^2\,\partial\ln r/\partial T$$

$$\Delta S = R\,\partial(T\ln r)/\partial T$$

The results for ReF_6 and TcF_6 at three temperatures are summarized in Table XXVIII. The thermodynamic properties of ReF_6 and TcF_6 calculated for the hypothetical molecules must then be corrected by these amounts. There is also a contribution from the change of zero-point energy for these molecules:

$$\Delta E_0 = -hc(\Delta_2 + \tfrac{3}{2}\Delta_5).$$

For ReF_6, this amounts to -442 cal mole^{-1}, and for TcF_6 to -432 cal mole^{-1}. To obtain values for ΔG and ΔH, these amounts must be appropriately added to the values given in Table XXVIII.

The preceding formulas are also appropriate to IrF_6 and to RhF_6. As might be expected, effects for these molecules are also negligible, e.g., all the corresponding entries for IrF_6 and RhF_6 in Table XXVIII would be zero. The value of ΔE_0 is -4.6 cal mole^{-1} for IrF_6 and -0.004 cal mole^{-1} for RhF_6.

TABLE XXVIII. Vibronic Effects on Thermodynamic Properties

T, °K	$-\Delta(G - E_0)/T$, cal deg^{-1} mole^{-1}	$\Delta(H - E_0)$, cal mole^{-1}	ΔS, cal deg^{-1} mole^{-1}
ReF_6			
200	0.103	18.5	0.196
298.15	0.136	20.2	0.203
900	0.175	12.4	0.189
TcF_6			
200	0.087	17.1	0.172
298.15	0.117	19.1	0.181
900	0.154	11.1	0.166

The value of ΔS for ReF_6 and TcF_6 is nearly constant over the range of temperatures considered in Table XXVIII. The magnitude of ΔS of about 0.2 cal deg^{-1} mole^{-1} is unfortunately very close to the uncertainties in the calculated values of S^0 and in the values of S^0 obtained in careful calorimetric measurements. At best, S^0 is determined experimentally with an uncertainty of a little better than 0.1 cal deg^{-1} mole^{-1}. An uncertainty of 1–2 cm^{-1} in the vibrational frequencies would introduce an uncertainty of similar magnitude in the calculated value of S^0. Thus, for the present time, the hope of an experimental verification of the effect of vibronic coupling on the entropy of these molecules is not very promising.

We have not undertaken similar calculations for OsF_6 ($J' = 2$) and for RuF_6 ($J' = 1$). For these molecules the energy levels involving ν_2 and ν_5 are much more complex than those for the

$J' = \frac{3}{2}$ ones. From the relative magnitudes of their vibronic parameters, we presume that the effects on the thermodynamic properties for OsF_6 and RuF_6 would be of the same magnitude as those for ReF_6 and TcF_6. Thus, fair estimates for the thermodynamic properties of OsF_6 and RuF_6 can be obtained in the usual way by using the frequencies in Table XXVI.

References

1. Born, M., and Oppenheimer, R., *Ann. Physik* [4] **84,** 457 (1927).
2. Herzberg, G., and Teller, E., *Z. Physik. Chem. Leipzig* **21,** 410 (1933).
3. Renner, E., *Z. Physik* **92,** 172 (1934).
4. Dressler, K., and Ramsay, D. A., *J. Chem. Phys.* **27,** 971 (1957).
5. Pople, J. A., and Longuet-Higgins, H. C., *Mol. Phys.* **1,** 372 (1958).
6. Dressler, K., and Ramsay, D. A., *Phil. Trans. Roy. Soc. London* **A251,** 553 (1959).
7. Longuet-Higgins, H. C., *Advan. Spectr.* **2,** 429 (1961).
8. Moffitt, W., and Liehr, A. D., *Phys. Rev.* **106,** 1195 (1957).
9. Moffitt, W., and Thorson, W., *Phys. Rev.* **108,** 1251 (1957).
10. Longuet-Higgins, H. C., Öpik, U., Pryce, M. H. L., and Sack, R. A., *Proc. Roy. Soc. London* **A244,** 1 (1958).
11. Child, M. S., and Longuet-Higgins, H. C., *Phil. Trans. Roy. Soc. London* **A254,** 259 (1961).
12. Jahn, H. A., and Teller, E., *Proc. Roy. Soc. London* **A161,** 220 (1937).
13. Jahn, H. A., *Proc. Roy. Soc. London* **A164,** 117 (1938).
14. Van Vleck, J. H., *J. Chem. Phys.* **7,** 61 (1939); **7,** 72 (1939).
15. Moffitt, W., Goodman, G. L., Fred, M., and Weinstock, B., *Mol. Phys.* **2,** 109 (1959).
16. Weinstock, B., and Chernick, C. L., *J. Am. Chem. Soc.* **82,** 4116 (1960).
17. Weaver, E. E., Weinstock, B., and Knop, C. P., *J. Am. Chem. Soc.* **85,** 111 (1963;) Malm, J. G., Sheft, I., and Chernick, C. L., *ibid.* **85,** 110 (1963); Slivnik, J., Brcic, B., Volavsek, B., Marsel, J., Vrscaj, V., Smalc, A., Frlec, B., and Zemljic, Z., *Croat. Chem. Acta* **34,** 187 (1962); Dudley, F. B., Gard, G., and Cady, G. H., *Inorg. Chem.* **2,** 228, (1963).
18. Glemser, O., *Symposium on Inorganic Fluorine Chemistry,* Argonne National Laboratory, Sept. 1963.
19. Weinstock, B., and Claassen, H. H., *J. Chem. Phys.* **31,** 262 (1959).
20. Weinstock, B., Claassen, H. H., and Malm, J. G., *J. Chem. Phys.* **32,** 181 (1960).
21. Claassen, H. H., Selig, H., and Malm, J. G., *J. Chem. Phys.* **36,** 2888 (1962).
22. Weinstock, B., Claassen, H. H., and Chernick, C. L., *J. Chem. Phys.* **38,** 1470 (1963).
23. Herzberg, G., *Infrared and Raman Spectra of Polyatomic Molecules,* Van Nostrand, New York, 1945.

24. Malm, J. G., Weinstock, B., and Claassen, H. H., *J. Chem. Phys.* **23,** 2192 (1955).
25. Yost, D. M., Steffens, C. C., and Gross, S. T., *J. Chem. Phys.* **2,** 311 (1934).
26. Euken, A., Ahrens, H., Bartholomé, E., and Bewilogua, L., *Z. Physik. Chem.* **B26,** 297 (1934).
27. Brockway, L. O., and Pauling, L., *Proc. Natl. Acad. Sci. U.S.* **19,** 68 (1933).
28. Braune, H., and Knoke, S., *Z. Physik. Chem. Leipzig* **21,** 297 (1933),
29. Redlich, O., Kurz, T., and Rosenfeld, P., *Z. Physik. Chem. Leipzig* **19,** 231 (1932).
30. Sachsse, H., and Bartholomé, E., *Z. Physik. Chem. Leipzig* **28,** 257 (1937).
31. Lagemann, R. T., and Jones, E. A., *J. Chem. Phys.* **19,** 534 (1951).
32. Gaunt, J., *Trans. Faraday Soc.* **49,** 1122 (1953).
33. Gaunt, J., *Trans. Faraday Soc.* **50,** 893 (1954).
34. Gruen, D. M., Malm, J. G., and Weinstock, B., *J. Chem. Phys.* **24,** 905 (1956).
35. Goodman, G. L., and Fred, M., *J. Chem. Phys.* **30,** 849 (1959); Hutchison Jr., C. A., and Weinstock, B., *J. Chem. Phys.* **32,** 56 (1960).
36. Tsujimura, S., Cohen, D., Chernick, C. L., and Weinstock, B., *J. Inorg. Nucl. Chem.* **25,** 226 (1963).
37. Bigeleisen, J., Mayer, M. G., Stevenson, P. C., and Turkevich, J., *J. Chem. Phys.* **16,** 442 (1948).
38. Braune, H., and Pinnow, P., *Z. Physik* **B35,** 239 (1937).
39. Bauer, S. H., *J. Chem. Phys.* **18,** 27 (1950).
40. Schomaker, V., and Glauber, R., *Nature* **170,** 290 (1952); Glauber, R., and Schomaker, V., *Phys. Rev.* **91,** 1182 (1953).
41. Burke, T. G., Smith, D. F., and Nielsen, A. H., *J. Chem. Phys.* **20,** 447 (1952).
42. Hawkins, N. J., Mattraw, H. C., and Carpenter, D. R., *Report No.* 1041 (Feb. 1954), Knolls Atomic Power Laboratory, Schenectady, New York.
43. Claassen, H. H., Weinstock, B., and Malm, J. G., *J. Chem. Phys.* **25**, 426 (1956).
44. Claassen, H. H. Private communication of unpublished work.
45. Hawkins, N. J., Mattraw, H. C., and Sabol, W. W., *J. Chem. Phys.* **23**, 2191 (1955).
46. Fermi, E., *Z. Physik* **71,** 250 (1931).
47. Weinstock, B., *Record Chem. Progr. Kresge-Hooker Sci. Lib.* **23,** 23 (1962).
48. Osborne, D. W., Weinstock, B., and Burns, J. H., Abstract, American Chemical Society Meeting, Division of Physical Chemistry ,April 1963, Los Angeles, California.
49. Claassen, H. H., and Weinstock, B., *J. Chem. Phys.* **33,** 436 (1960).
50. Child, M. S., *Mol. Phys.* **3,** 605 (1960).
51. Tanner, K. N., and Duncan, A. B. F., *J. Am. Chem. Soc.* **73,** 1164 (1951).

52. Weinstock, B., and Claassen, H. H. Unpublished data.
53. Mattraw, H. C., Hawkins, N. J., Carpenter, D. R., and Sabol, W. W., *J. Chem. Phys.* **23,** 985 (1955).
54. Westrum Jr., E. F., and Weinstock, B. To be published.
55. Gaunt, J., *Trans. Faraday Soc.* **50,** 209 (1954).
56. Chernick, C. L., Claassen, H. H., and Weinstock, B., *J. Am. Chem. Soc.* **83,** 3165 (1961).
57. Osborne, D. W., Schreiner, F., Malm, J. G., Selig, H., and Rochester, L., Abstracts of the N.Y. meeting of the American Chemical Society, Sept. 1963.
58. Runciman, W. A., and Schroeder, K. A., *Proc. Roy. Soc. London* **A265,** 489 (1962).
59. Griffith, J. S., *Theory of Transition Metal Ions,* Cambridge University Press, Cambridge, England, 1961.
60. Selig, H., Chernick, C. L., and Malm, J. G., *J. Inorg. Nucl. Chem.* **19,** 377 (1961).
61. Claassen, H. H., Selig, H., Malm, J. G., Chernick, C. L., and Weinstock, B., *J. Am. Chem. Soc.* **83,** 2390 (1961).
62. Schomaker, V., Weinstock, B., Smith, D. W., and Kimura, M. Unpublished work.
63. Moffitt, W., and Thorson, W., *Calcul des Fonctions d'Onde Moleculaire,* Daudel, R., Ed., Centre Nationale de la Recherche Scientifique, Paris, 1958, p. 141.
64. Thorson, W., *J. Chem. Phys.* **29,** 938 (1958).
65. Child, M. S., *Phil. Trans. Roy. Soc. London* **A255,** 31 (1962); *J. Mol. Spectr.* **10,** 357 (1963).
66. Liehr, A. D., *Ann. Rev. Phys. Chem.* **13,** 41 (1962).
67. Hougen, J. T., *J. Chem. Phys.* **37,** 1433 (1962).
68. Longuet-Higgins, H. C., *Mol. Phys.* **6,** 445 (1963).
69. Wigner, E., *Group Theory,* English Translation, Academic Press, New York, 1959.
70. Longuet-Higgins, H. C., and Goodman, G. L. Unpublished work.
71. Ballhausen, C. J., *Introduction to Ligand Field Theory,* McGraw-Hill, New York, 1962.
72. Tanabe, Y., and Sugano, S., *J. Phys. Soc. Japan* **9,** 753 (1954).
73. Wigner, E. *Gruppentheorie,* Vieweg-Verlag, Brunswick, 1931, p. 264.
74. Eckart, C., *Rev. Mod. Phys.* **2,** 305 (1930).
75. Edmonds, A. R., *Angular Momentum in Quantum Mechanics,* Princeton University Press, Princeton, New Jersey, 1957.
76. Rotenberg, M., Bivins, R., Metropolis, N., and Wooten, J. K. Jr., *the* 3-*j and* 6-*j symbols,* The Technology Press, M. I. T., Cambridge, Mass., 1959.
77. Racah, G., *Phys. Rev.* **62,** 448 (1942); *ibid.* **63,** 367 (1943).
78. Goodman, G. L. Unpublished results.
79. Kemble, E. C., *The Fundamental Principles of Quantum Mechanics,* McGraw-Hill, New York, 1937.
80. Malm, J. G., and Selig, H., *J. Inorg. Nucl. Chem.* **20,** 189 (1961).

81. Claassen, H. H., Malm, J. G., and Selig, H., *J. Chem. Phys.* **36,** 2890 (1962).
82. Ruff, O., and Tschirch, F. W., *Chem. Ber.* **46,** 929 (1913).
83. Weinstock, B., and Malm, J. G., *J. Am. Chem. Soc.* **80,** 4466 (1958).
84. *Noble Gas Compounds*, Hyman, H. H., Ed., The University of Chicago Press, Chicago and London, 1963.
85. Claassen, H. H., *J. Chem. Phys.* **30,** 968 (1959).
86. Gullikson, C. W., Nielsen, J. Rud, and Stair, A. T., Jr., *J. Mol. Spectr.* **1,** 151 (1957).
87. Osborne, D. W. Private communication.
88. Andrews, Frank C., *Equilibrium Statistical Mechanics,* John Wiley New York, 1963, p. 101.

ELECTRONIC CORRELATION IN ATOMS AND MOLECULES*

R. K. NESBET, *IBM San Jose Research Laboratory, San Jose, California*

CONTENTS

I. INTRODUCTION

This article will discuss the present status of the theory of electronic correlation in atoms and molecules. Only isolated stationary states will be considered, so excitation energies and transition probabilities will not be discussed. Effects due to degeneracy or near degeneracy of approximate wave functions will not be considered here. For this reason some of the special properties of the electronic continuum in metals lie outside the scope of the present paper. Nevertheless, many aspects of this argument are pertinent to the theory of the electronic properties of solids. Applications to nuclear theory could also be considered, but the detailed analysis given here is inappropriate for nuclear

* Most of the material given here was presented in a series of lectures at the Theoretical Chemistry Institute, University of Wisconsin, April, 1963.

theory unless the nuclear two-particle interaction can be represented by an integrable operator, i.e., one without an infinite repulsive core of finite extension.

The central role of the Hartree–Fock approximation in electronic correlation theory has been stressed in earlier papers.[1] Here this argument will be extended to a general statement of working hypotheses about the nature of electronic correlation, and the theoretical and empirical justification for these hypotheses will be discussed. A number of special cases in which the Hartree–Fock approximation and its relation to the correlation problem have been misunderstood in the general literature will be discussed, and it will be shown that several apparent counter-examples to the present working hypotheses arise solely from this misunderstanding.

A review article by Löwdin[2] surveys correlation theory from a point of view that differs from that taken here in its estimate of the importance of Hartree–Fock theory. A number of special methods not discussed here will be found in Löwdin's article. The present paper will emphasize the importance of Hartree–Fock theory in formulating a general theory of electronic systems in which practicable calculations can lead to results subject to experimental verification.

A particular approach to the correlation problem, based on Hartree–Fock theory, has been developed by Sinanoğlu and is reviewed in a recent article.[3] Some details of this work will be discussed here. The recent literature on formal many-particle perturbation theory has been reviewed in books by Thouless and by Kumar, each giving an extensive bibliography.[4]

The working hypotheses about electronic correlation to be considered here are stated in Section II. Sections III and IV present a summary of the Hartree–Fock theory and of correlation theory, respectively. Some of the evidence supporting the proposed working hypotheses is given in Section V. A number of apparent counter-examples that can be refuted by careful consideration of the theory presented here are discussed in Section VI.

II. WORKING HYPOTHESES

The conclusions of the argument presented here can be summarized in three working hypotheses. First, electronic properties of a non-degenerate atomic or molecular stationary state are determined primarily by the Hartree–Fock approximation in *all* circumstances. It is essential to understand the detailed implications of the Hartree–Fock approximation before an intelligent consideration of correlation effects is possible.[1] Second, except for degeneracies or near degeneracies (usually a consequence of symmetry, and dealt with by group-theoretical methods) correlation energies are additive among localized pairs of electrons. Analysis of correlation effects should start from a consideration of localized transforms of occupied Hartree–Fock orbitals.[5,6] In a closed-shell configuration, correlation energy is divided into pair correlation energies for each localized electron pair, nearly constant and insensitive to environment, and smaller interpair correlation energies equivalent to London dispersion forces.[7] The effect of structure and molecular geometry is relatively much greater for interpair correlation energies, but the latter are much smaller than the intrapair energies. The third hypothesis is that electronic correlation has a very small effect on electronic properties described by one-particle operators,* e.g., static electric moments and electric or magnetic polarizabilities, but *not* operators such as L^2 or S^2.

III. SUMMARY OF HARTREE–FOCK THEORY†

Any N-electron Schrödinger wave function can be expressed as a linear combination of the complete orthonormal set of Slater determinants constructed from all subsets of N orbitals (one-particle wave functions) ϕ_i chosen from an arbitrary complete orthonormal set. Let some determinant (normalized) from this set, denoted by

$$\Phi_0 = \det \phi_1(1) \ldots \phi_N(N) \tag{1}$$

be selected as reference state. Let the complete set of orbitals (assumed to be denumerable) be divided into two subsets, N orbitals ϕ_i, $i \leq N$, occupied in Φ_0, and orbitals ϕ_a, $a > N$, the

* A recent discussion has been given by Goodisman and Klemperer.[8] The principal theorem used to justify this hypothesis was proved in reference 9.

† The argument here follows reference 1. Various alternative forms of the Hartree–Fock approximation are discussed in the Appendix, below, where the terminology used here is defined.

remainder of the complete set, unoccupied in Φ_0. Then a typical Slater determinant of the complete set of determinants is

$$\Phi_{ijk\cdots}^{abc\cdots} = \det \phi_1 \ldots \phi_a \ldots \phi_b \ldots \phi_c \ldots \phi_N \tag{2}$$

if

$$\Phi_0 = \det \phi_1 \ldots \phi_i \ldots \phi_j \ldots \phi_k \ldots \phi_N \tag{3}$$

with the convention that

$$i < j < k < \ldots \leq N < a < b < c < \ldots \tag{4}$$

It should be emphasized that the choice of orbitals and reference state is completely arbitrary. The orbitals are functions of all variables, continuous or discrete (spin or isotopic spin), that describe a single particle. Integrations over these variables imply summation over discrete variables. Many theorems which are usually proved in the context of a particular choice of orbital basis are in fact identities valid for general bases, and they will be proved here as such. The Hartree–Fock theory can be thought of as a physically motivated procedure for choosing the basis orbitals appropriate to a particular problem.

The N-particle Hamiltonian is assumed to be constructed from Hermitian one- and two-particle operators, in the form

$$H(1, \ldots, N) = \sum_i K(i) + \sum_{ij} Q(ij) \tag{5}$$

where the indices refer to particle variables. The single index i is summed over $1, \ldots, N$ and the index pair ij is summed over $i < j \leq N$.

In this determinantal basis for the N-particle problem, phase relations and matrix elements are identical with those derived in the notation of second quantization with state vectors

$$\mathbf{\Phi}_0 = \eta_1^* \ldots \eta_i^* \ldots \eta_j^* \ldots \eta_N^* \,|\, \text{vac}\rangle \tag{6}$$

$$\mathbf{\Phi}_{ij\cdots}^{ab\cdots} = \eta_1^* \ldots \eta_a^* \ldots \eta_b^* \ldots \eta_N^* \,|\, \text{vac}\rangle \tag{7}$$

$$= \eta_a^* \eta_b^* \ldots \eta_j \eta_i \mathbf{\Phi}_0 \tag{8}$$

where η_a^*, η_i are, respectively, creation and annihilation operators for particles in orbital states a and i. The Hamiltonian in second quantization is

$$\mathbf{H} = \sum_i \sum_j (i|K|j) \eta_i^* \eta_j + \tfrac{1}{2} \sum_i \sum_j \sum_k \sum_l (ij|Q|kl) \eta_i^* \eta_j^* \eta_l \eta_k \tag{9}$$

where the summations are over *orbital* indices and run over the complete orbital basis set. Dirac notation is used for the matrix elements of K and Q. The second quantization formalism is very convenient for proving general results, such as the linked-cluster expansion theorem, derived to several orders of perturbation theory by Brueckner[10] and to arbitrary order by Goldstone.[11]

The general rules for matrix elements of the many-particle Hamiltonian in a basis of Slater determinants can be derived most easily by using the fact that a normalized determinant Φ can be expressed as

$$\Phi = \det \Pi = (N!)^{-\frac{1}{2}} \mathscr{A} \Pi \tag{10}$$

where Π is a serial product,[12] as in Eq. (1), of the N occupied orbitals of Φ, and $\mathscr{A}$ is the total antisymmetrizing operator

$$\mathscr{A} = 1 - \sum P_{ij} + \sum P_{ijk} - \cdots \tag{11}$$

summed over all permutations P of the particle indices. Since

$$\mathscr{A}^2 = N!\mathscr{A} \tag{12}$$

the operator $\mathscr{A}$ is proportional to a projection operator. It follows from this, since Π is normalized to unity, that[13]

$$(\Phi_\alpha, H\Phi_\beta) = (\Pi_\alpha, H\mathscr{A}\Pi_\beta) \tag{13}$$

Matrix elements between determinants in the complete set and an arbitrary reference state Φ_0 are[14]

$$(\Phi_{ijk}^{abc}, H\Phi_0) = 0, \text{ more than two substitutions} \tag{14}$$

$$(\Phi_{ij}^{ab}, H\Phi_0) = (ab|Q(1 - P_{12})|ij) = (ab|R|ij) \tag{15}$$

$$(\Phi_i^a, H\Phi_0) = (a|K|i) + \sum_j (aj|R|ij) = (a|\mathscr{H}_0|i) \tag{16}$$

$$\begin{aligned}(\Phi_0, H\Phi_0) &= \sum_i (i|K|i) + \sum_{ij} (ij|R|ij) \\ &= \tfrac{1}{2}\sum_i [(i|K|i) + (i|\mathscr{H}_0|i)]\end{aligned} \tag{17}$$

Summations over indices $ijk \ldots$ here always denote the range $i < j < k < \ldots \leq N$. The definition of the auxiliary operator R is clear from Eq. (15). The operator $\mathscr{H}_0$ is the Hartree–Fock one-particle Hamiltonian for the reference state Φ_0,

$$\mathscr{H}_0 = K + \sum_{i(\text{occ.})} (i|R|i) \tag{18}$$

All matrix elements in the complete determinantal basis can be deduced from these formulas and expressed in terms of R and $\mathscr{H}_0$. For an arbitrary determinant, matrix elements $(\Phi_\nu, H\Phi_\mu)$ are expressed at first in terms of the operator

$$\mathscr{H}_\mu = K + \sum_{i(\mu)} (i(\mu)|R|i(\mu)) \tag{19}$$

where $i(\mu)$ denotes an orbital occupied in Φ_μ. Then, since $\mathscr{H}_\mu - \mathscr{H}_0$ is a sum of operators like $(i|R|i)$, the result can be written immediately in terms of $\mathscr{H}_0$ and R.

A useful formal result that follows immediately from this analysis is that diagonal energy differences between Slater determinants can be expressed entirely in terms of matrix elements that involve only the orbitals occupied in one determinant but not in the other. If diagonal elements are denoted by

$$D_{ijk\ldots}^{abc\cdots} = (\Phi_{ijk\ldots}^{abc\cdots}, H\Phi_{ijk\ldots}^{abc\cdots}) \tag{20}$$

then

$$\begin{aligned} D_{ijk\ldots}^{abc\cdots} - D_0 = {} & (a|\mathscr{H}_0|a) + (b|\mathscr{H}_0|b) + \ldots - (i|\mathscr{H}_0|i) \\ & - (j|\mathscr{H}_0|j) - \ldots + (ab|R|ab) + \ldots + (ij|R|ij) + \ldots \\ & - (ai|R|ai) - \ldots \end{aligned} \tag{21}$$

Thus the energy difference is given by the obvious difference of one-particle energies, corrected by adding all interactions between pairs of new orbitals $(ab|R|ab)$, by adding all interactions between all pairs of old orbitals $(ij|R|ij)$, and by subtracting all mixed pair interactions. It should be emphasized that this formula is an identity, completely independent of the choice of reference state and orbital basis. In particular, the operator $\mathscr{H}_0$ is in general not diagonalized in this orbital basis, so $(a|\mathscr{H}_0|a)$ is a mean value not an eigenvalue. Familiar examples of Eq. (21) are

$$D_i^a - D_0 = (a|\mathscr{H}_0|a) - (i|\mathscr{H}_0|i) - (ai|R|ai) \tag{22}$$

$$\begin{aligned} D_{ij}^{ab} - D_0 = {} & (a|\mathscr{H}_0|a) + (b|\mathscr{H}_0|b) - (i|\mathscr{H}_0|i) - (j|\mathscr{H}_0|j) \\ & + (ab|R|ab) + (ij|R|ij) - (ai|R|ai) - (aj|R|aj) \\ & - (bi|R|bi) - (bj|R|bj) \end{aligned} \tag{23}$$

In atomic and molecular (and nuclear) problems, the correction terms (the R integrals here) are not at all unimportant. One of the difficulties with the usual many-particle perturbation theory[9,11] is that the zeroth-order Hamiltonian is taken to be a sum of one-particle operators, so that these R integrals are absent from the energy denominators of the theory. To anyone familiar with atomic and molecular computations, this appears to be a serious limitation. The perturbation theory ought to be formulated in terms of the corrected denominators indicated here, which are much better approximations to the actual excitation energies of a many-electron system. The linked-cluster expansion as derived by Goldstone[11] cannot be used with these denominators. In a recent calculation on the Be atom, apparently the first application of the Goldstone formalism to an atomic problem, it was found necessary to introduce these corrected denominators by summing infinite series of diagrams.[15]

Among the identities indicated by Eq. (21) are the well-known results

$$D^a - D_0 = (a|\mathscr{H}_0|a); \quad D_i - D_0 = -(i|\mathscr{H}_0|i) \tag{24}$$

sometimes referred to (misleadingly) as Koopmans' theorem. The contribution of Koopmans' paper[16] was not these identities, which have no physical content, but a theorem to the effect that if the reference state Φ_0 has stationary energy and if the orbitals ϕ_a or ϕ_i are eigenfunctions of the operator $\mathscr{H}_0$, then the determinants Φ^a, Φ_i formed from Φ_0 by adding ϕ_a to the occupied set or by removing ϕ_i, respectively, also have stationary energy with respect to further variations involving these orbitals. It follows that all these determinants are approximations to eigenfunctions of the $N + 1$, N, and $N - 1$ particle systems, and the energy differences given by the identities of Eq. (24) can be interpreted as electron affinities or ionization potentials, respectively.

To put some physical content into the formal results derived above, consider a stationary state eigenfunction Ψ of the N-particle Hamiltonian and expand it in the complete set of Slater determinants indicated above. It is convenient to pick Φ_0 to be the determinant of greatest weight in this expansion, and to relabel the orbitals and determinants accordingly. If the coefficient of Φ_0 is set equal to unity, thus normalizing in the sense that

$$(\Phi_0, \Psi) = 1 \tag{25}$$

then

$$\Psi = \Phi_0 + \sum_{\mu}{}' \Phi_\mu c_\mu \tag{26}$$

where μ denotes the sets of indices in Eq. (2). The sum excludes $\mu = 0$. The advantage of not fixing an orbital basis in advance is clear here, since by construction all

$$|c_\mu| \leq 1 \tag{27}$$

with advantages from the point of view of perturbation theory. This crucial point has been overlooked in discussions of the molecular-orbital theory at large internuclear separations, and consideration of its implications leads to a much more satisfactory theory.[17]

If there are several determinants of equally large weight in the expansion of Ψ, as in some open-shell configurations or in some systems with a very large number of particles, this corresponds to a situation of near degeneracy in the perturbation theory, which lies outside the scope of the present article. Generally the matrix of the Hamiltonian must be diagonalized over the set of nearly degenerate determinants, to determine a modified reference state expressed as a linear combination of determinants.

For a stationary state,

$$H\Psi = E\Psi \tag{28}$$

so that by Eqs. (15) and (16), taking the Φ_0 component of Eq. (28),

$$E = (\Phi_0, H\Phi_0) + \sum_{\mu}{}'(\Phi_0, H\Phi_\mu)c_\mu \tag{29}$$

$$= D_0 + \sum_i\sum_a (i|\mathscr{H}_0|a)c_i^a + \sum_{ij}\sum_{ab}(ij|R|ab)c_{ij}^{ab} \tag{30}$$

If H contains only one- and two-particle operators, Eq. (30) is an exact result, independent of the orbital basis. If Φ_0 is chosen carefully enough that Eq. (27) holds, this is already a very useful formula. As will be shown below, the one-particle terms can be removed completely by an appropriate choice of the basis orbitals. The matrix elements $(ij|R|ab)$ are inherently small quantities, double-exchange integrals over four mutually orthogonal orbitals.[1]

At this point it is convenient to choose basis orbitals that simplify Eq. (30) as much as possible. The one-particle terms can be removed either by requiring all the coefficients c_i^a to vanish, which will be discussed below, or by requiring that

$$(i|\mathscr{H}_0|a) = 0, \quad i \leq N < a \tag{31}$$

This equation, which will be called the "Hartree–Fock condition", separates the complete set of orbitals (a Hilbert space) into two orthogonal subspaces. The choice of basis in either subspace is still unspecified.

It can be shown[1] that Eq. (31) is equivalent to any of the following statements:

$$\delta D_0 = 0 \tag{32}$$

for all variations of orbitals preserving normalization of the single determinant Φ_0;

$$\text{Re}\,(\Phi_i^a, H\Phi_0)d\alpha^* = 0, \quad i \leq N < a,\ d\alpha \text{ arbitrary} \tag{33}$$

$$(\Phi_i^a, H\Phi_0) = 0, \quad i \leq N < a \tag{34}$$

$$(a|\mathscr{H}_0|i) = 0, \quad i \leq N < a \tag{35}$$

$$\mathscr{H}_0\phi_i = \sum_j \phi_j \varepsilon_{ji}, \quad i \leq N \tag{36}$$

The sum over j is restricted to occupied orbitals of Φ_0, while these orbitals are determined up to a unitary transformation by any of these equivalent conditions. The statement that Eq. (32) is equivalent to Eq. (36) is the general result of the unrestricted Hartree–Fock theory.* $\mathscr{H}_0$ is invariant under unitary transformations of the occupied orbitals of Φ_0, due to its definition in Eq. (18) as a formal trace of a Hermitian quadratic form. Because $\mathscr{H}_0$ is Hermitian if the operators K and Q are, it follows that the matrix of coefficients ε_{ji} in Eq. (36) is Hermitian and can be diagonalized by a unitary transformation of the N occupied orbitals. This produces a canonical form of the Hartree–Fock equations,

$$\mathscr{H}_0\phi_i = \varepsilon_i\phi_i, \quad i \leq N \tag{37}$$

in the form of a time-independent Schrödinger equation.

An approximation to this equation was first used by Hartree,[18]

* See the Appendix, p. 358.

and later extended by Fock[19] to include exchange terms, essentially by proving the equivalence of Eqs. (32) and (36). Some additional constraints on the orbital basis were imposed for open-shell atomic configurations by Fock's original analysis.* Hartree's self-consistent field without exchange was analyzed by Gaunt[20] and Slater[21] in terms of a condition analogous to Eq. (34), but of course not satisfied exactly. This analysis was extended by Brillouin[22] and by Møller and Plesset[9] to the equations of Fock, also derived explicitly for a single Slater determinant of unspecified shell structure by Dirac.[23] The equivalence between Eqs. (32) and (34) is usually referred to as Brillouin's theorem.

These theorems hold in the same form in a truncated orbital Hilbert space, if operators are defined by matrix elements in that space.[1] This remark is relevant to computations using finite orbital basis sets. All of the earlier papers cited here proved Brillouin's theorem under the requirement that all orbitals should satisfy Eq. (37). This is unnecessarily restrictive, since the condition given by Eq. (36) requires only that the unoccupied orbitals ϕ_a be orthogonal to the set of occupied orbitals.[1] The importance of this remark is that it makes it unnecessary to introduce an explicit continuum in atomic and molecular perturbation calculations. Criteria for the specific choice of the orbital basis will be discussed below in connection with the correlation problem.

An immediate consequence of Eq. (34) and of Eq. (14) is that in the first order of perturbation theory the wave function Ψ is approximated by a linear combination of Φ_0 and the doubly substituted determinants Φ_{ij}^{ab}. It follows from this that the mean value of any one-particle operator, like $\Sigma_i K(i)$ in Eq. (5), which has no matrix elements between Φ_0 and any Φ_{ij}^{ab}, differs from the Hartree–Fock value by second-order terms only, quadratic in the coefficients c_{ij}^{ab} estimated for the first-order wave function.[9] More explicitly, if

$$\Psi = \Phi_0 + \sum_{ij}\sum_{ab} \Phi_{ij}^{ab} c_{ij}^{ab} \tag{38}$$

$$= \Phi_0 + \sum_{\mu} \Phi_\mu c_\mu \tag{39}$$

* See the Appendix, p. 358.

such that $F_{\mu 0} = 0$ for any one-particle operator F, the mean value of F is

$$\frac{(\Psi, F\Psi)}{(\Psi, \Psi)} = F_{00} + \left[\frac{\sum_{\mu}|c_{\mu}|^2(F_{\mu\mu} - F_{00}) + \sum\sum_{\mu\nu \neq \mu} c_{\mu}^{*}c_{\nu}F_{\mu\nu}}{1 + \sum_{\mu}|c_{\mu}|^2}\right] \tag{40}$$

For many operators, unless c_{μ} is negligible the difference $F_{\mu\mu} - F_{00}$ is small and of indefinite sign, while the off-diagonal terms are always of indefinite sign, so that the correction to F_{00} can be much smaller than might be suggested by the magnitude of the coefficients c_{μ}.

This result applies only to the *unrestricted* Hartree–Fock approximation,* which for open-shell atomic configurations implies that symmetry operators such as $\mathbf{S}^2$ or $\mathbf{L}^2$ do not necessarily have precise values for the determinant Φ_0. Since these are two-particle operators, there is no conflict with the present result.

IV. CORRELATION THEORY

A. General Remarks

The many-particle wave function for fermions must be totally antisymmetric, which implies that it vanishes when two particles with identical spin and isotopic spin quantum number have identical space coordinates. This so-called Fermi hole prevents any strong interaction between electrons of the same spin, even in the Hartree–Fock approximation. However, two electrons of opposite spin in general have finite probability of being in the same volume element in the Hartree–Fock approximation. The two-particle interaction can influence the Hartree–Fock wave function only indirectly, through the mean-value integrals $(i|R|i)$ indicated in Eq. (18). A strongly repulsive or attractive two-particle potential, singular as the relative coordinates r_{ij} vanish, modifies the many-particle wave function near these singularities and introduces a correlation between each pair of particles that cannot be described by the single determinant

* See Appendix, p. 358.

wave function of the Hartree–Fock approximation. This correlation is characterized by the fact that the joint probability of finding particle i in volume element $d\tau_1$ and of finding particle j with opposite spin in volume element $d\tau_2$ is not equal to the product of the absolute probability of finding i in $d\tau_1$ and the absolute probability of finding j in $d\tau_2$, while this relationship is required by the structure of a determinantal wave function.

The Coulomb interaction between electrons leads also to a long-range correlation effect, described by London[24] as the dispersion force due to mutually induced instantaneous dipole moments. This long-range correlation is not affected by the Fermi hole, and is essentially independent of spin. A very similar effect occurs when an electron occupies an orbital that surrounds and lies outside of an inner atom or ion, as does the series electron in an alkali atom. Then the *internal* dipole potential (proportional to R^{-2}, where R is the mean radius of the series electron) due to virtual excitation of this orbital interacts with the induced dipole moment of the inner ion. In the second-order perturbation theory for the energy, the formalism is identical with that of London's theory of the dispersion force (where the interaction integral, arising from external dipole fields, varies as R^{-3}), and the energy is proportional to R^{-4} and to the polarizability of the inner atom.[25] This "polarization potential" is the dominant long-range interaction in electron scattering by a neutral atom, just as the London potential, proportional to R^{-6}, is the dominant long-range interaction between neutral atoms. These long-range correlation effects are of considerable physical importance, but they involve quite small energies and can be described satisfactorily for many purposes by second-order perturbation theory. The usual theory of these effects uses the approximation of pure dipole interactions, which could introduce much larger errors than does the use of second-order perturbation theory.[26]

The treatment of short-range correlation is a much more formidable problem. Since there are few interesting properties of atoms or molecules that depend on two-electron operators, other than the energy itself, the present argument will be primarily concerned with correlation energy. When there is no important near degeneracy, so that some single Slater determinant Φ_0 is the largest component in the true wave function, the correlation energy is most usefully defined as the difference between the exact non-relativistic energy E and the unrestricted Hartree–Fock energy D_0.* Obviously this would not be a useful definition if the occupied orbitals of Φ_0 were artificially constrained to have symmetry properties inappropriate to the electronic state Ψ in question.[17] This is why the unrestricted Hartree–Fock approximation is the most reasonable starting point for correlation theory.

With this choice of the reference state Φ_0, Eq. (35) holds, and the exact correlation energy given by Eq. (30) is[27]

$$E - D_0 = \sum_{ij} \sum_{ab} (ij|R|ab) c_{ij}^{ab} \tag{41}$$

This is formally additive over occupied pairs of orbitals, so that the correlation energy of pair ij can be defined as

$$E_{ij} = \sum_{ab} (ij|R|ab) c_{ij}^{ab} \tag{42}$$

With this formula, the theory of correlation energy reduces to the problem of estimating or computing the coefficients c_{ij}^{ab}.[27] Unless Φ_0 is inappropriately chosen, the magnitude of each of these coefficients is inherently less than or equal to unity. The unoccupied orbitals ϕ_a are not yet specified, except by the orthogonality condition

$$(\phi_a, \phi_i) = 0, \quad i \leq N < a \tag{43}$$

The formulas given above can also be written in terms of normalized two-particle antisymmetric functions

$$\psi_{ij}(1, 2) = \phi_{ij}(1, 2) + \chi_{ij}(1, 2) \tag{44a}$$

where

$$\chi_{ij}(1, 2) = \sum_{ab} \phi_{ab}(1, 2) c_{ij}^{ab} \tag{44b}$$

and

$$\begin{aligned} \phi_{ab}(1, 2) &= \det \phi_a(1)\phi_b(2) \\ \phi_{ij}(1, 2) &= \det \phi_i(1)\phi_j(2) \end{aligned} \tag{44c}$$

* See Appendix, p. 358.

if *det* includes the normalizing factor $2^{-\frac{1}{2}}$. Then Eqs. (41) and (42) are

$$E - D_0 = \sum_{ij} E_{ij} \tag{45}$$

where

$$E_{ij} = (\phi_{ij}, Q(1, 2)\chi_{ij}) \tag{46}$$

The condition given by Eq. (43) becomes

$$\int \phi_k^*(1)\chi_{ij}(1, 2)\, d\tau_1 = 0, \quad k = 1, \ldots, N \tag{47}$$

referred to as one-particle orthogonality to the occupied orbitals of Φ_0.

Because the orthogonality conditions are somewhat simpler, it is often more convenient to carry out general derivations in an arbitrary orbital basis. The results can always be transcribed into explicit equations for functions analogous to the χ_{ij} defined above. This does not imply that an orbital expansion is necessarily the most satisfactory way of computing the functions χ_{ij} or the corresponding coefficients c_{ij}^{ab}.

The simplest useful estimate of these coefficients is given by Rayleigh–Schrödinger perturbation theory, taken to first order in the wave function and to second order in the energy, where the non-diagonal part of the matrix of the many-particle Hamiltonian in a basis of Slater determinants which includes the Hartree–Fock function Φ_0 is treated as a first-order perturbation.[1] Then

$$c_{ij}^{ab} = -(ab|R|ij)/(D_{ij}^{ab} - D_0) + \cdots \tag{48}$$

with the denominator given by Eq. (21). The standard formal perturbation theory, given originally by Møller and Plesset,[9] retains the two-particle terms in Eq. (21) as diagonal elements of the first-order perturbing matrix and uses the canonical Hartree–Fock orbitals of Eq. (37), to get

$$c_{ij}^{ab} = -(ab|R|ij)/(\varepsilon_a + \varepsilon_b - \varepsilon_i - \varepsilon_j) + \cdots \tag{49}$$

For finite systems, the difference in the denominators between these two formulas is significant unless the excitation energy is rather large. If large diagonal elements are retained in the perturbing matrix they will come in to all orders of perturbation theory. For this reason Eq. (48) should lead to more rapid convergence of the theory.[1,15] The standard requirement that canonical Hartree–Fock basis orbitals be used is not necessary. Perturbation theory could equally well be developed in terms of diagonal elements $(a|\mathscr{H}_0|a)$, etc., in Eq. (49), with explicit nondiagonal elements of $\mathscr{H}_0$ retained in the formalism in higher orders.

The freedom of choice that remains for the orbital basis can be used to simplify the perturbation theory. In particular, if the occupied Hartree–Fock orbitals are transformed to a localized basis, as suggested by Lennard-Jones,[28] many of the two-electron integrals become extremely small, since they vanish exponentially or as high inverse powers of the separation distance between such orbitals. In the Hartree–Fock approximation, the condition that the sum of all exchange integrals $(ij|Q|ji)$, $i \neq j$, shall be minimized makes the classical or Hartree part of the invariant Hartree–Fock energy as large as possible, and leads to an intuitive picture of localized chemical bonds and lone-pair electrons in molecules, with classical electrostatic interactions.[29] This same kind of transformation, if also applied to the unoccupied orbitals, tends to make individual terms in the second-order correlation energy as large as possible, and should decrease the number of terms that have to be considered.[1] Criteria for localizing unoccupied orbitals have been suggested by Boys,[30] and calculations on the formaldehyde molecule[31] and on water[32] have been reported in which the use of localized orbitals both simplifies the calculations and leads to a description of the wave function in close correspondence to chemical intuition. Since any large effects of correlation are due to the singularity of the two-particle interaction as relative coordinates vanish, the use of a localized orbital basis to evaluate these effects is an intuitively very reasonable procedure.

There are three proposals for improving the Hartree–Fock approximation that deserve serious consideration, since they all start from an improvement upon second-order perturbation theory. The first two methods are variational in nature, using exact energy expressions for well defined trial wave functions. The third method, based on the Bethe–Goldstone equation,[33] is most easily derived by a selective summation of high-order terms in perturbation theory and will be discussed in some detail after the more straightforward variational methods have been described. The variational methods were both formulated in terms of localized orbitals in their original presentation, while the very extensive literature on the Bethe–Goldstone equation, based as it is on the perturbation theory that goes back to Møller and Plesset,[9] almost entirely ignores the physical significance of the use of localized orbitals as developed by Lennard-Jones and his collaborators.[28] As a consequence of this, the dependence of the total correlation energy on the number, N, of particles is treated in a most artificial manner. In particular, for an extended system the canonical Hartree–Fock orbitals are in general spread equally over all equivalent sites. Thus, all of the pair correlation energies E_{ij} of Eq. (42) are numbers of similar magnitude, and the number of them is proportional to N^2. Since the total correlation energy is generally proportional to N, it follows that the individual E_{ij} for delocalized orbitals must vary as N^{-1}, so that the individual terms become negligible as N increases while the number of important terms increases quadratically. This is obviously an awkward situation. When localized orbitals are used, E_{ij} for a localized pair refers to the correlation energy of a fixed unit, an electron pair in a chemical bond or in an atomic inner shell, and on the basis of chemical experience it can be expected to be independent of the total number of atoms in the system and hence to be independent of N. When orbitals i and j are localized on different atoms or in different chemical bonds, the correlation energy E_{ij} describes a dispersion force or van der Waals interaction[7,24] which is important only for a small number of near neighbors in a molecule or solid.

The two proposed variational methods also have characteristic weaknesses, and in concluding this section a compromise method will be proposed, which amounts to the use of the Bethe–Goldstone equation in a localized orbital basis.

B. Paired-Electron Wave Function

The first variational method to be considered here was proposed by Hurley, Lennard-Jones, and Pople[5] as a direct generalization of the work on localized orbitals cited above. The trial wave function is taken to be an antisymmetrized product of antisymmetric electron-pair functions, analogous to those of Eqs. (44),

$$\Psi = \det \psi_{12}(1, 2)\psi_{34}(3, 4) \ . \ . \ . \ \psi_{N-1,N}(N-1, N) \tag{50}$$

where det implies a normalization constant multiplying the total antisymmetrizing operator. The electron-pair functions are a generalization of doubly occupied spatial orbitals, and the assumed wave function Ψ describes the inner closed shells of an atom or the singlet ground state of a molecule. In order to derive a tractable variational formula for the total energy, the pair functions are required to satisfy the condition of "strong" orthogonality,

$$\int \psi_{ij}^*(1, 2)\psi_{kl}(1, 3)\, d\tau_1 = 0, \quad ij \neq kl \tag{51}$$

The mean value of the Hamiltonian is then derived and is found to be very similar to the Hartree–Fock energy formula, Eq. (17). Later papers discussing these functions have retained the strong orthogonality constraint and the energy formula, but have discussed some of the implications of the special form assumed for Ψ.[34] No *ab initio* calculations have been reported that use this method, although the approximate wave function for Be obtained by Watson[35] by the straightforward method of superposition of configurations has been analyzed and shown to be very closely approximated by a function of the form assumed in Eq. (50).[36]

In the papers cited above it has been shown that Eq. (51) is equivalent to the requirement that a complete orbital basis be divided into subsets with no common members, each containing a pair of orbitals ϕ_i, ϕ_j and all other orbitals that occur in the two-particle expansion of the pair function ψ_{ij}, analogous to Eqs. (44). This constraint on the pair functions is reasonable only if the orbitals ϕ_i, ϕ_j are a localized pair (spatially identical with different spin factors). The dominant part of the correlation energy is expected to come from such localized pairs, and is described by a Ψ of the assumed form.

Hurley, Lennard-Jones, and Pople pointed out that the pair functions have a simple canonical form (for singlet pairs)

$$\psi_{i\alpha,i\beta} = \phi_{i\alpha,i\beta} + \sum_a \phi_{a\alpha,a\beta} c_{i\alpha,i\beta}^{a\alpha,a\beta} \tag{52}$$

Since the general two-particle expansion of an electron-pair function would contain one-particle excitations, this canonical form implies that the orbitals ϕ_i are modified somewhat from localized Hartree–Fock orbitals. Brillouin's theorem, Eq. (34), implies that these modifications are of second order or higher in the many-particle perturbation theory. It can be seen from Eq. (52) that in the modified orbital basis *interpair* excitations, described by coefficients c_{ik}^{cd} where ij and kl are different localized pairs, are not present in the wave function. Thus Ψ' of Eq. (50) does not describe interpair correlations. Since the pair functions are independent except for the constraint of strong orthogonality, four-particle and higher excitations, described by Slater determinants $\Phi_{ijkl\ldots}^{abcd\cdots}$, occur in Ψ' with coefficients given by products of the pair excitation coefficients. Thus,

$$c_{ijkl}^{abcd} = c_{ij}^{ab} c_{kl}^{cd} \tag{53}$$

If the occupied orbitals ϕ_i, redefined for the reference state Φ_0 implied by Eq. (52), are localized and if the unoccupied orbitals ϕ_a are also localized, then the perturbation formulas, Eq. (48) or (49), imply that the coefficients c_{ij}^{ab} should be quite small unless a and b are localized in the same spatial region as are i and j.[1] The strong orthogonality constraint is simply an extreme form of this, requiring a and b to be absent from the expansion of any pair function other than ψ_{ij}. Equation (53) is reasonable for localized orbitals, since it becomes exact for weakly interacting two-electron systems with strong internal correlations, such as two He atoms, as the separation between localized pairs increases.

C. Superposition of Two-Particle Excitations

The second variational method that has been proposed is a direct extension of first-order perturbation theory for the wave function (second-order for the energy). If a Hartree–Fock orbital basis is used, the first-order estimate of coefficients c_i^a in Eq. (30) is zero, since by Eq. (34) the matrix elements $(\Phi_i^a, H\Phi_0)$ vanish for Hartree–Fock orbitals. The first-order wave function is of the form[1]

$$\Psi' = \Phi_0 + \sum_{ij} \sum_{ab} \Phi_{ij}^{ab} c_{ij}^{ab} \tag{54}$$

The proposed method is to use this as a variational trial function.[27] The resulting equations for the coefficients c_{ij}^{ab} have been shown to be closely analogous to the Bethe–Goldstone equation.[27] Since these equations are valid in a localized orbital basis they are a step toward incorporation of the localization concept of Lennard-Jones into the formal theory of strong correlations. The empirically established additivity of nuclear energies in the alpha-particle model can be understood in terms of an assumed wave function of this form, in a basis of localized orbitals.[6] Since there are four independent spin, isotopic spin states for a given nucleonic spatial orbital function, the pair-function wave function proposed by Hurley *et al.*, Eq. (50), is inappropriate, while Eq. (54) can describe strong correlations among four particles as long as the Hamiltonian contains only one- and two-particle operators. In the electronic case, this function describes correlations between electrons of parallel spin as well as between those of antiparallel spin and is not limited to singlet closed shells. Interactions between different localized pairs ij and kl are included, since terms ik or jl occur as well as ij or kl. The principal disadvantage of Eq. (54) is that it ignores the four-particle and higher excitations inherent in the many-particle wave function, as indicated by Eq. (53), when there are strong correlations within independent pairs. A comparison of the variational equations for the coefficients c_{ij}^{ab} in Eq. (54) with the Bethe–Goldstone equation shows that this difficulty (omitting four-particle and higher excitations) will not be important until the total correlation energy $E - D_0$ becomes a significant fraction of the two-particle energies.[27] However, since $E - D_0$ increases in proportion to the number of particles present, Eq. (54) is at best an approximation valid for relatively small systems.

In analogy to the transformation of basis orbitals that leads to Eq. (52) for the electron-pair wave function, orbitals for a completely general wave function can be chosen by the condition

$$c_i^a = 0, \quad i \leq N < a \tag{55}$$

replacing the Hartree–Fock condition given by Eq. (35). Equation (55) will be called the "Brueckner condition" for determining the basis orbitals, since it reduces to the modified Hartree–Fock equations proposed by Brueckner[37] when certain approximations are made.[27] Equation (55) has been called the "exact self-consistent field theory" by Löwdin.[38] If Eq. (55) is satisfied, the wave function of Eq. (54) includes the reference state and all double excitations, and single excitations have been made to vanish. In these circumstances, when $(\Psi, H\Psi)/(\Psi, \Psi)$ is stationary, Eq. (55) reduces to[27]

$$\begin{aligned}(a|\mathscr{G}_0|i) = (a|\mathscr{H}_0|i) &+ \sum_j \sum_{bc} (aj|R|bc)c_{ij}^{bc} \\ &+ \sum_{jk} \sum_b (jk|R|ib)c_{jk}^{ba} + \sum_j \sum_b (j|\mathscr{H}_0|b)c_{ij}^{bc} \\ = 0, \quad i \leq N < a &\end{aligned} \tag{56}$$

The summations, as before, include all values of indices such that $j < k \leq N < b < c$. By definition, the coefficients c_{ij}^{ab} are separately antisymmetric in either the upper or lower index set. These equations define a Brueckner one-particle Hamiltonian $\mathscr{G}_0$ and are similar to the Hartree–Fock condition, Eq. (35), with corrections to second order in the many-particle perturbation theory. In the philosophy of the Brueckner theory, the energy mean value for the wave function of Eq. (54) is made stationary for a fixed choice of orbital basis, then Eq. (55) or (56) is used to redetermine the orbital basis, and the process is iterated until both conditions are satisfied. This will be called "Brueckner self-consistency" since the usual Hartree–Fock self-consistency, the solution of Eq. (35), is replaced by solution of Eq. (56), but this must be repeated at each iteration of the Brueckner process. This procedure has its origin in the fact that Eq. (35) is meaningless for the hard-core potential which is generally used to describe the two-particle interaction in nuclei, because matrix elements become infinite, while Eq. (56) effectively modifies the two-particle interaction so as to remove the non-integrable singularity. For an integrable interaction, such as the Coulomb interaction between electrons, it is unlikely that Brueckner self-consistency is worth the added computational effort since single-particle as well as three-particle or higher excitations can be brought into the wave function either by going to higher-order perturbation theory or by directly introducing such terms into a variational trial function.

The variational equations for a trial function identical to Eq. (54), expressed in terms of the two-particle functions defined by Eq. (44) rather than in an orbital expansion, have recently been derived by Szasz.[39] Comparison of this work with the orbital expansion method[27] shows the great formal advantage of the latter, which allows the derivation to be presented very compactly. In the orbital expansion method all equations are derived directly from the Schrödinger equation, Eq. (28), by taking components of the entire equation in the basis of Slater determinants given by Eq. (2). If the set of determinants is truncated, as in Eq. (54), the same equations hold as matrix eigenvalue equations, equivalent to the usual variational equations. Thus, for an arbitrary set of determinants $\{\Phi_0; \Phi_\mu, \mu \neq 0\}$, the components of Eq. (28) are, if c_0 is set equal to unity,

$$E = D_0 + \sum_{\mu} H_{0\mu} c_\mu \tag{57}$$

which, in a Hartree–Fock orbital basis, is just Eq. (41), and

$$E c_\mu = D_\mu c_\mu + H_{\mu 0} + \sum_{\nu \neq \mu} H_{\mu\nu} c_\nu, \quad \mu \neq 0 \tag{58}$$

The summations are over the truncated set of determinants. If all the c_{ijk}^{abc} and coefficients of higher excitations are zero, by truncation of the determinantal basis set, the Brueckner condition in the form of Eq. (56) is obtained from Eq. (58) for the coefficients c_i^a by setting all these coefficients equal to zero. The corresponding equations for c_{ij}^{ab} are derived in the same way and have been given in full in reference 27.

D. The Bethe–Goldstone Equation

The Bethe–Goldstone equation[33] is the time-dependent Schrödinger equation for two particles of an N-particle system, where the interaction with the remaining $N - 2$ particles is represented by a self-consistent field analogous to that of the Hartree–Fock theory and by an orthogonality constraint, analogous to Eq. (47), which results from the Pauli exclusion principle. The integral equation form of this equation was developed by Brueckner and collaborators[37] from the multiple-scattering formalism of Watson.[40] The differential equation form was derived by Bethe and Goldstone for the case of infinite uniform nuclear matter, for which all canonical Hartree–Fock orbital pairs near the Fermi surface are equivalent and the two-particle equation reduces to a single equation in the relative coordinates of the two particles. Solutions for nuclear matter have been discussed by several authors.[33,41] For finite systems, the equations for different orbital pairs are not equivalent, and the Bethe–Goldstone equation for each pair is at least as difficult to solve as is the Schrödinger equation for the helium atom, since reduction to relative coordinates is not possible.

It has been pointed out by Szasz[42] that earlier work by Fock, Wesselow, and Petrashen[43] on the alkali-earth elements had included a variational derivation of the equation for a two-particle wave function, for the two valence electrons, that reduces to the Bethe–Goldstone equation except for the fact that the remaining $N - 2$ particles occupy orbitals that satisfy modified Hartree–Fock equations. Szasz shows that if unmodified Hartree–Fock orbitals are used for the $N - 2$ particle wave function a form of the Bethe–Goldstone equation can be derived. It is

$$[\mathscr{H}_{ij}(1) + \mathscr{H}_{ij}(2) + (1 - \Omega_{ij}(1, 2))Q(1, 2)]\psi_{ij}(1, 2) = \varepsilon_{ij}\psi_{ij}(1, 2) \tag{59}$$

where

$$\mathscr{H}_{ij} = \mathscr{H}_0 - (i|R|i) - (j|R|j) \tag{60}$$

in analogy to the notation used above. The projection operator $(1 - \Omega_{ij})$ removes all components containing occupied orbitals ϕ_k, $k \neq i, j$, from the orbital expansion of an arbitrary two-particle function. The pair function ψ_{ij} is constrained to be orthogonal to the $N - 2$ occupied orbitals in the sense of Eq. (47), equivalent to the application of the projection operator. Thus ψ_{ij} is of the form of Eqs. (44) except that one-particle excitations, such as ϕ_{ib} or ϕ_{aj} in the notation used there, also occur unless the Brueckner condition, Eq. (55), is used to determine the orbitals ϕ_i, ϕ_j.

The Brueckner theory actually takes an important step beyond the earlier work of Fock, Wesselow, and Petrashen by proposing that the Bethe–Goldstone equation be applied to *every* pair of particles, thus treating each pair independently of the others except for the requirement that the occupied orbitals ϕ_i be drawn from a common set for all pairs. These orbitals are determined by the Brueckner condition, Eq. (55), rather than by the Hartree–Fock condition, Eq. (35), and this formally excludes single excitations from the wave function. Correlation energy is computed by Eq. (41). In the theory of nuclear matter all pairs are assumed to be equivalent, which introduces a number of simplifications, and the orbitals are taken to be plane waves. Hence the integrals $(i|R|i)$ and $(j|R|j)$ are negligibly small, so $\mathscr{H}_{ij}$ can be replaced by $\mathscr{H}_0$ and the indices ij are no longer required. These simplifications reduce Eq. (59) to the usual Bethe–Goldstone equation.[33]

It has been shown by Goldstone[11] that the Bethe–Goldstone equation corresponds to a partial summation of diagrams in the many-particle perturbation theory. Estimates of the neglected diagrams of lowest order, for the case of nuclear matter, indicate that they have only a quite small effect on the correlation energy.[33] This analysis is incomplete and perhaps not directly relevant to the application of Eq. (59) to finite systems. For example, Eq. (58) for the coefficients c_{ij}^{ab} contains terms that cannot be expressed in the form of the integral equation assumed by Brueckner.[27] These terms can be identified as hole–hole or hole–particle interactions in the perturbation theory and ought in principle to be included, as shown by Brout.[44] The attempt by Brout to derive these equations variationally is not valid for a finite system, since Brout's argument neglects the constraint conditions shown by Hurley, Lennard-Jones, and Pople[5] to be necessary if the variations of the pair functions are to be independent. Recent work by Sinanoğlu[3,45] is very similar to the paper by Brout, but an adequate discussion of the constraint conditions is not given by either of these authors. A rigorous variational derivation of the Bethe–Goldstone equation is probably not possible.

If Hartree–Fock basis orbitals are used and single-particle excitations are neglected, or if orbitals are determined by the Brueckner condition so that single-particle excitations are absent, the Bethe–Goldstone equation for pair ij is just the variational equation for a trial wave function

$$\Psi = \Phi_0 + \sum_{ab} \Phi_{ij}^{ab} c_{ij}^{ab} \tag{61}$$

Since this is just Eq. (54) with all terms omitted which do not refer to the occupied pair ij, the variational equations for the coefficients c_{ij}^{ab} are obtained from those derived in reference 27 by simply omitting all occupied orbital indices other than ij. The energy is $D_0 + E_{ij}$ since only a single pair is considered. The equations derived in this way are

$$
\begin{aligned}
(D_0 + E_{ij} - D_{ij}^{ab})c_{ij}^{ab} &= (ab|R|ij) + \sum_{cd}{}'(ab|R|cd)c_{ij}^{ab} \\
&+ \sum_{c}{}'\{(ja|R|jc)c_{ij}^{bc} - (jb|R|jc)c_{ij}^{ac} + (ia|R|ic)c_{ij}^{bc} - (ib|R|ic)c_{ij}^{ac} \\
&\qquad + (b|\mathscr{H}_0|c)c_{ij}^{ac} - (a|\mathscr{H}_0|c)c_{ij}^{bc}\} \qquad (62)
\end{aligned}
$$

$$
= (ab|R|ij) + \sum_{cd}{}'(ab|R|cd)c_{ij}^{cd} + \sum_{c}{}'\{(b|\mathscr{H}_{ij}|c)c_{ij}^{ac} + (a|\mathscr{H}_{ij}|c)c_{ij}^{cb}\} \qquad (63)
$$

Here the primed summation over c includes all values $N < c \neq a, b$ and the primed summation over cd includes all pairs $N < c < d$ except the specific pair ab. By definition the coefficients c_{ij}^{ab} are antisymmetric in the indices ab, and this fact is used in simplifying Eq. (62). It can easily be seen that for canonical Hartree–Fock basis orbitals, Eqs. (63) and (59) are identical except for the single-particle excitation effects which have been neglected. Actually Eq. (63) is a slight generalization of the various forms of the Bethe–Goldstone equations already in the literature. It is valid in a basis of localized orbitals since it is not assumed in the derivation that the occupied orbitals are eigenfunctions of the one-particle operators $\mathscr{H}_0$, for the Hartree–Fock condition, or $\mathscr{G}_0$, for the Brueckner condition, but only that equations analogous to Eq. (36) are satisfied.

A proper justification of the Bethe–Goldstone equations requires two things. First, they must be shown to be part of a well defined computational process that leads eventually to an exact solution of the many-particle Schrödinger equation. Second, the later stages of this process must be carried out to the extent that corrections to the Bethe–Goldstone equations are computed accurately or estimated and found to be negligible. In principle the many-particle perturbation theory provides such a procedure, but Eqs. (63) in their general form, applicable to localized orbitals, have not yet been derived directly from perturbation theory. Another approach[27] is to use the many-particle Schrödinger equation, Eq. (28), to develop the hierarchy of coupled equations given by Eq. (58) for the coefficients of Slater determinants in the expansion of the exact wave function. Alternatively, these equations can be expressed in terms of the many-particle functions $\psi_{ijk\ldots}$ defined in analogy to the two-particle functions ψ_{ij} of Eqs. (44). In the latter form this approach has been used by Brenig[46] to show that the Bethe–Goldstone equations can be derived by introducing the "two-particle approximation" into this hierarchy of equations. This is equivalent to assuming equations like Eq. (53) for all the three-particle and higher excitations. It is justified physically by Brenig's argument that ψ_{ij} can be replaced by the simpler function ϕ_{ij} of Eq. (44) except in the neighborhood of the singularity of the two-particle interaction operator, and that higher-order excitations are a consequence of the two-particle excitations, which are influenced directly by this operator. The equivalence between the hierarchy of coupled equations considered by Brenig and the corresponding coupled equations derived by an orbital expansion[27] has been demonstrated by Kumar,[4,47] who has used this relationship to discuss corrections to the two-particle approximation. The formal results derived by Brout[44] are very similar to those of Brenig.

These theoretical arguments suggest that the Bethe–Goldstone equations, given by Eqs. (63) in a basis of localized orbitals, should provide a fairly accurate approximation to the correlation energy for systems of arbitrary size, if Eq. (41) is used to compute the correlation energy. Although requiring abandonment of a straightforward variational development, the Bethe–Goldstone equations have an advantage over the method proposed by Hurley, Lennard-Jones, and Pople in that all possible pairs are treated in a similar manner. The inherent restriction of the two-particle excitation function of Eq. (54) to systems with a relatively small number of particles is removed by use of the Bethe–Goldstone equations, which would be exact for a gas of weakly interacting He atoms. However, analysis of the variational equations for Eq. (54) shows that interactions not described by the Bethe–Goldstone equations occur between the excitations of different occupied orbital pairs, even in the limit of small systems for which $E - D_0$ is negligible.[27] A satisfactory evaluation of the usefulness of the Bethe–Goldstone equations requires analysis of these coupling terms, which in a sense have been omitted more because they are inconvenient than because they are known to be small, as well as solution of the Bethe–Goldstone equations themselves for non-trivial examples.

Equations similar to the Bethe–Goldstone equations have recently been proposed by Sinanoğlu.[45] The comments made above on Brout's paper[44] and on the use of canonical Hartree–Fock orbitals in the general perturbation theory apply also to this work. In addition there is an error in Sinanoğlu's analysis that accounts for the apparent difference between the proposed equations and the Bethe–Goldstone equations as given by Eq. (59). This error arises from the definition by Sinanoğlu of the "fluctuation potential"

$$m_{ij}(ij) = Q(ij) + (ij|R|ij) - (i|R(ij)|i) - (j|R(ij)|j) \tag{64}$$

Here $(i|R(ij)|j)$ is an operator in variable space j, obtained by integrating the two-particle operator $\phi_i^*(i)R(ij)\phi_i(i)$ over variable space i. The peculiarity of Eq. (64) can be seen by expressing this operator in the general particle variables 1, 2,

$$m_{ij}(1, 2) = Q(1, 2) + (ij|R|ij) - (i(1)|R(1, 2)|i(1)) - (j(2)|R(1, 2)|j(2)) \tag{65}$$

which obviously is not symmetrical in the variable space indices 1, 2, contrary to the form expected in a many-fermion theory for a Hamiltonian containing the symmetrical interaction $Q(1, 2)$. This difficulty can be traced to the fact that the perturbing Hamiltonian in many-particle theory, as originally derived by Møller and Plesset,[9] cannot be expressed entirely as a two-particle operator, a form tacitly assumed by Sinanoğlu. Following Møller and Plesset, if the zeroth-order Hamiltonian is taken to be

$$H^{(0)} = \sum_i \mathscr{H}_0(i) - \sum_{ij}(ij|R|ij) \tag{66}$$

then the perturbing operator is obtained by subtracting $H^{(0)}$ from Eq. (5) to get

$$V = \sum_{ij}[Q(ij) + (ij|R|ij)] + \sum_i[K(i) - \mathscr{H}_0(i)] \tag{67}$$

$$= \sum_{ij}[Q(ij) + (ij|R|ij)] - \sum_i\sum_k(k(2)|R(i, 2)|k(2)) \tag{68}$$

Here $\Sigma_k(k(2)|R(i, 2)|k(2))$ is a one-particle operator in the variable space i, while k is a dummy index of summation denoting occupied orbitals in the Hartree–Fock reference state. These two indices are unrelated to each other and, in particular, the term with $k = i$ occurs in Eq. (68). The definition of an effective two-particle interaction as in Eq. (64) rests on an artificial identification of variable space indices with orbital indices. Even if this is done, some formal one-particle operator must remain in Eq. (68), since the terms with $k = i$ do not vanish. These terms have apparently been omitted in Sinanoğlu's analysis. It must be remembered that the summation over indices ij, for the two-particle operator $Q(ij)$, excludes the diagonal element $i = j$, and that $(ii|R|ii)$ vanishes identically.

Although it is a purely arbitrary definition, a formally correct expression for Eq. (68) is

$$V = \sum_i V_i(i) + \sum_{ij} m_{ij}(ij) \tag{69}$$

where

$$V_i(1) = (i(2)|R(1, 2)|i(2)) \tag{70}$$

and

$$m_{ij}(1, 2) = Q(1, 2) + (ij|R|ij) - (i(1)|R(1, 2)|i(1)) - (j(2)|R(1, 2)|j(2)) \tag{71}$$

The artificiality of these definitions is clear from the fact that the operators in Eq. (69) have indices that depend on the particle variables. If the equivalent form, Eq. (67), is used, it is obvious that all particles are identical. The sum of the integrals $(ij|R|ij)$ is just a number, used to shift the energy of the reference state.

V. SUPPORTING EVIDENCE

Any judgement about the validity of approximate methods must depend ultimately on the numerical results of computations. Despite the very large number of papers dealing with formal correlation theory, there are in the literature only a few quantitative results that supply information about the validity of the approximations that have been proposed. Under these circumstances, simply because adequate calculations have not been carried out, there is a tendency to over-emphasize details of elaborate methods that represent only very small corrections to simpler methods, while neglecting careful quantitative work with the latter.

These remarks are pertinent to the historical development of the Hartree–Fock theory of atoms and molecules. The theorem discussed above in connection with Eq. (40), first stated by Møller and Plesset,[9] has had very little quantitative application. There are few one-particle properties of atoms that can be directly measured. Moreover, most atoms have open-shell ground-state configurations, and the theorem does not apply directly to the traditional Hartree–Fock wave functions obtained for such states.[1] The self-consistent-field method of Hartree involves integro-differential equations that are virtually insoluble for molecules. Progress in this area has had to wait for the development of the matrix Hartree–Fock method* and for the application of electronic digital computers to the evaluation of the very large number of complicated integrals needed in this method. This work has been carried far enough for reasonable approximations to a ground-state molecular Hartree–Fock wave function to have been obtained in a few cases.[48] Since the electric dipole moment of a diatomic molecule is directly measurable, the dipole moments computed with these wave functions give, almost for the first time, an indication of the quantitative validity of the theorem of Møller and Plesset. These results are summarized in Table I. The error in all cases is considerably less than one per cent, in agreement with results for various one-particle operators in the He atom.[8,49]

Many physical properties of atoms and molecules are of the nature of polarizabilities, essentially the second-order energy effect of a perturbing field acting on a non-degenerate state of an unperturbed system. Formal methods for the analysis of such effects have recently been reviewed by Dalgarno.[50] The conceptually most simple method, called by Dalgarno the coupled Hartree–Fock method, is an application of the unrestricted Hartree–Fock approximation* to the perturbed system. Equations for the first-order wave function and second-order energy are given by a perturbation expansion of the Hartree–Fock equations.[51] Because

* See the Appendix, p. 358, for references and definitions.

TABLE I. Dipole Moments μ of Diatomic Molecules in Debye Units. Polarity is given as positive if the effective charge associated with the lighter nucleus is positive. Calculated values are from approximate molecular Hartree–Fock calculations

Molecule	HF	LiH	LiF
μ, calc.	1.827[a]	−5.888[c]	6.297[e]
μ, obs.	1.818[b]	−5.882[d]	6.284[f]

[a] Nesbet, R. K., *J. Chem. Phys.* **36,** 1518 (1962).
[b] Weiss, R., *Phys. Rev.* **131,** 659 (1963).
[c] Kahalas, S. L., and Nesbet, R. K., *J. Chem. Phys.* **39,** 529 (1963).
[d] Wharton, L., Gold, L. P., and Klemperer, W., *J. Chem. Phys.* **33,** 1255 (1960); *ibid.* **37,** 2149 (1962).
[e] McLean, A. D., *J. Chem. Phys.* **39,** 2653 (1963).
[f] Wharton, Klemperer, Gold, Strauch, Gallagher, and Derr, *J. Chem. Phys.* **38,** 1203 (1963); Vidale, G., *J. Phys. Chem.* **64,** 314 (1960).

of computational difficulties this method has until recently been applied only to light atoms or atomic ions[50] and to H_2.[52] Careful calculations by this method on the LiH molecule have recently been reported by Stevens, Pitzer, and Lipscomb.[53] In general the computed values of magnetic susceptibility and magnetic screening have the same degree of accuracy as the static quantities (first-order effects) discussed above, when computed with the best available molecular Hartree–Fock functions. Recent calculations of electric dipole polarizabilities of atoms[54] have been carried out by a simplified method called by Dalgarno the uncoupled Hartree–Fock method.[50] For light atoms or positive ions the agreement with known experimental results is comparable to that of the other Hartree–Fock results mentioned above, but there appear to be significant discrepancies for heavier atoms and for negative ions. It would be premature to attribute these discrepancies to correlation effects until the coupled Hartree–Fock method has been applied. The negative ions may be a special case of near degeneracy, since their small binding energy makes excitation into the continuum possible at very low energies.

To summarize the Hartree–Fock results, it appears that physical properties of atomic and molecular ground states have been calculated within an accuracy of approximately one per cent whenever an adequate approximation to the Hartree–Fock wave function has been used. Computational difficulties have made this possible so far only for atoms and for a few small molecules. Since these errors are so small, it is reasonable to expect that except in the case of near degeneracies, which require a preliminary configuration interaction matrix diagonalization, the simple application of the lowest relevant order of perturbation theory for correlation corrections should reduce these errors to a fraction of one per cent. Only in exceptional cases, when extremely high accuracy is needed from a computed result, is it likely that more sophisticated methods will be required. The important applications of such methods will be to the direct calculation of correlation energies.

The question of the correlation energy has also had a curious history, due to computational difficulties. The self-consistent-field method used by Hartree[18,19] to solve the Hartree–Fock equations leads to orbital radial functions which are expressed as numerical tables. The number of significant figures in total energy, computed from such functions obtained before the advent of digital computers, is quite small and is not adequate for the purpose of computing correlation energies. Hartree–Fock functions that make more accurate energy computations possible have recently been obtained by the matrix technique[55] and a survey of the relativistic and correlation energies has been carried out.[56]

As a result of this work, in connection with the molecular Hartree–Fock calculations mentioned above, the net correlation contribution to molecular binding is known to within a few tenths of an electron volt for several molecules.[48] From these results has emerged the remarkable fact that the correlation energy of a localized pair of electrons is very insensitive to the environment of the pair. This result has made possible the analysis of the net correlation energy in molecules in terms of the known correlation energy of separated atomic ions with similar electronic structure,[57,58] or in terms of empirical bond correlation energies and of the united atom limit.[58] Some results are given in Table II.

TABLE II. Net Correlation Energy Contribution ΔE_c to Molecular Dissociation Energies D_e from Approximate Molecular Hartree–Fock Calculations (SCF), Compared with Empirical Estimates $E_c(\mathrm{U}) - E_c(\mathrm{S})$ and $E_c(\mathrm{I}) - E_c(\mathrm{S})$. $E_c(\mathrm{U})$, $E_c(\mathrm{I})$, $E_c(\mathrm{S})$ are, respectively, correlation energies[a] of the united atom, the separated ions with polarity indicated by dipole moment, and the separated atoms in their ground states. The units are electron volts. Dissociation energies D_e are observed values plus zero-point energy

<table>
<tr><th>Molecule</th><th>H_2</th><th>HF</th><th>HF</th><th>LiH</th><th>LiF</th><th>CH_4</th></tr>
<tr><td>D_e(exp.)[b]</td><td>4.75</td><td colspan="2">6.08</td><td>2.52</td><td>5.99</td><td>18.20</td></tr>
<tr><td>D_e(SCF)</td><td>3.64[c]</td><td>4.05[d]</td><td>4.11[e]</td><td>1.44[f]</td><td>4.03[g]</td><td>11.98[h]</td></tr>
<tr><td>$E_c(\mathrm{S})$</td><td>0.00</td><td colspan="2">−8.82</td><td>−1.23</td><td>−10.05</td><td>−4.30</td></tr>
<tr><td>$E_c(\mathrm{U})$</td><td>−1.14</td><td colspan="2">−10.69</td><td>−2.57</td><td>−12.22</td><td>−10.69</td></tr>
<tr><td>$E_c(\mathrm{I})$</td><td>+1.08</td><td colspan="2">−10.83</td><td>−2.27</td><td>−12.01</td><td>−9.61[i]</td></tr>
<tr><td>$E_c(\mathrm{U}) - E_c(\mathrm{S})$</td><td>−1.14</td><td colspan="2">−1.88</td><td>−1.34</td><td>−2.17</td><td>−6.39</td></tr>
<tr><td>$E_c(\mathrm{I}) - E_c(\mathrm{S})$</td><td>−1.08</td><td colspan="2">−2.01</td><td>−1.03</td><td>−1.96</td><td>−5.30</td></tr>
<tr><td>ΔE_c</td><td>−1.11</td><td>−2.03</td><td>−1.97</td><td>−1.08</td><td>−1.96</td><td>−6.22</td></tr>
</table>

[a] Clementi, E., *J. Chem. Phys.* **38,** 2248 (1963); *ibid.* **39,** 175 (1963); Clementi, E., and McLean, A. D., *Phys. Rev.* **133,** A419 (1964).

[b] Data from Herzberg, G., *Diatomic Molecules*, Van Nostrand, Princeton, 1955, 2nd ed., or from Gaydon, A. G., *Dissociation Energies*, Chapman and Hall, London, 1953, revised ed.

[c] Kolos, W., and Roothaan, C. C. J., *loc. cit.*, ref. 48.

[d] Clementi, E., *loc. cit.*, ref. 48.

[e] Nesbet, R. K., *loc. cit.*, ref. 48.

[f] Kahalas, S. L., and Nesbet, R. K., *loc. cit.*, ref. 48.

[g] McLean, A. D., *loc. cit.*, ref. 48.

[h] Krauss, M., *loc. cit.*, ref. 48.

[i] Extrapolated E_c for C^{4-} obtained from the data of reference *a* by Carlson, K. D., and Skancke, P. N., *loc. cit.*, ref. 57.

Although the theory of correlation energy based on the Bethe–Goldstone equation for localized pairs, or on the second-order perturbation theory when it is adequate, indicates that correlation energy for localized orbitals should be insensitive to environment, the degree of this insensitivity indicated by the data in Table II goes beyond what might have been expected from the theory. This is a quantitative result that could not in principle be derived by purely formal arguments, without computation. These results are compatible with earlier work on two-electron atomic ions of the He series, discussed by Löwdin.[2,49] There is a great deal of empirical evidence for the additivity of bond energies in molecules, to an accuracy of the order of one tenth of an electron volt.[59] This was part of the motivation behind the Lennard-Jones work on localized orbitals,[28] which established the additivity of bond energies in the Hartree–Fock approximation.[60] Since the net correlation energy is a large fraction of a typical molecular binding energy, this empirical evidence cannot be explained solely by the Hartree–Fock approximation. A similar situation holds for the alpha-particle model as used to describe nuclear binding energies.[6]

Except for the rather special case of the helium atom there have been only a few attempts to compute correlation energies. Earlier work, mainly concerned with near-degeneracy situations in atoms, is reviewed by Hartree.[61] The calculations by Boys[62] represented a new and more ambitious approach. Using the method of superposition of configurations constructed from an arbitrary orthogonalized set of atomic basis orbitals (exponential functions multiplied by powers of r and by spherical harmonics), Boys and collaborators were able to compute energies roughly halfway between the Hartree–Fock energies and the experimental values (omitting relativistic terms in the electronic Hamiltonian). Most of the published results were obtained before digital computers became available. The amount of computation in this work was estimated by Boys to be no greater than that involved in Hartree's self-consistent-field calculations. The informal perturbation method used in this work to determine orthonormal basis orbitals was systematized in later calculations with larger basis sets, by using the matrix Hartree–Fock method. Later calculations using the same method, but including more basis orbitals and more configurations, obtained 97.4% of the correlation energy of He[63] and 89.3% of the correlation energy of Be.[35] The numbers of configurations used were 20 and 37, respectively, and a single diagonalization of matrices of this size can be accomplished in a few minutes on contemporary computers.

A fact of considerable importance stressed in the papers by Boys, and confirmed by later work, is that the configurations contributing most strongly to the correlation energy must be constructed from orbitals which are as similar as possible to the occupied orbitals which they replace in the reference state ("root function" in Boys' work). In an orthonormal set these orbitals have spatial extension similar to that of the occupied orbitals, but they must have additional nodes in regions of high density in order to be orthogonal to the occupied orbitals. These perturbing orbitals are quite different from the functions which describe Rydberg-series electrons in atoms, since the latter spread out very rapidly as the number of nodes increases, while the former are confined to the same spatial extension as the occupied orbitals. The additional nodes in regions of high density cause these perturbing orbitals to have large kinetic energy, while the potential energy is comparable to that of the occupied orbitals. For this reason the energy of the perturbing orbitals lies in the continuum of the energy spectrum of the Hartree–Fock one-electron Hamiltonian, and the perturbing orbitals must be thought of as wave packets in the continuum. Such functions, which occur naturally in discrete orbital basis sets orthogonalized to occupied Hartree–Fock orbitals, are a natural by-product of the matrix Hartree–Fock method. They occur in an atomic calculation if basis orbitals are included that have exponents as large as those characteristic of a given shell, but that have higher powers of r and higher-order spherical harmonic factors. Criteria for the choice of localized perturbing orbitals in molecules, when the occupied orbitals of the reference state are localized, have been discussed by Boys.[30]

The very accurate electronic wave function for H_2 obtained by Kolos and Roothaan[64] has been analyzed by Davidson and Jones[65] in the diagonalized quadratic form discussed by Hurley, Lennard-Jones, and Pople.[5] This is the "natural" orbital expansion of Löwdin,[66] in the special case of a two-particle wave function. Davidson and Jones show that 89% of the correlation energy is obtained if this expansion is truncated at the first five natural spatial orbitals (counting the degenerate π-orbitals as two). The remaining error is 0.1246 electron volts, an accuracy almost within the limits of additivity of empirical bond energies.[59] As pointed out by Davidson and Jones, this set of orbitals includes the nodeless occupied orbital σ_g together with all the four others that correspond to independent ways of introducing a single nodal surface. Orbitals π_{ux} and π_{uy} correspond to longitudinal nodal planes, σ_u to a transverse nodal plane, and σ_g to an internal closed nodal surface. Very similar results are obtained for He.[67] It can be concluded that the perturbing orbitals most important in calculating the correlation energy of a doubly occupied localized spatial orbital might be expected always to have the same topological description: three independent transverse nodal planes and an internal closed nodal surface superimposed on the original orbital. Furthermore, from the apparent constancy of the correlation energy of localized orbital pairs, inclusion of perturbing orbitals of this kind should in general account for some 90% of the correlation energy of a localized pair.

The calculation reported on He[63] went to a higher stage of this same topological pattern, including all combinations of three or less nodal surfaces for orbitals perturbing an occupied $1s^2$ pair. It can be conjectured from the result of this calculation that inclusion of this many nodal surfaces would in general lead to some 97% of the correlation energy of a localized pair. This is an accuracy well within the limits of the empirical additivity of chemical bond energies. If this is so, the prospects for a useful approximate solution of the Bethe–Goldstone equations by direct orbital expansion are very favorable, since the general characteristics of the perturbing orbitals are known from the argument given here. If a similar degree of accuracy is not attained for the total correlation energy when such perturbing orbitals are included for each occupied localized orbital in solving the Bethe–Goldstone equations, this would imply that the error is due to formal corrections to these equations, such as the coupling terms discussed in Section IV.

A recent calculation by Kelly[15] on Be used the formalism of the standard many-particle perturbation theory.[11] In particular, the continuum eigenfunctions of the ground-state Hartree–Fock one-particle Hamiltonian were used explicitly as unoccupied orbitals in the perturbation formulas. Energy denominators were originally taken to be of the form given in Eq. (49), but this left large terms in all higher orders of the perturbation expansion, in the form of a geometric series. This series was summed and shown to be equivalent to changing the denominators to the form of Eq. (48). In a more recent paper[68] this and other partial summations found to be necessary in the calculation are shown to be implicit in the coupled equations derived variationally[27] for the trial wave function given by Eq. (54). Kelly's computed correlation energy is 96.3% of the empirical value.[56] Since this calculation included the entire continuum for *s*, *p*, and *d* orbitals, perturbing orbitals with two nodal surfaces for each occupied orbital can be represented by wave packets in this continuum, and the accuracy of the computed correlation energy is compatible with the accuracy expected from consideration of calculations with discrete basis sets. This calculation gives probably the most accurate evaluation of an interpair correlation energy, found to be −0.135 electron volts for the 1*s*–2*s* interaction. This is, of course, the sum of four independent 1*s*–2*s* pair correlations, for the four different orbital spin assignments.

The advantage of working directly with localized orbitals is obvious for large systems, since, as shown above, the number of important orbital pairs increases linearly with N for localized orbitals while increasing quadratically with N for canonical Hartree–Fock orbitals that extend over the entire system. This distinction does not arise for He, with only one pair, or for Be, where the canonical orbitals are already separated into distinct inner and outer shells. The next most simple case with a closed-shell structure is the ^1S ground state of Ne. Because of the near degeneracy between all eight canonical orbitals in the outer shell, $2s^2 2p^6$, and because these orbitals are not separated spatially from each other, the coupling terms omitted in using the Bethe–Goldstone equations in this orbital basis are likely to be especially large. The localized orbitals for Ne are four doubly occupied tetrahedral equivalent orbitals[28,69] and the coupling terms should not be any larger for such orbitals than they are for the orbitals of Be. For this reason, calculations on Ne will be a more meaningful test of proposed methods than are the calculations on Be.

Additivity of local bond or lone-pair energies in molecules is not just a question of the formal additivity indicated by Eq. (41) but also requires local correlation energies to be insensitive to structural changes of the molecular environment.[59] The Bethe–Goldstone equations in a localized orbital basis, which treat each pair separately, are compatible with this kind of additivity. Since there has been no quantitative calculation of corrections to these equations, which include terms that couple the equations for different orbital pairs, it is premature to assert that these equations account for the observed additivity of chemical-bond energies. It would be more correct to invert this statement, since the observed chemical additivity provides evidence for the hypothesis that solution of the Bethe–Goldstone equations will give a very useful estimate of total correlation energies.

It has been proposed by Sinanoğlu[70] that the Bethe–Goldstone equations should be solved in a canonical Hartree–Fock basis, and that the solutions should then be transformed to a localized basis to evaluate quantities of chemical interest, such as bond or lone-pair correlation energies. Since the theory can be developed in terms of localized Hartree–Fock orbitals from the outset,[1,27] this indirect approach is not necessary. It would be cumbersome for large systems for the reasons given above. If coupling terms connecting the Bethe–Goldstone equations for different pairs are important when the orbitals are not spatially separated, this procedure would be less accurate for systems such as Ne than would be the direct use of the Bethe–Goldstone equations in a localized orbital basis.

VI. COUNTER-EXAMPLES

A. Open Shells

One of the reasons that the power of the Hartree–Fock theory has been under-estimated is that Hartree's self-consistent-field calculations on atoms with open-shell configurations do not satisfy the conditions of the unrestricted Hartree–Fock approximation.[1] The theorem of Møller and Plesset,[9] stating that errors in one-particle properties are second-order perturbations, holds only in the unrestricted theory. A striking improvement in computed magnetic hyperfine structure interactions in Li[71,72] and N[73] has been achieved by taking into account the single-particle excitations which result from the arbitrary constraints inherent in the traditional Hartree–Fock method. In Li and N the unrestricted inner-shell orbitals are spin-polarized, having different radial functions for orbitals of opposite spin, as a result of the dependence on spin of the exchange potential due to the unfilled outer shell.[1]

B. The Role of the Continuum

Some early configuration interaction calculations on atoms used perturbing orbitals characteristic of the Ryberg series of the atom, and found that this series gave only a very small fraction of the total effect expected. This was interpreted to imply that continuum states would have to be included explicitly. Moreover, the standard many-particle perturbation theory is formulated in terms of eigenfunctions of the Hartree–Fock one-particle Hamiltonian, which necessarily includes an explicit continuum. This conclusion was shown to be unnecessary by demonstrating that Brillouin's theorem can be stated in general by Eq. (31), which means that localized orbitals can be used for the unoccupied set. The only condition is that these orbitals must be orthogonal to the occupied set.[1] The work of Boys,[62] discussed above, showed clearly that the perturbing orbitals in correlation theory must be thought of as localized wave packets in the continuum, not closely related to the much more diffuse Rydberg-series orbitals. In the matrix Hartree–Fock method a discrete set of orbitals is used from the beginning. For correlation calculations it is only necessary to include in this set orbitals which, when orthogonalized to localized occupied orbitals, give perturbing orbitals of the kind discussed in Section V. Orbitals of this kind, often expressed as polynomials in local coordinates multiplied into occupied orbitals, have also been found to represent the first-order changes of occupied orbitals in recent polarizability calculations.[53,54,74] It does not appear to be essential to introduce an explicit continuum.

C. Molecular-Orbital Theory at Large Internuclear Distances

The standard molecular-orbital theory of the covalent bond describes the bond as a doubly occupied spatial orbital at all internuclear distances. This necessarily goes to a combination of ionic states as two atoms are separated. The point overlooked in this analysis is that the Hartree–Fock approximation for the separated atoms in their ground-state configuration is also a Hartree–Fock approximation for a molecule made up of these atoms. Thus the ionic configuration is not the Hartree–Fock state of lowest energy at large internuclear distances, and it should not be used as the reference state. The character of the reference state must change at some intermediate distance. These ideas have been used to analyze the N_2 molecule at all internuclear distances[17] and to develop a theory of the spin-dependent Heisenberg exchange interaction between atoms with open-shell configurations.[75] If the proper configuration is chosen, correlation effects are inherently small at all internuclear distances and are described by the theory discussed in the present paper. This work on the Heisenberg exchange interaction also provides an example of the use of localized orbitals in dealing with correlation energy in solid-state theory.

APPENDIX: PROPOSED HARTREE–FOCK TERMINOLOGY

A. Traditional Hartree–Fock Method

This refers to the method used in Hartree's atomic calculations, referred to by Hartree as the "self-consistent field with exchange". The variational equations used, first derived by Fock,[76] are integro-differential equations for radial factors $R_{nl}(r)$ of the occupied orbitals in a Slater determinant, or a linear combination of Slater determinants that in the non-relativistic case is an eigenfunction of $\mathbf{L}^2$ and $\mathbf{S}^2$. The radial factors are constrained to be independent of the quantum numbers m_l and m_s (equivalence restriction) and the orbitals are assumed to have definite angular momentum quantum numbers l (symmetry restriction). Hartree has given a detailed review of this method.[77]

B. Matrix Hartree–Fock Method

This is a computational technique that can be used in connection with any of the formal variants of the Hartree–Fock approximation. Occupied orbitals are represented as linear combinations of a fixed set of basis orbitals, and the matrix representation of the Hartree–Fock equations in this basis gives a set of non-linear matrix equations to be solved iteratively in analogy to Hartree's method. The matrix method was proposed in a form applicable to closed-shell states of molecules by Roothaan.[78] A very similar procedure had been used earlier by Coulson[79] and the same method was proposed independently by Hall.[80] Roothaan refers to this procedure as the "expansion method".[81] A version of this method equivalent to the traditional method for open shells has been developed by Roothaan.[82]

C. Spin-Polarized Hartree–Fock Method

Slater[83] pointed out that the traditional method involved a constraint by requiring orbitals of opposite spin to have identical spatial factors in configurations with singly occupied spatial orbitals. It was proposed to drop this equivalence restriction on m_s. The matrix form of the spin-polarized theory was given by Pople and Nesbet[84] and by Berthier.[85]

D. Unrestricted Hartree–Fock Method

This terminology refers to a variational calculation in which the trial function is a single Slater determinant with no *a priori* constraints on the form of the occupied orbitals. The first complete discussion of these constraints, in the form of symmetry or equivalence restrictions in the traditional method, was given by Nesbet,[86] although this form of the theory had been tacitly assumed in the much earlier theoretical discussions of Dirac[87] and of Møller and Plesset.[88] It should be noted that Brillouin[89] commented explicitly that what is now known as Brillouin's theorem did not apply exactly to Fock's equations for open-shell configurations.

E. Truncated Hartree–Fock Method

This is a modified version of the unrestricted method in which symmetry and equivalence restrictions are applied by simply omitting terms from the matrix equations that couple different orbital symmetry species and by averaging over the subspecies index (m_l or m_s) to obtain equations independent of the subspecies.[86] This method was used for the first atomic matrix Hartree–Fock calculations,[86,90] and it is equivalent to the traditional method whenever all occupied shells of the same symmetry species (l) are occupied in the same way.[91] This method was called the "method of symmetry and equivalence restrictions" in earlier papers.

F. Extended Hartree–Fock Method

This refers to a variational method proposed by Löwdin[92] in which the trial function is the projection onto an eigenstate of $\mathbf{L}^2$ and $\mathbf{S}^2$ of a Slater determinant of open-shell configuration. Except in trivial cases the projected function is a linear combination of Slater determinants constructed from orbitals which do not form an orthonormal set. In general the energy of such a function includes part of the correlation energy as defined in the present paper.

References

1. Nesbet, R. K., *Proc. Roy. Soc. London* **A230,** 312 (1955); *Rev. Mod. Phys.* **33,** 28 (1961).
2. Löwdin, P. O., *Advan. Chem. Phys.* **2,** 207 (1959).
3. Sinanoğlu, O., *Advan. Chem. Phys.* **6,** 315 (1963).
4. Thouless, D. J., *The Quantum Mechanics of Many-Body Systems,* Academic Press, New York, 1961; Kumar, K., *Perturbation Theory and the Nuclear Many Body Problem,* Interscience Publishers, New York, 1962.
5. Hurley, A. C., Lennard-Jones, J. E., and Pople, J. A., *Proc. Roy. Soc. London* **A220,** 446 (1953).
6. Nesbet, R. K., *Phys. Rev.* **100,** 228 (1955).
7. Pitzer, K. S., *Advan. Chem. Phys.* **2,** 59 (1959); McWeeny, R., *Rev. Mod. Phys.* **32,** 335 (1960).
8. Goodisman, J., and Klemperer, W., *J. Chem. Phys.* **38,** 721 (1963).
9. Møller, C., and Plesset, M. S., *Phys. Rev.* **46,** 618 (1934).
10. Brueckner, K. A., *Phys. Rev.* **100,** 36 (1955).
11. Goldstone, J., *Proc. Roy. Soc. London* **A239,** 267 (1957).
12. Boys, S. F., *Proc. Roy. Soc. London* **A206,** 489 (1951).
13. Löwdin, P. O., *Phys. Rev.* **97,** 1509 (1955).
14. Condon, E. U., and Shortley, G. H., *The Theory of Atomic Spectra,* Cambridge University Press, New York, 1951, pp. 169–174.
15. Kelly, H. P., *Phys. Rev.* **131,** 684 (1963).
16. Koopmans, T., *Physica* **1,** 104 (1933).
17. Nesbet, R. K., *Phys. Rev.* **122,** 1497 (1961).
18. Hartree, D. R., *Proc. Cambridge Phil. Soc.* **24,** 89, 111 (1928).
19. Fock, V., *Z. Physik* **61,** 126 (1930); *ibid.* **62,** 795 (1930).
20. Gaunt, J. A., *Proc. Cambridge Phil. Soc.* **24,** 328 (1928).

21. Slater, J. C., *Phys. Rev.* **32**, 339 (1928).
22. Brillouin, L., *Actualités sci. et ind.* No. 159, Hermann et Cie., Paris, 1934.
23. Dirac, P. A. M., *Proc. Cambridge Phil. Soc.* **26**, 376 (1930).
24. London, F., *Z. Physik* **63**, 245 (1930).
25. Mayer, J. E., and M. G., *Phys. Rev.* **43**, 605 (1933); Van Vleck, J. H., and Whitelaw, N. G., *Phys. Rev.* **44**, 551 (1933).
26. Margenau, H., *Rev. Mod. Phys.* **11**, 1 (1939); Nesbet, R. K., *Mol. Phys.* **5**, 63 (1962).
27. Nesbet, R. K., *Phys. Rev.* **109**, 1632 (1958).
28. Lennard-Jones, J. E., *Proc. Roy. Soc. London* **A198**, 1, 14 (1949); and later papers with Hall, G. G., and Pople, J. A.
29. Lennard-Jones, J. E., and Pople, J. A., *Proc. Roy. Soc. London* **A202**, 166 (1950); Edmiston, C., and Ruedenberg, K., *Rev. Mod. Phys.* **35**, 457 (1963).
30. Boys, S. F., *Rev. Mod. Phys.* **32**, 296 (1960).
31. Foster, J. M., and Boys, S. F., *Rev. Mod. Phys.* **32**, 300, 303 (1960).
32. McWeeny, R., and Ohno, K. A., *Proc. Roy. Soc. London* **A255**, 367 (1960).
33. Bethe, H. A., and Goldstone, J., *Proc. Roy. Soc. London* **A238**, 551 (1957). This is a formal development based on Brueckner's work, reviewed in detail by Bethe, H. A., *Phys. Rev.* **103**, 1353 (1956).
34. Parks, J. M., and Parr, R. G., *J. Chem. Phys.* **28**, 335 (1958); Kapuy, E., *Acta Phys. Acad. Sci. Hung.* **9**, 237 (1958); **10**, 125 (1959); **11**, 409 (1960); **12**, 185 (1960); Shull, H., *J. Chem. Phys.* **30**, 1405 (1959); Arai, T., *J. Chem. Phys.* **33**, 95 (1960); Löwdin, P. O., *J. Chem. Phys.* **35**, 78 (1961).
35. Watson, R. E., *Phys. Rev.* **119**, 170 (1960).
36. Sinanoğlu, O., *Proc. Natl. Acad. Sci. U.S.* **47**, 1217 (1961); *J. Chem. Phys.* **36**, 706 (1962); Allen, T. L., and Shull, H., *J. Phys. Chem.* **66**, 2281 (1962).
37. Brueckner, K. A., Levinson, C. A., and Mahmoud, H. M., *Phys. Rev.* **95**, 217 (1954); Brueckner, K. A., *Phys. Rev.* **96**, 508 (1954); **97**, 1353 (1955); Brueckner, K. A., and Levinson, C. A., *Phys. Rev.* **97**, 1344 (1955); Brueckner, K. A., and Wada, W., *Phys. Rev.* **103**, 1008 (1956).
38. Löwdin, P. O., *J. Math. Phys.* **3**, 1171 (1962).
39. Szasz, L., *Phys. Rev.* **132**, 936 (1963).
40. Watson, K. M., *Phys. Rev.* **89**, 575 (1953).
41. Gomes, L. C., Walecka, J. D., and Weisskopf, V. F., *Ann. Phys. N.Y.* **3**, 241 (1958).
42. Szasz, L., *Z. Naturforsch.* **14a**, 1014 (1959).
43. Fock, V., Wesselow, M., and Petrashen, M., *Zh. Eksperim. i Teor. Fiz.* **10**, 723 (1940). Later references are given by Szasz, ref. 42.
44. Brout, R., *Phys. Rev.* **111**, 1324 (1958).
45. Sinanoğlu, O., *J. Chem. Phys.* **36**, 706, 3198 (1962).
46. Brenig, W., *Nucl. Phys.* **4**, 363 (1957).
47. Kumar, K., *Nucl. Phys.* **21**, 99 (1960).

48. H_2: Kolos, W., and Roothaan, C. C. J., *Rev. Mod. Phys.* **32,** 219 (1960); HF: Nesbet, R. K., *Rev. Mod. Phys.* **32,** 272 (1960); *J. Chem. Phys.* **36,** 1518 (1962); Clementi, E., *J. Chem. Phys.* **36,** 33 (1962); LiH: Kahalas, S. L., and Nesbet, R. K., *J. Chem. Phys.* **39,** 529 (1963); LiF: McLean, A. D., *J. Chem. Phys.* **39,** 2653 (1963); CH_4: Krauus, M., *J. Chem. Phys.* **38,** 564 (1963).
49. Bethe, H. A., and Salpeter, E. E., *The Quantum Mechanics of One- and Two-Electron Atoms,* Springer-Verlag, Berlin, 1957, p. 164.
50. Dalgarno, A., *Advan. Phys.* **11,** 281 (1962).
51. Peng, H., *Proc. Roy. Soc. London* **A178,** 499 (1941).
52. Sinha, S. K., and Mukherji, A., *J. Chem. Phys.* **32,** 1652 (1960); Aleksandrov, I. V., *Soviet Phys. "Doklady"* (English Transl.) **3,** 325, 729 (1958).
53. Stevens, R. M., Pitzer, R. M., and Lipscomb, W. N., *J. Chem. Phys.* **38,** 551 (1963); Stevens, R. M., and Lipscomb, W. N., *J. Chem. Phys.* **40,** 2238 (1964).
54. Yoshimine, M., and Hurst, R. P., *Phys. Rev.* **135,** A612 (1964).
55. Clementi, E., Roothaan, C. C. J., and Yoshimine, M., *Phys. Rev.* **127,** 1618 (1962); Clementi, E., *J. Chem. Phys.* **38,** 996, 1001 (1963); *ibid.* **41,** 295, 303 (1964).
56. Clementi, E., *J. Chem. Phys.* **38,** 2248 (1963); **39,** 175 (1963).
57. Nesbet, R. K., *J. Chem. Phys.* **36,** 1518 (1962); McLean, A. D., *J. Chem. Phys.* **39,** 2653 (1963); Carlson, K. D., and Skancke, P. N., *J. Chem. Phys.* **40,** 613 (1964).
58. Clementi, E., *J. Chem. Phys.* **38,** 2780 (1963); *ibid.* **39,** 487 (1963).
59. A recent discussion citing earlier references is given by Allen, T. L., and Shull, H., *J. Chem. Phys.* **35,** 1644 (1961).
60. Hall, G. G., *Proc. Roy. Soc. London* **A205,** 541 (1951).
61. Hartree, D. R., *The Calculation of Atomic Structures,* John Wiley and Sons, Inc., New York, 1957.
62. Boys, S. F., *Proc. Roy. Soc. London* **A201,** 125 (1950); *ibid.* **A127,** 136, 235 (1953); Bernal, M. J. M., and Boys, S. F., *Phil. Trans. Roy. Soc. London* **A245,** 139 (1952); Donath, W. E., *J. Chem. Phys.* **35,** 817 (1961).
63. Nesbet, R. K., and Watson, R. E., *Phys. Rev.* **110,** 1073 (1958).
64. Kolos, W., and Roothaan, C. C. J., *Rev. Mod. Phys.* **32,** 219 (1962).
65. Davidson, E. R., and Jones, L. L., *J. Chem. Phys.* **37,** 2966 (1962).
66. Löwdin, P. O., *Phys. Rev.* **97,** 1474 (1955).
67. Löwdin, P. O., and Shull, H., *Phys. Rev.* **101,** 1730 (1956); Shull, H., and Löwdin, P. O., *J. Chem. Phys.* **30,** 617 (1959).
68. Kelly, H. P., and Sessler, A. M., *Phys. Rev.* **132,** 2091 (1963).
69. Linnet, J. W., and Poe, A. J., *Trans. Faraday Soc.* **47,** 1033 (1951).
70. Sinanoğlu, O., *J. Chem. Phys.* **37,** 191 (1962).
71. Nesbet, R. K., *Quarterly Progress Report, Solid State and Molecular Theory Group,* MIT, July 15, 1956, p. 3; October 15, 1956, p. 47 (unpublished); *Phys. Rev.* **118,** 681 (1960).
72. Sachs, L. M., *Phys. Rev.* **117,** 1504 (1960).

73. Bessis, N., Lefebvre-Brion, H., and Moser, C. M., *Phys. Rev.* **124,** 1124 (1961).
74. Earlier references are cited in reference 50 and reference 53.
75. Nesbet, R. K., *Ann. Phys. N.Y.* **4,** 87 (1958); *Phys. Rev.* **119,** 658 (1960); *ibid.* **122,** 1497 (1961).
76. Fock, V., *Z. Physik* **61,** 126 (1930); *ibid.* **62,** 795 (1930).
77. Hartree, D. R., *The Calculation of Atomic Structures,* John Wiley and Sons, Inc., New York, 1957.
78. Roothaan, C. C. J., *Rev. Mod. Phys.* **23,** 69 (1951).
79. Coulson, C. A., *Proc. Cambridge Phil. Soc.* **34,** 204 (1938).
80. Hall, G. G., *Proc. Roy. Soc. London* **A205,** 541 (1951).
81. Roothaan, C. C. J., and Bagus, P., in *Methods in Computational Physics,* Vol. II, Academic Press, New York, 1963.
82. Roothaan, C. C. J., *Rev. Mod. Phys.* **32,** 179 (1960).
83. Slater, J. C., *Phys. Rev.* **82,** 538 (1951).
84. Pople, J. A., and Nesbet, R. K., *J. Chem. Phys.* **22,** 571 (1954).
85. Berthier, G., *J. Chim. Phys.* **51,** 363 (1954).
86. Nesbet, R. K., Ph.D. Dissertation, University of Cambridge, 1954 (unpublished); *Proc. Roy. Soc. London* **A230,** 312 (1955); *Rev. Mod. Phys.* **35,** 552 (1963).
87. Dirac, P. A. M., *Proc. Cambridge Phil. Soc.* **26,** 376 (1930).
88. Møller, C., and Plesset, M. S., *Phys. Rev.* **46,** 618 (1958).
89. Brillouin, L., *Actualités sci. et ind.* No. 159, Hermann et Cie., Paris, 1934.
90. Nesbet, R. K., and Watson, R. E., *Phys. Rev.* **110,** 1073 (1958); *Ann. Phys. N.Y.* **9,** 260 (1960); Nesbet, R. K., *J. Chem. Phys.* **32,** 1114 (1960); Watson, R. E., *Phys. Rev.* **119,** 170 (1960).
91. Nesbet, R. K., *Rev. Mod. Phys.* **33,** 28 (1961).
92. Löwdin, P. O., *Phys. Rev.* **97,** 1509 (1955).

DIAMOND SYNTHESIS

R. H. WENTORF, JR., *General Electric Research Laboratory, Schenectady, New York*

CONTENTS

I. INTRODUCTION

Lavoisier's discovery that diamond is essentially pure carbon may be regarded as the first of many steps which scientists took in their irregular march toward the goal of diamond synthesis. A surprisingly large number of workers have given some serious attention to this problem, although frequently they did not report their efforts, especially if they were unsuccessful. Bundy *et al.*[1] list many of the early papers in this field. The culmination of these efforts to achieve reproducible diamond synthesis was announced by the General Electric Company in 1955. A few days later a Swedish group announced that they had succeeded in synthesizing diamond.

Figure 1 shows the arrangements of carbon atoms in the lattices of graphite and diamond. Note that in diamond the sheets of puckered hexagonal rings are arranged in an ABCABC sequence, but in perfect crystalline graphite the sheets are arranged in an ABAB sequence. Thus the diamond lattice is not obtained from a simple compression of the graphite lattice; some of the sheets must be shifted as well. This is the nub of the problem.

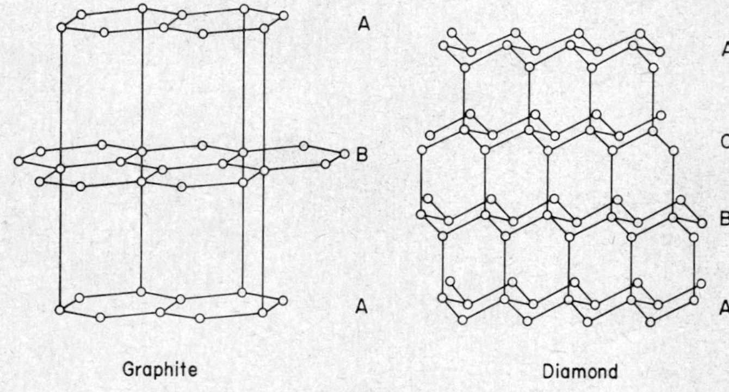

Fig. 1. The arrangement of carbon atoms in graphite and in diamond.

Large pieces of perfectly crystalline graphite are relatively rare; in many natural and in most artificial graphites the perfect regions are rather small and are separated from each other by stacking faults (some of which provide the ABCABC sequence of rhombohedral graphite) or disordered or amorphous regions, particularly in artificial graphites.[2] Nevertheless, the strong bonding between carbon atoms, even in imperfectly crystalline carbons, makes carbon a refractory element which is reluctant to change its form even on a small scale. Bridgman[3] compressed graphite to an estimated 400 kb at 25°C, but no permanent change was noted in the graphite and no diamond was found.

The fact that diamond expands to graphite when strongly heated in vacuum suggested to the early workers that pressure would assist the transformation to diamond. The pressures required for diamond stability were estimated to 1200°K by

Rossini and Jessup[4] from a knowledge of the heats of combustion, specific heats, compressibilities, and thermal expansions of diamond and graphite. Berman and Simon[5] subsequently refined this estimate and made a plausible extrapolation of the equilibrium line between graphite and diamond which has been found to agree

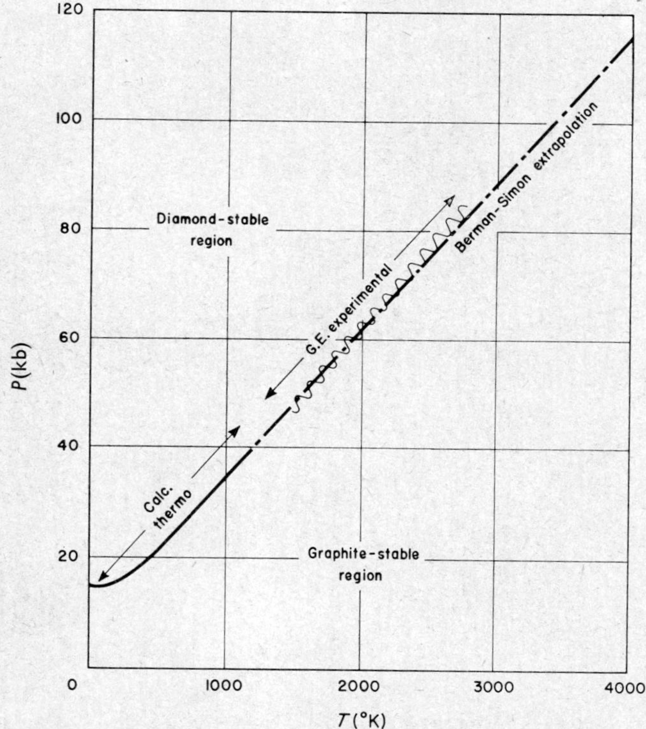

Fig. 2. Part of the phase diagram for carbon.

well with laboratory findings. Figure 2 shows a portion of the phase diagram for carbon in terms of pressure and temperature. The solid line represents the course of the equilibrium line as calculated from thermodynamic data, and the wavy portion was determined experimentally by the growth or dissolution of diamond or graphite, according to Bundy *et al.*[6]

It is sometimes possible for a thermodynamically unstable

phase to form instead of the stable phase. For example, phosphorus vapor commonly condenses to white phosphorus even though the red or black forms are more stable, or white tin may be grown by reducing a solution of a tin salt under conditions of pressure and temperature appropriate for the stability of gray tin. It might be expected that diamond might similarly form where graphite is stable, and much effort has been directed along these lines.[1] However, before certain recent work, to be discussed later, no-one had succeeded in reproducibly making diamond by "low" (i.e. sub-equilibrium) pressure processes. (Perversely enough, it is found that graphite may sometimes form where diamond is thermodynamically stable.) So far nearly all reproducible syntheses have been conducted at pressures and temperatures suitable for the stability of diamond, and a respectable body of knowledge has been accumulated on this subject.

The investigations of the transformation processes from graphite to diamond are particularly interesting, quite apart from their commercial or industrial importance, because pressure is a significant variable and because the change in atomic bonding is pronounced. Analogous examples are not common; the two closest are the transformation of silica to the rutile-like stishovite[7] and the transformation of hexagonal boron nitride to the cubic diamond-like form.[8] The former change involves an increase in the silicon coordination number from 4 to 6. It occurs at pressures ranging from 90 to at least 150 kilobars[9] (1 kilobar = 987 atmospheres) and temperatures of about 1000–2000°C and is not yet very easily accessible for experiments. The transformation in boron nitride superficially resembles that in carbon, except that the cubic boron nitride is stable at somewhat lower pressures and higher temperatures, the catalysts, when used, are different, and the transformation has been observed to take place at 25°C at pressures of about 130 kb,[10] conditions which produce no permanent change in graphite.

Another stimulus for the study of diamond formation is provided by the occurrence in nature of many kinds of diamonds not yet produced in the laboratory. A knowledge of the conditions necessary for the genesis of natural diamonds would shed light on the history of the earth and the composition of its interior.

Even though considerable study has been given to the processes by which diamonds may form, these processes do not always appear to be simple ones. More work is indicated in order to achieve a good understanding of them. Nevertheless, the work performed so far has many interesting, and sometimes puzzling, aspects.

II. THE DIRECT TRANSFORMATION TO DIAMOND

Probably the most straightforward way to make diamond is to expose graphite or other forms of substantially pure carbon to such a severe combination of pressure and temperature that diamond is the product. Although simple in principle, and often attempted in the past,[1] this method was among the last to yield diamond in the laboratory.

A. Explosive Shock Methods

Early indications that elemental carbon, despite its refractoriness, might yield and become diamond were obtained by Alder and Christian[11] from dynamic studies of graphite which had been briefly compressed and heated by the passage through it of a shock wave driven by high explosive. It was observed that when the shock pressure was in the range of 180–400 kb (with corresponding temperatures of about 500–800°K), the free surface velocity of the carbon block fell as the thickness of the block was increased. This behavior implied that a time-dependent, energy-absorbing change was occurring in the compressed carbon. The most reasonable interpretation was that diamond was being formed in times of the order of a few microseconds. Unfortunately no recovery of the shocked carbon was possible in order to examine it directly for the presence of diamond. Experiments which produced peak pressures between 400 and 600 kb (thereby compressing the graphite to densities about that of diamond) indicated that virtually complete transformation to diamond took place in less than a microsecond. At peak pressures between about 600 and 850 kb, with concomitant temperatures above about 1300°K, a marked displacement of the shock Hugoniot curve indicated that the diamond formed by shock had changed into another form which, by analogy with germanium and silicon, was presumed to be a metallic, close-packed liquid.

In 1961 DeCarli and Jamieson[12] announced the results of some work by which they had recovered diamonds from carbon which had been subjected to explosive shock. The momentum of the shocked carbon was transferred to wood and water. It was estimated that the graphite was exposed to pressures of about 300 kb for times of a few microseconds. The recovered carbon was found to contain about 10% of diamond, in the form of small (0.01–0.1 mm) lumps made up of tiny crystals whose size was estimated to be in the range 1–10 microns. The diamonds were separated from unchanged graphite by bromoform, in which diamond sinks but graphite floats. The diamonds were identified by X-ray and electron diffraction tests. Interestingly enough, it was found that highly perfect, crystalline natural graphite usually did not produce diamond under these conditions of pressure, temperature, and time.

The equations of state of the various forms of graphite carbon are not well known and the phase changes complicate dynamic measurements on the compressed material, so that it is difficult to estimate the temperature of the compressed carbon behind the shock wave in such experiments. According to the estimates of Alder and Christian,[11] perfectly crystalline graphite would reach a temperature of about 600°K on being shock-loaded to 300 kb; higher temperatures, perhaps over 2000°K, might be reached in the more compressible, less perfectly crystalline, less dense artificial graphites. By analogy, one would expect the less dense portions of a piece of imperfectly crystalline graphite to be briefly heated hotter than neighboring denser regions. Furthermore, graphite is most compressible along its *c*-axis, so that some crystallites may be heated hotter than others, depending upon their orientations relative to the shock front. Added to these thermal inhomogeneities is the extensive shearing which takes place in the material at a plane shock front, which could facilitate the transformation by bringing small sheets of graphite into the diamond stacking sequence.

The carbon exists in the compressed, agitated state for only a few microseconds—time enough for many atomic vibrations or changes in chemical bonding. Ultimately it expands to its final density and becomes cooler in the process. It is difficult to estimate the temperature of the carbon after expansion, for we do not know how much irreversible work has been done on it. Evidently its average temperature is well below the temperature at which diamond turns partially to graphite, i.e., about 1500°K, because one would expect such extremely small diamond crystallites to graphitize rather easily. The rarefaction wave, like the compression wave, must produce some local shear as it passes through the material, and this shearing tends to break up large crystalline aggregates. The stresses of bringing the shocked material to rest also do their share of fragmentation.

The X-ray diffraction patterns of the diamond produced by this shock-loading method indicated that the diamond crystallite size was of the order of a few microns; the K-alpha doublet could not be resolved for the (331) reflection with copper radiation. The diamonds were black in color, which indicates that the crystals were rich in defects. It is likely that in the short times available for the reaction only a rather imperfect lattice could be constructed.

Perhaps no diamond was observed to form in the highly crystalline graphite used by DeCarli and Jamieson because the shock temperatures attained in this material were too low for the transformation to take place in the time intervals available, or because the micro-crystallites of graphite which existed in the shocked material were too large to be re-stacked in the diamond sequence.

B. Static High-Pressure Methods

Bundy[13] has described the formation of diamond directly from other kinds of carbon at pressures of about 120 kb and higher, and temperatures in the range of 3000–4100°K. In these experiments small bars of carbon, about $0.5 \times 1 \times 2$ mm, were buried in insulating stone or ceramic and were compressed in a double-opposed-piston pressure device shown in cross-section in Fig. 3. The pressures were estimated by reference to the transition in iron at 131 kb. After the desired pressure had been reached, the bar was heated to the desired temperature in a few milliseconds by a pulse of electric current from a bank of capacitors. The temperature of the bar remained fairly constant for about 10 milliseconds before the dying current surge permitted cooling. The current and voltage were displayed on an oscilloscope and

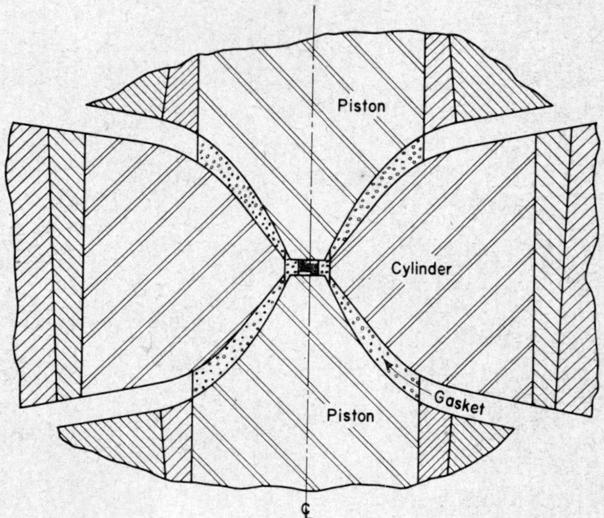

Fig. 3. High-pressure apparatus.

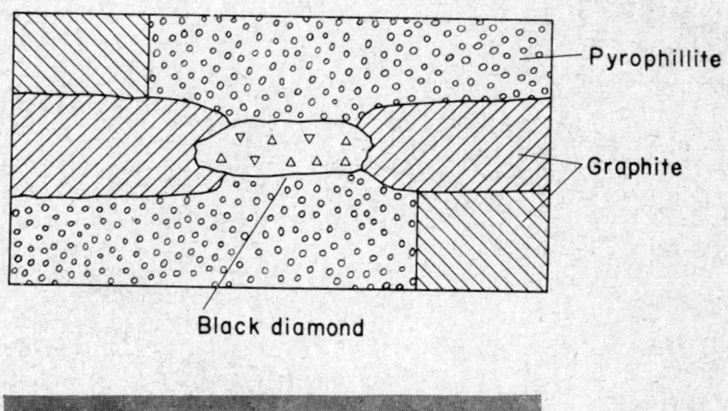

Fig. 4. Diamond formed in carbon bar.

recorded photographically, so that fairly accurate estimates could be made of the temperature and electrical resistance of the carbon. The heating and cooling periods were relatively long compared with those obtainable in shock experiments, but yet were short enough to prevent extensive chemical reaction between the hot carbon and its containers.

The diamond formed in these experiments was recovered as dark polycrystalline masses. Figure 4 shows a graphite bar whose center region has been converted to diamond. Various kinds of starting carbon, such as spectroscopic graphite, natural graphites, pyrolytic carbon, or lampblack, were all found to produce diamond, in contradistinction to the shock-wave method. Some small differences in heating and diamond-forming behavior were noted among the various carbons as the result of differing electrical resistances or anisotropy; the diamonds formed from highly crystalline graphites showed preferred orientations. Ordinarily the diamond produced is a good electrical insulator, but when a graphite containing 0.2–0.3% boron was used, semiconducting diamond was formed.

A summary of these experiments is represented graphically in Fig. 5. It will be noted that the threshold pressure for diamond formation during these brief heating times increases somewhat as the temperature falls. It is difficult to obtain kinetic data on the rate of transformation to diamond in these experiments because the formation rate is a steep function of temperature, and when a small patch of diamond forms, its higher electrical resistance to the heating current raises the temperature around it so that more diamond forms, etc., in runaway fashion.

These experiments, coupled with similar experiments at lower pressures on the melting of graphite, have permitted an estimate of the location of the triple point between diamond, graphite, and molten carbon. This triple point is shown in Fig. 5 at about 4100°K and 125 kb, well within the range expected from the work of other experimenters.

Thus it is plain that it is not necessary for the carbon to be molten in order that diamond should form; the transformation occurs in the hot solid matter. However, an interesting result was obtained when carbon was heated to about 4500°K at a pressure of about 175 kb. These conditions are well above the triple point, and one would expect that first diamond would form, then melt, and then, on cooling, freeze back to diamond again, so that diamond would be the material recovered from the experiment. But the product which was obtained consisted

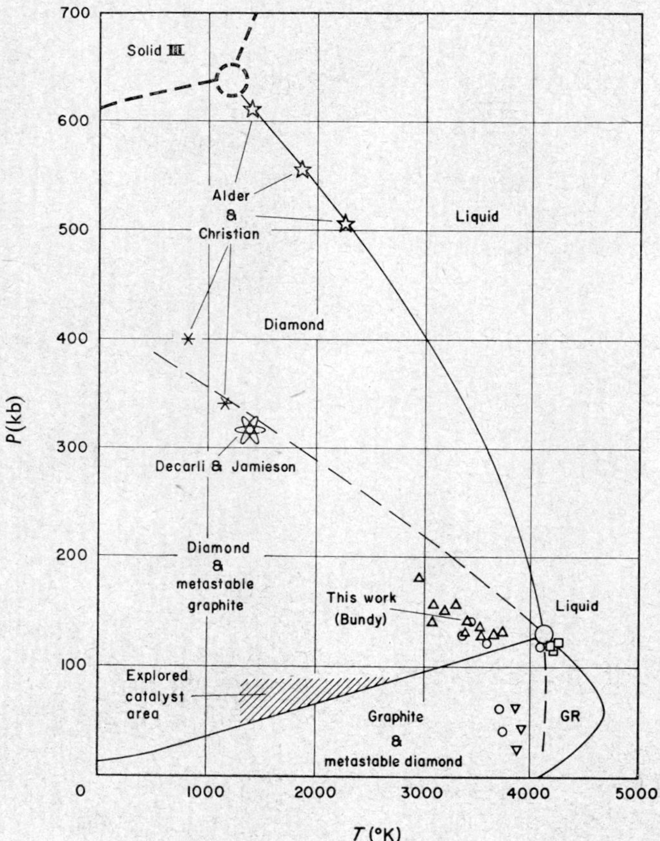

Fig. 5. Phase diagram for carbon.

of an outer diamond skin containing a core mixture of graphite and diamond. The skin was diamond formed from solid graphite; the core was produced from freezing carbon above the triple-point pressure. It appears that both diamond and graphite may nucleate and grow from the liquid, and that all of the graphite formed did not have an opportunity to transform to diamond during the few milliseconds available while cooling. On the other hand, no diamond was ever observed to have formed out of molten carbon below about 120 kb.

Further studies[14] of the transformation of various carbons to diamond have been performed at lower temperatures held for longer times than in the flash-heating experiments. The same pressure apparatus was used to expose the carbons to various pressures in the range 115–175 kb, but the heating was done by means of a heavy alternating current. The carbons were contained in small titanium tubes in order to reduce considerably the local overheating resulting from diamond formation. It was found that at temperatures in the range 1500–2500°K, most carbons turned at least partially into diamond. At the lower temperatures, heating times of about an hour were necessary to produce a diamond content of about 10%, but at about 2500°K most of the carbon was changed to diamond in a few minutes at 150 kb. It was found that some carbons formed diamond with greater readiness than others. A highly refined natural graphite powder yielded much more diamond in shorter times and at lower temperatures than, for example, spectroscopic carbon. Furthermore, the rate of diamond formation did not follow a simple first-order reaction rate but instead was extremely high at first, as if many proto-diamond nuclei were present which grew rapidly during the first moments of heating, and afterward grew only slowly.

X-ray diffraction analyses of many of the diamond-bearing products of these experiments displayed two strong but not sharp lines at spacings of 3.1 and 2.19 A which arise from a new phase of carbon which is evidently intermediate in structure between diamond and graphite. Traces of this structure were also noted in several of the flash-heated carbons. So far it has not been possible to isolate this kind of carbon in pure form. It is associated with diamond formation from the more perfectly crystalline graphites but does not appear to be a necessary intermediate since it does not invariably accompany diamond formation from all kinds of carbon.

Many hydrocarbons, when thermally decomposed at about 2000°C and 150 kb, yielded diamond as the crystalline product. However, naphthalene and anthracene yielded graphite and no diamond even after several minutes of heating. This is one of many examples in which graphite forms and persists at thermodynamic conditions appropriate for the stability of diamond; more such examples will be described below.

III. THE CATALYST METHOD

A. General

The use of a catalyst permits diamond to form from non-diamond carbons at pressures and temperatures significantly lower than those found necessary for the direct conversion processes. Thus pressures and temperatures of the order of 60 kb and 1500°C suffice for rapid diamond formation in the presence of effective catalysts such as nickel, iron, etc.

The details of the General Electric method[15] were published in 1959. Although there are differences in viewpoint concerning the nature of the diamond-forming reaction, this method consists in subjecting carbon with metal to temperatures high enough to provide an active (usually molten) metallic phase at pressures high enough for diamond to be stable. Diamond thereupon forms and is preserved by cooling the system before reducing the pressure. This method has been repeated in many laboratories all over the world and finds use in the manufacture of diamonds for industrial purposes.

Following the publication of the early papers,[15,16] many new findings have been made in this field of diamond synthesis. Although much work has been devoted to the problem it has so many facets that it is not yet possible to say that this method is completely understood in all its details. Some of these findings will be discussed in this paper in an attempt to make the picture of this interesting phase transformation more complete.

The work described here was performed in the "Belt" high-pressure, high-temperature apparatus.[17] It is shown in cross-section in Fig. 6. Two conical pistons of cemented tungsten carbide, supported by steel rings, are driven into a corresponding short conical cylinder of cemented tungsten carbide, also supported by steel rings. A compressible gasket, made of a pyrophyllite and steel "sandwich", seals the gaps between pistons and the cylinder. The reaction zone is heated by passing a heavy current through it via the pistons; pyrophyllite or other refractory materials provide thermal and electrical insulation for the reaction zone. The pressure in the chamber is estimated from the force applied to the pistons; the forces required to produce certain reference transitions in bismuth (25.3 kb, 88 kb) and barium (59 kb) are determined so as to provide a pressure calibration. The temperature may be estimated by means of thermocouples

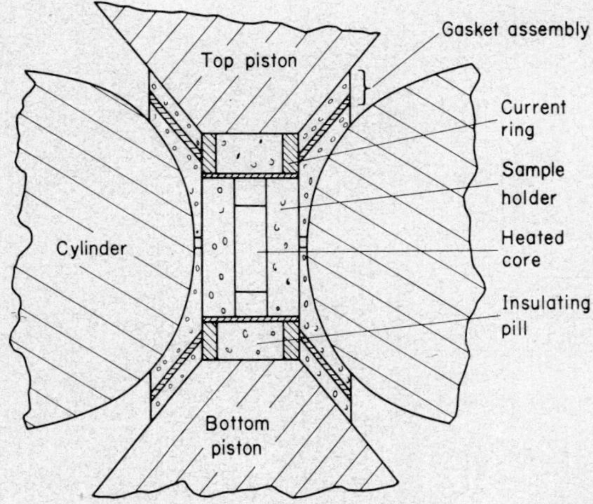

Fig. 6. "Belt" high-pressure, high-temperature apparatus.

led into the chamber, or, once a given cell configuration has been calibrated, by means of the heating power dissipated in the sample. The problems of constructing and operating very high pressure equipment are discussed at length in many publications (see, for example, *Very High Pressure Techniques*, Butterworths, London, 1961) and need not be dealt with here. The "Belt" apparatus is capable of maintaining diamond synthesis conditions for periods of many hours.

Figure 7, taken from Wentorf and Bovenkerk,[18] shows a pressure–temperature plot of part of the equilibrium line between graphite and diamond as well as the melting line of a typical catalyst metal, here nickel, and the melting line of the metal–carbon eutectic. Diamond formation takes place in the shaded region where diamond is stable and the metallic phase is molten.[6]

At the low-temperature end of this region cubes tend to form; at the higher temperature octahedra are favored. Higher temperatures favor more clear and colorless crystals, and, as the synthesis temperature is reduced, the crystals tend to become yellowish, greenish, and finally black at the lowest temperatures, which are still above the melting temperature of the metallic phase.

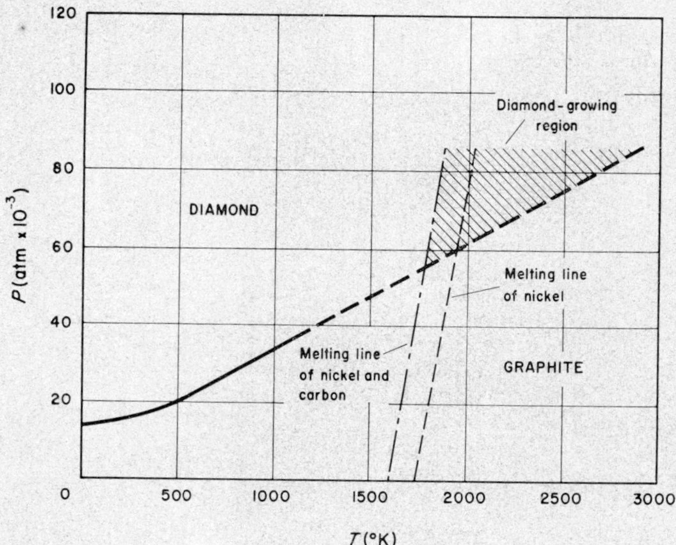

Fig. 7. Diamond synthesis conditions.

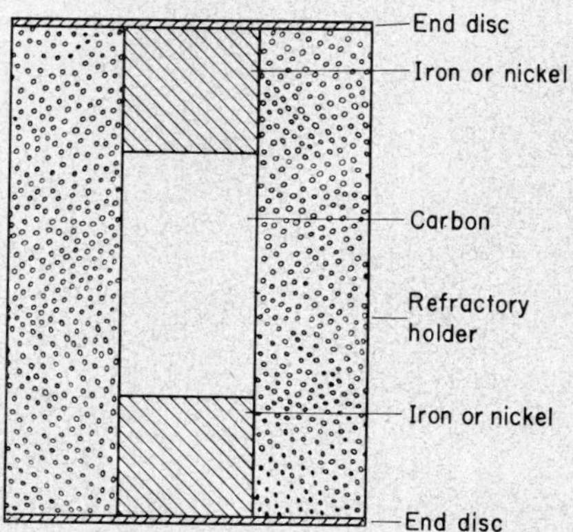

Fig. 8. Reaction cell for diamond making.

Figure 8, taken from Bovenkerk *et al.*,[15] shows a typical arrangement of carbon and catalyst metal in a reaction cell. As the temperature is increased at high pressures, the metal next to the carbon melts and diamond forms out of it. As time goes by, a thin metallic layer, about 0.1 mm thick, advances into the carbon and leaves diamond behind it. The rate of advance of the film depends upon the pressure, temperature, and other factors which will be discussed later, but it may be as high as 0.5 mm/sec. This rate would correspond to a diffusion constant of about 0.01 cm^2/sec for carbon in a liquid metallic phase, which is in the range expected for these circumstances.

B. Diamond vs. Graphite Formation

(1) *Metal–Carbon Systems*

In all the metal–carbon systems studied so far, with the possible exception of tantalum, it appears that the metallic phase must

Fig. 9. Freshly grown cube face of diamond.

be molten in order for diamond to form. The metallic phase may be a carbide or a eutectic mixture of metal and carbide, or of metal and carbon, or combinations of these.

An illustration of the action of the molten phase is furnished by Fig. 9, showing a cube face of a diamond crystal freshly grown with nickel. This photomicrograph was made by E. H. Hull of a crystal grown by H. M. Strong at the General Electric Research Laboratory. The regular surface pattern may be explained as follows. The fit between the crystal lattices of diamond and nickel is close; both are cubic, with an a_0 of 3.56 and 3.52 A, respectively. It was evident in the early days of diamond synthesis that crystals of nickel forming near a diamond

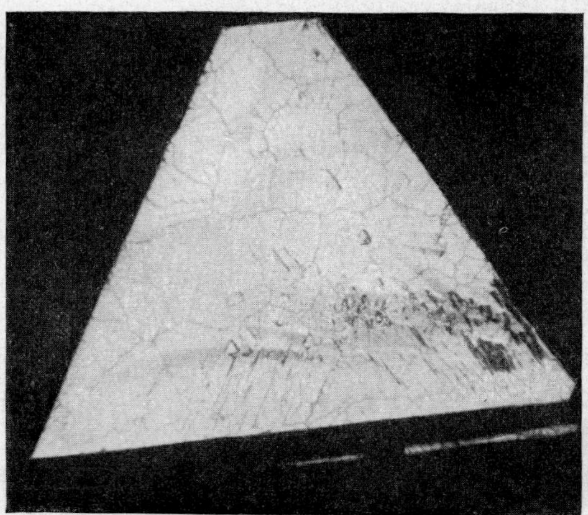

Fig. 10. Freshly grown octahedral face of diamond.

were often trapped in the diamond. These inclusions of nickel could be detected in various ways, such as by microscopic examination, X-ray diffraction,[19] or by the ferromagnetic properties of the diamonds. (They have the same Curie temperature as pure nickel.) Thus, as the nickel–carbon melt begins to freeze, the nickel forms dendrites of solid metal oriented on the diamond face. Where the metal is solid, diamond no longer forms, but where it is still molten, diamond continues to grow. Thus, if the rate of cooling is slow enough for this partially frozen state to exist for a few seconds, a network of shallow ditches forms on the crystal face, corresponding to the dendrites of nickel. The ditches are about 1 micron deep and are arranged with the square symmetry of the cube face.

Figure 10 is a similar photomicrograph of the octahedral face of the same diamond which shows this effect in a different way.

Here the face was nearly covered by grains of nickel, oriented parallel to the diamond, which began to freeze simultaneously at several places on the face. Underneath these frozen portions very little diamond growth occurred. However, where these grains joined one another the fit was not quite perfect, and there existed grain boundaries richer in impurities and having slightly lower melting temperatures than the grains themselves. Diamond grew briefly at these boundaries until they, too, froze, and so irregular ridges of diamond are found on the octahedral face corresponding to the grain boundaries. (One can photograph the same diamond face before and after the metal film is removed to establish this correspondence.)

Such surface patterns could be the result of the expected variations in the diffusion rate of carbon through the various media of molten, solid, or grain-boundary material. On the other hand, it is frequently observed that carbon may be transported by diffusion through frozen nickel–carbon mixtures at diamond-stable pressures but that it is always deposited as graphite without the formation of any diamond along with the graphite. It is evidently not a matter of the rate of carbon transport or the mere precipitation of carbon from solid (or liquid) solution: diamond formation from iron or nickel appears to be connected with certain molten metallic phases.

So far it has not been possible to establish the composition of these molten phases during diamond growth on account of their inaccessibility inside the high-pressure, high-temperature apparatus and the small volume of the active diamond-forming region. However, some information may be gleaned from the metallographic examination of quenched metal–carbon mixtures.

To date the nickel–carbon system has received the most study. At cooling rates of 1000°C/min or slower, the formerly molten metallic phase is found to consist of graphite flakes or nodules scattered through nickel. The carbon is always observed to have separated initially as graphite, even at pressures as high as 150 kb, and it remains graphite unless the pressure is very high (e.g., 130–150 kb, based upon the transition in iron at 131 kb) and the graphite is given time to transform directly to diamond in the solid state in the manner described earlier. When the pressure is below about 50 kb, a nickel melt saturated with carbon and quenched at about 20,000°C/sec shows nickel grains surrounded by a network of graphite. When the pressure is above about 60 kb, i.e., high enough for diamond to form, the quenched melt shows a eutectic structure of nickel and a carbide. H. M. Strong has obtained good evidence that this carbide decomposes during slow cooling to yield graphite and nickel. It is not yet possible to say whether this carbide is present as a separate species in the molten nickel alloy during diamond formation; electrolysis experiments, described briefly below, have been found to be unexpectedly complex and difficult to interpret.

When molten iron–carbon systems from which diamond is forming are cooled at about 200°C/sec, they are found to contain diamond, Fe_3C, and some graphite. Platinum–carbon melts from which diamond is forming, when similarly cooled, yielded diamond, platinum, and flakes of graphite.

The presence of small amounts of graphite is noteworthy. This graphite is usually found as tiny flakes scattered through the metallic phase, and tends to be more concentrated near the cool end of a temperature gradient. Thus, not quite all of the carbon present in a specimen can be converted to diamond, even when no metal carbides are recovered, as when nickel and carbon are used, and when nickel is melted in contact with diamond only, at diamond-stable pressures, part of the diamond dissolves in the nickel and is recovered as graphite on quenching. H. M. Strong found that even the nickel inclusions which are sometimes found in diamonds grown in the presence of nickel contain flakes of graphite. Thus, at pressures and temperatures appropriate for the stability of diamond, and in the intimate presence of diamond, graphite forms nevertheless. The evidence is that often graphite forms from the frozen metallic phase as it cools and the solubility of carbon in it falls. Graphite has also been observed to have crystallized, concurrently with diamond, out of molten metallic systems such as nickel–carbon or iron–carbon, particularly when the pressure and temperature were close to the graphite–diamond equilibrium but definitely still on the diamond side.

(2) *Carbon Solvents Which Deposit Graphite*

There exist other solvents for carbon besides the metals of Groups VI, VII, and VIII of the periodic table. According to

Bever and Floe,[20] the solubility of carbon in molten copper varies from about 0.0001% to 0.006% by weight in the temperature range 1200–1800°C.

Other known solvents for carbon at one atmosphere include cuprous oxide and kimberlite, the rock in which diamond occurs in the South African pipes.

In the course of experimentation at high pressures and temperatures at the General Electric Research Laboratory, it was observed that substances such as pyrophyllite (a hydrated alumino-silicate) or serpentine (a hydrated magnesium silicate) at temperatures of the order of 1600°C were capable of transporting carbon from hot regions to cooler regions.[21] The carbon was deposited as graphite even though the pressures and temperatures were suitable for diamond stability. A number of other substances were then examined as potential carbon solvents at high pressures and temperatures, using a reaction cell arrangement as shown in Fig. 11.

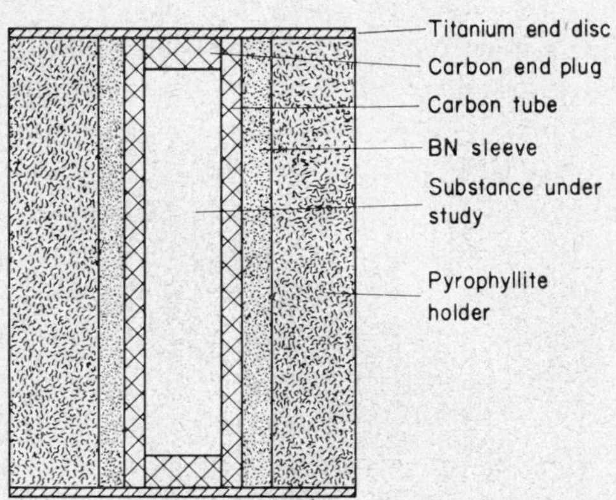

Fig. 11. Carbon solubility test cell.

In these experiments the carbon tube was heated by passing a heavy current through it. The tube was buried in packed hexagonal boron nitride in order to avoid any entrance of pyrophyllite through a crack into the carbon tube, where it might combine with or change the carbon-solvent powers of the substance in the tube. (For example, KCl alone exhibits scarcely any carbon-solvent powers at 1700°C and 60 kb, but a mixture of KCl and pyrophyllite may transport carbon in significant amounts.) The greater heat loss from the ends of the tube provided a temperature gradient; the midpoint of the tube was several hundred degrees hotter than the ends. Thus if carbon were somewhat soluble in the material in the tube, and more soluble in hotter material, carbon would be dissolved from the tube wall near the midlength and deposited near the ends. In this way rather small solubilities could be detected but not accurately measured.

It was found that CdO, ZnO, Cu_2O, CuCl, AgCl, serpentine, kimberlite, pyrophyllite, and pyrophyllite mixed with KCl were effective in dissolving, transporting, and precipitating carbon at 65 kb and temperatures in the range 1400–1800°C. Undoubtedly other carbon solvents exist. After heating periods of from 5–20 min the precipitated carbon appeared as black, shiny plates of generally hexagonal outline ranging in size from 20–200 microns, or as black spheroidal masses having diameters of 50–100 microns. Only graphite was observed to precipitate from these solvents under the above conditions. When measured diamond-seed crystals were placed in these carbon solvents they remained unchanged in size but acquired a coating of graphite. By comparison with the solvent activity of nickel under similar conditions, the solubility of carbon in the above non-metallic substances was estimated to be between 0.01 and 1%. CdO and Cu_2O were partly reduced to the metal, although Cd or Cu metal alone was not observed to be an effective carbon solvent in similar tests. It is possible that some of the carbon was transported in the form of CO: the decomposition of CO to carbon and CO_2 is favored at lower temperatures.

It is probably worth mentioning that the metals of Groups VI, VII, and VIII form various compounds with Co, H, etc., in accordance with a rule by which the electrons necessary to form the noble-gas grouping about the metal atom are furnished by the adducts. Moist CuCl is also observed to form a carbonyl, $CuCl \cdot H_2O \cdot CO$; silver sulfate in H_2SO_4 absorbs a mole of CO per mole of sulfate. The analogies between carbonyl formation and carbon-solvent powers suggest that carbon atoms in solution may donate some of their electrons to the solvent atoms, so that some positively charged carbon ions might be present. This view is supported by the concentration-cell experiments of Sanbongi and Ohtani[22] in their study of the electromotive forces of solutions of carbon in molten iron; they found the solution containing the most carbon took on a negative charge. The average charge per carbon atom was estimated to be about 2.3 electronic units.

It is possible to use direct current for heating carbon and metal samples similar to the one shown in Fig. 8, and interesting effects have been observed in such experiments. These effects are the consequences of the potential gradient associated with the direct current; usually the voltage drop across the heated zone is of the order of 1 or 2 volts. It appears that various kinds of carbon ions are present and drift in the applied field so that diamond and/or graphite may be formed preferentially in certain regions. In spite of considerable work with these phenomena, they are still not well understood. For example, with steep potential gradients, diamond formation is preferred from carbon transported as positively charged ions; with smaller potential gradients the opposite appears to be true. Sometimes the local formation of diamond will increase the local potential gradient (diamond is a poor electrical conductor under the prevailing conditions) so that instabilities and rapid fluctuations ensue. Studies of these systems are continuing in order to determine the charges, lifetimes, and compositions of the various species present in the active melts and their relationships to the formation of diamond and/or graphite.

C. Driving Forces for the Graphite–Diamond Reaction

The experimental evidence accumulated so far indicates that the main driving force for the graphite–diamond reaction is the free energy difference between non-diamond carbon and diamond at the synthesis pressure and temperature. A gram of graphite at diamond-stable conditions has a higher free energy than a gram of diamond, so that graphite will then tend to be more soluble than diamond in the same solvent. In this way the solution will become supersaturated with respect to diamond, and, if the proper catalytic effects operate, diamond will crystallize from the solution.

One may estimate the free energy differences between graphite and diamond at various diamond synthesis conditions from the course of the diamond–graphite equilibrium line[6] and the probable differences in specific volume between the two phases. Thus, at constant temperature,

$$\left(\frac{\partial \Delta G}{\partial p}\right)_T = \Delta V \tag{1}$$

or

$$\Delta G = \int_{p_1}^{p_2} V\, dp \cong \Delta V \cdot (p_2 - p_1) \tag{2}$$

At pressures and temperatures around 60 kb and 2000°K, the atomic volume of diamond may be taken to be about 12/3.52 = 3.41 cm^3; the atomic volume of graphite, 12/2.25 = 5.33 cm^3. (The small effects of compression and thermal expansion tend to cancel each other.) Thus ΔV is about 1.92 cm^3, and a pressure of 1 kb above the equilibrium line would increase the free energy difference between graphite and diamond by about 47 calories/g-atom.

In reality some of the crystal faces of diamond have slightly higher free energies than others, depending upon the impurities adsorbed on the various faces, their intrinsic surface energies, and the temperature. Thus the equilibrium line is not perfectly sharp, and the diamond crystals may have different habits, depending upon the relative rates of nucleation and growth on the various faces.

Furthermore, it is plain that the free energies of the various carbonaceous starting materials may vary. A carbon which had been converted to highly crystalline graphite by exposure to very high temperatures (e.g., 3500°C) would be expected to have, on the average, a slightly lower free energy than a carbon prepared by graphitization at lower temperatures (e.g., 2400°C). The less perfectly crystalline carbons prepared at lower temperatures contain significant amounts of more or less amorphous carbon mixed in with small crystalline regions;[2] such amorphous regions would have higher free energy contents than the crystalline parts.

Thus the driving force for diamond formation would be expected to be greater when less perfectly crystalline graphitic carbon is used as the carbon source, and one might expect that some of this amorphous material, given sufficient opportunity, would dissolve and recrystallize as more perfectly crystalline graphite if it did not immediately change to diamond. The net driving force for diamond formation would not be expected to remain constant with time at constant pressure and temperature but would be expected to fall if an inhomogeneous graphitic carbon were the source material. The driving forces for diamond formation from highly amorphous carbons such as lamp black, charcoal, etc. would be expected to be high, but such materials often contain large amounts of adsorbed impurities which may react with the metal-solvent catalyst or interfere with diamond formation. These impurity effects often easily overpower the structural effects.

An obvious way to create a free energy difference favoring the formation of diamond from graphite is to use a temperature change, either in space or in time, so that graphite dissolves in hot material and diamond grows from cooler material. However, in practice, it appears that free energy differences created by temperature differences are not as significant as the isothermal free energy differences which exist between diamond and non-diamond carbon at diamond-stable pressures. At only moderate diamond-stable overpressures, diamond is found to form from graphite, via the usual metallic film, whether the graphite be hotter or cooler than the diamond which is being formed.

Figure 12 shows a cross-section of a reaction zone assembly in which a strong temperature gradient was produced. The top half of the cell was heated by passing a heavy current through it from the top and then radially outward at the midlength to the surrounding pressure vessel. A temperature gradient of about 300°C/mm was produced in the bottom half of the cell by thermal conduction from the hot midlength to the cool bottom end; no heating current flowed through the bottom half of the cell. Pieces of carbon and catalyst metal could be arranged in different ways in the cell. In the arrangement shown in the figure, the (molten) metal is hotter than the graphite, but at pressures of the order of 60–65 kb it was found that diamonds formed at the metal–graphite interface just as when the temperature gradient was essentially zero, or reversed in direction. The structural free energy difference between graphite and diamond overpowered

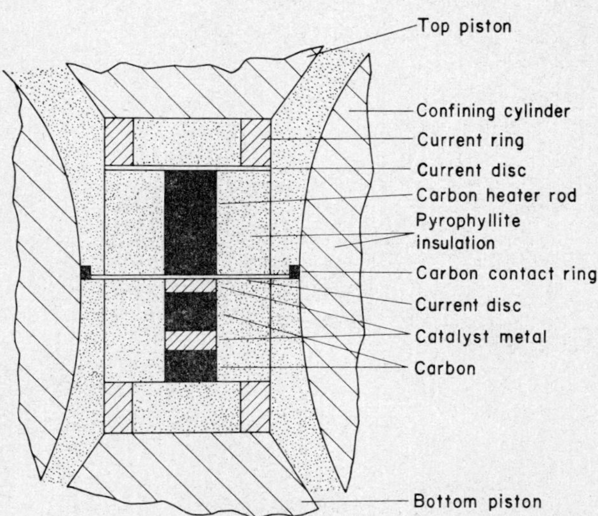

Fig. 12. Reaction cell for studying the effects of temperature gradient on diamond formation.

that produced by the steep temperature gradient. The metallic film separating diamond and graphite was about 0.1 mm thick, so that the diamond was about 30°C hotter than the graphite out of which it formed in this experiment.

D. Diamond Nucleation and Growth Phenomena

(1) *Quasi-solution of Carbon*

As mentioned earlier, at high temperatures and high over-pressures the nucleation rate is very rapid and many small diamonds form with an octahedral habit. Twinning on the octahedral face becomes very common. Cyclic five-fold octahedral face twins may appear; one is shown in Fig. 13. It is likely that such a symmetrically grown crystal did not grow segment by segment, but rather grew from a five-fold twin nucleus. Such a nucleus would be common in the spectroscopic carbon used as the carbon source, for this material is not perfectly crystalline and would be expected to contain many of the five-membered ring systems so common in carbon compounds. Thus, one might look to the source carbon for some diamond nuclei, and suppose that at rapid growth and nucleation rates the carbon is not entirely dispersed as atoms in the metallic solution but may retain part of its solid structure.

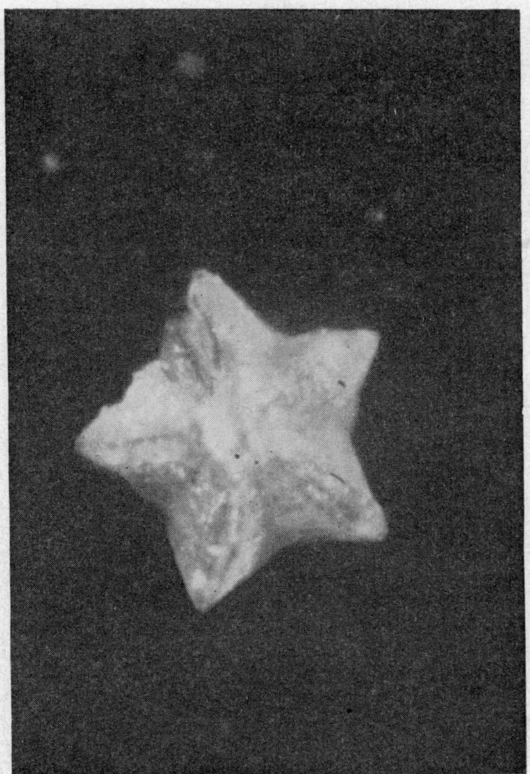

Fig. 13. Cyclic five-fold octahedral face twin of diamond.

Other evidence which favors such a view is furnished by experiments in which highly crystalline natural graphite was converted to diamond by the use of catalyst metals in a reaction cell as shown in Fig. 14. Thin discs of Ticonderoga or Madagascar graphite were placed between two pieces of catalyst metal, e.g., nickel, manganese, or iron. Each disc of graphite was cut from a sheet of graphite so that the graphite *c*-axis coincided with the axis of the cell. The *a*-axes of successive graphite discs could be

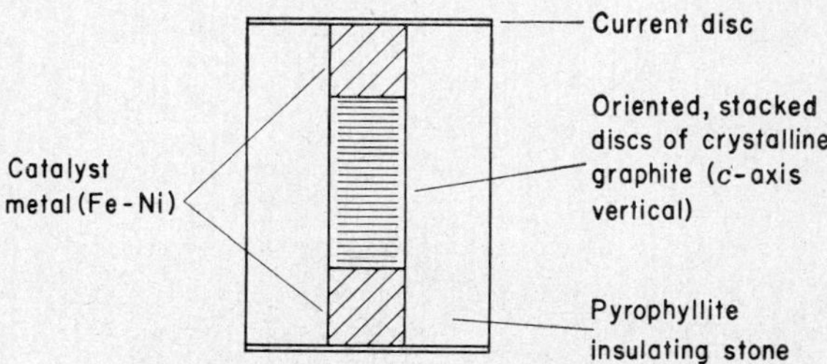

Fig. 14. Reaction cell using oriented graphite as carbon source.

aligned parallel to each other by observation of the characteristic slip lines on the flat surfaces of the discs. These slip lines run at 60° to one another on a single face; their orientation relative to the hexagonal rings of graphite is shown in Fig. 15. This was determined from an X-ray diffraction analysis of the Ticonderoga graphite by W. L. Roth of the General Electric Research Laboratory.

The cells were exposed to pressures of about 65 kb and heated for times of the order of 30 sec to temperatures of about 1700°C so that not all of the graphite was transformed to diamond. Upon removal from the pressure apparatus, the diamondiferous portions broke away easily from the remaining graphite, though frequently a thin layer of graphite remained attached to parts of the diamonds.

It was evident from microscopic examination that nearly all of the diamonds at a particular interface were oriented parallel to each other, with two octahedral faces parallel to the sheets of carbon atoms in the original graphite. A photomicrograph of the region of such a freshly grown diamond mass is shown in Fig. 16. Here one may see the triangular outline of the octahedral face of a diamond. Stuck to this diamond face is a thin sheet of graphite which shows the characteristic slip lines. From the orientation of these slip lines and the orientation of the edges of the diamond octahedral face, one may determine, as shown in Fig. 15, that the diamond formed as if by collapse of the graphite without any significant net rotation of the sheets of carbon atoms.

This effect was clearly not an oriented nucleation followed by regular growth, or an orientation of the metallic catalyst film on existing diamond, because when part of the Ticonderoga graphite

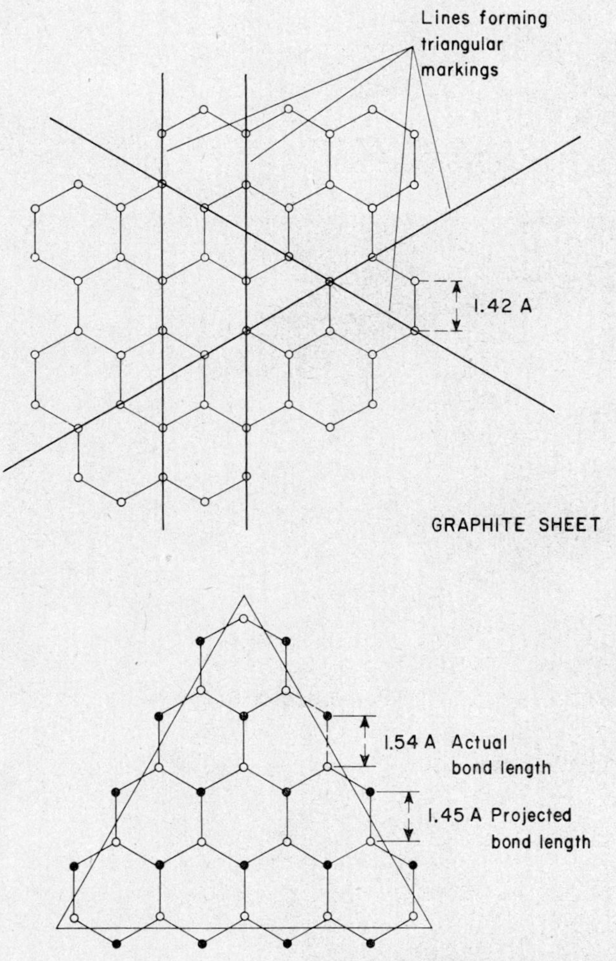

Fig. 15. Schematic drawing of the relationships between (a) the sheets of atoms in graphite and the slip lines visible on Ticonderoga graphite, and (b) the sheets of carbon atoms in diamond and the boundaries of an octahedral (111) face.

was replaced by spectroscopic carbon, the diamonds grown from the spectroscopic carbon were oriented in random fashion while the growth from the oriented graphite was oriented as described above.

When the pressure and temperature of the growth conditions were lowered so that the diamonds grew in the black, cube habit,

Fig. 16. Photomicrograph of interface between Ticonderoga graphite and diamond freshly grown from it.

H. P. Bovenkerk found that the diamonds were randomly oriented even when grown from oriented graphite. The rate of diamond growth is much slower under these conditions than when octahedra form, and apparently the detailed mechanisms of growth are not quite the same.

Thus it appears that complete solution into atoms of carbon, while it probably occurs, is not essential for the formation of diamond; evidently the diamonds may be rapidly built up, at least in part, from blocks of atoms. One would not expect such blocks to always fit together perfectly at their peripheries, nor would the atoms of catalyst always have an opportunity to escape from underneath or around the blocks as they are laid down. It is found that rapidly grown diamonds contain detectable amounts of metal, graphite, or carbides.

(2) *Solution of Carbon*

Cannon[23] has performed some experiments on the isotopic composition of a spectroscopic graphite and the diamonds made from it using nickel or nickel-manganese catalyst at about 58 kb and 1700°K. The graphite and the diamonds were burned in oxygen and the mass ratio 45:44 of the produced CO_2 was measured. It was found that the diamonds tended to be about 3% richer in ^{13}C than the starting graphite, and the exteriors of the diamond crystals were somewhat richer in ^{13}C than their interiors. These results imply that much of the carbon which was transformed to diamond in these experiments existed as monatomic carbon in the liquid metal; otherwise hardly any isotopic separation effects would be expected.

(3) *Growth Habits*

Bovenkerk[24] has given a detailed paper on the growth habits of synthetic diamonds.

So far growth spirals are only rarely found on the octahedral faces of synthesized diamonds; instead one sees an accumulation of layers parallel to the octahedral faces. Each octahedral layer then apparently must be nucleated afresh before growth can spread out across the face, and the conditions necessary for steady growth are very similar to those required for initial nucleation. This has the practical effect of making the growth of large crystals difficult to control because new layers may be nucleated before the old ones have been completed.

At lower synthesis pressures, in the 45–50 kb range, the diamonds tend to have the cube habit, but the octahedral faces may eventually predominate after long slow growth. Growth terraces or even coarse spirals may appear on the cube surfaces.[25] The growth mechanism appears to be different from that which produces octahedra, for material is being added mainly to the cube faces instead of to the octahedral faces. At somewhat higher pressures, cube, octahedral, and higher order faces may appear on the same crystal. At still higher synthesis pressures,

60 kb and over, and at higher temperatures the octahedral habit is predominant.

A possible explanation for this change in habit with pressure may be sought in the nature of the cubic and octahedral surfaces of diamond. The octahedral faces exhibit the lowest surface density of available carbon bonds (the octahedral planes are the planes of easy cleavage). The density of available carbon bonds is much higher on cube faces. Accordingly, less pressure would be

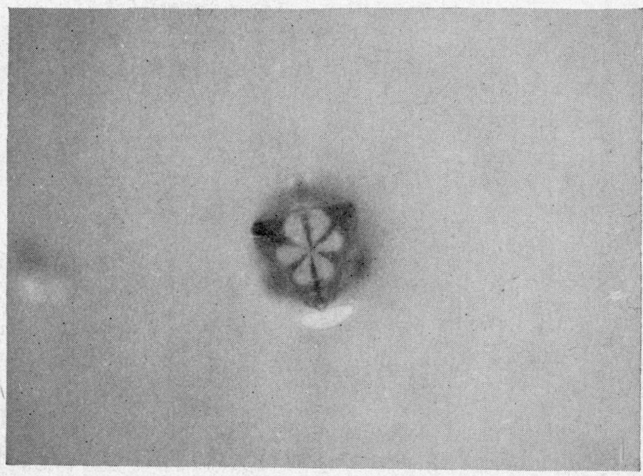

Fig. 17. Synthetic diamond octahedron showing strings of impurities along cube axes.

required to hold oncoming material in the diamond configuration to a cube face than to an octahedral face; not only diamond but also other impurities would tend to adhere to cube faces, and the cube faces should be the most likely regions for nucleation of fresh growing surfaces even when octahedra are being formed.

This view is consistent with the formation of dark, impure cubes at low temperatures and pressures and also with the appearance of octahedra such as the one shown in Fig. 17. In such a crystal the cube axes are delineated by strings of impurities. If the growth rate of octahedra is increased by increasing the pressure, the carbon may not be supplied to the growing faces as fast as new layers are nucleated and hoppered crystals may appear, often strung out atop each other along the cube axis which points toward the source of non-diamond carbon. Such crystals are shown in Fig. 18. During the growth of such crystals the leading edges of new octahedral layers grow away from the source of carbon and eventually starve. The linear growth rate may be as high as 0.5 mm/sec.

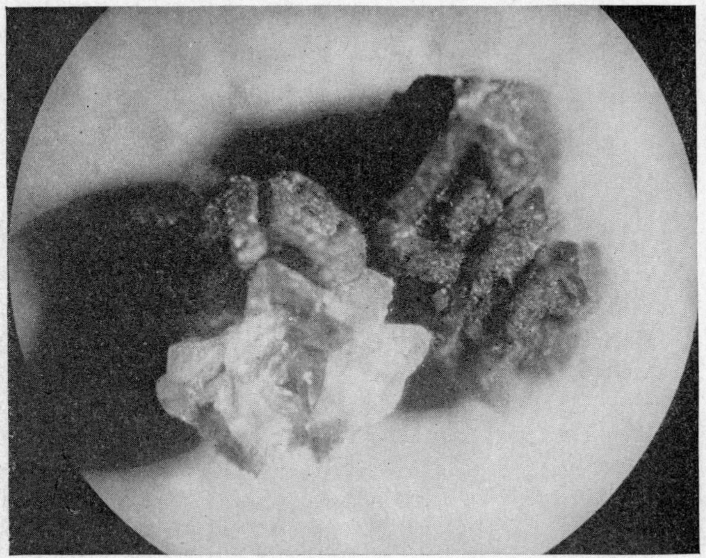

Fig. 18. Cluster of diamond octahedra grown rapidly along a cube axis.

At higher pressures, then, the growing diamonds tend to nucleate fastest along their cube directions, but cube faces rarely appear because the spreading rate of octahedral layers exceeds the spreading rate of cube-face layers. At lower pressures the octahedral faces do not tend to spread as rapidly, presumably because their lower surface concentration of available carbon bonds results in a lower net rate of acquisition of new material. At higher pressures the cube faces spread more slowly than octahedral faces because impurities (e.g., metal) must be rejected from the cube faces before new material can spread over them.

The "stickiness" of cube faces, which permits their growth at low pressures, may also hinder their growth relative to other faces. At extreme low temperatures the desorption of impurities from cube faces may proceed so slowly that their growth is virtually halted, and very little diamond will form unless the pressure is high enough to permit the growth and spreading of octahedral surfaces.

It is also found that the crystal habit may be affected by the kind of metal catalyst used; for example, iron favors octahedra, and nickel-chromium favors twinned octahedra.

Thus one may regard the variations of crystal habit and color with synthesis temperature and pressure as the result of the interplay of the forces of pressure and chemisorption upon the different crystal faces, although, due to the difficulties of measurement, these views have not yet been checked quantitatively and so may require revision. Probably the keys to the understanding of the specific catalytic powers required for the formation of diamond instead of graphite lie among factors such as those just considered above.

(4) *Formation of Unstable Phases and Catalytic Processes*

The formation of a phase outside its field of stability has been observed in other systems[26] and is sometimes referred to as "Ostwald's rule".[27] The explanation of the effect rests upon the condition that the unstable phase represents a level of free energy intermediate between the high level occupied by the substance in its dissolved or dispersed state and the low level occupied by the substance in its stable phase. In leaving the high level, the substance may become trapped in the intermediate, or metastable, level on its way to the low level. The permanence of this trapped or metastable phase depends on the magnitude of the excitation energy required to make it labile; because carbon is a refractory element, the excitation energy may be considerable. The function of the metallic liquid phase thus resembles that of a catalyst, for through it the desired reaction, that of diamond formation, is favored, while the undesired reaction, the formation of graphite, is repressed, although not always completely. The solid metallic phase, however, appears to act as a catalyst for graphite formation and in some way hinders the formation of diamond.

The above discussion of energy levels may, of course, be unsatisfactory to some-one who seeks a more detailed description of the processes of diamond and graphite formation. In this search one faces complexities of catalytic activity similar to those noted in other chemical systems, such as those involving enzymes, acid-catalyzed reactions, finely divided platinum, metal–hydrocarbon complexes, etc., and in addition it is much more difficult to make direct observations upon the active systems involved in diamond growth. The exact mechanisms by which the formation of diamond or graphite is directed are still objects of research.

However, it immediately appears unlikely that strong metal–carbon epitaxial factors are involved; i.e., the metal probably does not form a template for diamond. For first, it is noted that metal atoms of widely differing sizes are effective in diamond formation: e.g., Ni has an $a_0 = 3.52$ A; Pt has $a_0 = 3.91$ A; both are f.c.c. Secondly, the metallic phase must apparently be molten, not crystalline, in order for diamond to form. It is possible that next to a growing diamond surface the catalyst metal atoms are not so free as in the bulk liquid metal, and in this way a tenuous epitaxy may exist between metal and diamond, so that further diamond growth is favored. The arrangement of the metal atoms in this special diamond-directing layer could not be quite the same as in the bulk solid metal because graphite crystallizes from the solid.

As mentioned earlier, the crystallization of diamond may be strongly directed by the source graphite, both at moderate pressures in the presence of a catalyst or at pressures of 120 kb and over in the absence of a catalyst. And in the pressure range 115–160 kb there are observed to be marked differences in the tendencies of various forms of carbon to form diamond in the absence of a catalyst. Thus, graphite or some crystalline phase or phases present in normal graphite may serve as nuclei for diamond formation. Accordingly, there are indications that some of the intermediates involved in the formation of diamond may be extremely rich in carbon, and at pressures in the range 40–70 kb one of the functions of the catalyst metal is to permit the formation of these intermediates. Such intermediates may exist in the source carbon or be formed there by diffusion of metal atoms into the carbon, or they may be built up from carbon atoms in solution. The kinds of metal atoms necessary for the construction of these postulated intermediates may vary somewhat among the metals known to be catalysts, so that metal mixtures are also effective catalysts. However, the bonding of the metal atoms in the intermediates is necessarily not strong, otherwise diamond could not form from them, and by the same token the metal atoms will desert the intermediates in favor of the solid metal lattice should the system become too cool.

The existence of such intermediates has not been proven directly so far on account of the inaccessibility of the reaction zone at high pressures, but no satisfactory explanation of catalyzed diamond formation seems to be possible without postulating them, and it is difficult to find evidence which disproves their existence. One might expect their structures to be like hydrogen-poor polymeric forms of known transition metal–hydrocarbon complexes such as iron dicyclopentadiene, chromium dibenzene, etc. Studies of the growth of crystals other than diamond have suggested that colloidal or microcrystalline aggregates are involved in the deposition of new material on growing crystal faces.[28]

IV. METASTABLE DIAMOND FORMATION

Much effort has been devoted to the formation of diamond at pressures where graphite is stable, but until the past few years very little success had been obtained.[29]

Eversole[30] has been granted U.S. Patent No. 3030188 for a process for making diamond which involves passing gas containing a methyl group, e.g., methane, ethane, propane, methyl chloride, or acetone, over diamond powder at about 1000°C. The gas pressure is about 0.2 mm Hg. In a few hours the diamonds become covered with a black coating which is predominantly graphite; this black coating may be removed by treatment with sulfuric-chromic acid or by heating at about 1000°C for a few hours in hydrogen at 10–50 atmospheres pressure. After such a treatment, 0–1 micron diamond powder shows a weight gain of about 1–4%; electron-diffraction analyses of the treated surfaces show that they are diamond. A repetition of the process produces further gains in weight. So far the process has not been described in a scientific publication nor is it known whether it is used commercially.

Takasu[31] has recently reported the formation of diamond at low pressures during the decomposition of methyl chlorosilane on the (111) faces of silicon crystals. The process is carried out at about 600–1400°C at a pressure of 2–4 atmospheres. On the silicon there first forms an epitaxial layer of β-SiC, then an intermediate described as Si_2C, and finally, thin patches of diamond about 30–100 A thick and several dozen microns in diameter. The growth of diamond seems to halt after a few hours. Transmission electron-diffraction patterns of the diamond platelets show rings corresponding to the expected reflections for diamond. The platelets burn in oxygen. No diamond is observed when SiC is used instead of silicon for the substrate.

V. SEMICONDUCTING DIAMONDS

A. Natural Semiconducting Diamond

In 1952 Custers[32] published a paper describing the behavior of semiconducting natural diamonds which were classified as type II-b. Since then such diamonds have been studied by many workers,[33,34] and they have been determined to be *p*-type semiconductors with activation energies for conduction of about 0.35 eV, carrier concentrations of about 2×10^{13} cm^{-3}, and mobilities of about 1500 cm^2 V^{-1}. Corresponding to the conduction activation energy of 0.35 eV, there appears a strong, temperature-dependent, infrared absorption at about 3.57 μ. The crystals are frequently blue in color.

B. Growth of Semiconducting Diamond

It was found that semiconducting diamonds could be prepared in the laboratory by growing new diamond crystals in the presence of small amounts of boron, beryllium, or aluminum, which were added to the synthesis mixture.[35] Rather small amounts of boron (0.1% or less) sufficed to produce *p*-type crystals having an intense blue color. The crystals prepared in the presence of aluminum or beryllium tended to be colorless. All of the crystals prepared by these techniques were *p*-type, as determined from measurements of the thermoelectric power. Thus, it appears that the impurity atoms substitute for carbon atoms in the diamond lattice and behave as electron acceptors.

The electrical conductivity of a crystal increases with increasing concentration of impurity atoms in the synthesis mixture, although not all of the conductivity may arise from simple acceptor sites. Resistivities of 10^5 ohm-cm are common and resistivities as low as 10^2 ohm-cm have been measured. The conductivity of a crystal may not be uniform throughout the crystal; this condition is also frequently observed in natural type II-b diamonds.

The activation energy for conduction appears to depend somewhat on the particular impurity incorporated. From a plot of the logarithm of the resistivity vs. reciprocal temperature in the range 0–440°C, the activation energies were estimated as follows:

Al-doped, Fe catalyst:	0.32 eV
B-doped, Fe-Ni catalyst:	0.17–0.18 eV
Be-doped, Ni catalyst:	0.2–0.35 eV

Work is proceeding slowly on measurements of the Hall effect and absorption spectra of such semiconducting diamonds, but their small size (of the order of 0.5–1 mm) and non-uniformity present difficulties.

One might expect that diamonds containing nitrogen as an impurity might be *n*-type semiconductors. Many natural diamonds are known to contain significant amounts of nitrogen,[36] of the order of 10^{17}–10^{20} atm/cm^3, but their electrical resistivity is high, in the range 10^{13} ohm-cm. Electron spin resonance measurements made by H. H. Woodbury, G. W. Ludwig, C. M. Huggins, and P. Cannon at the General Electric Research Laboratory indicated that many synthetic diamonds also contain significant amounts of nitrogen.[37] However, neither these crystals nor crystals grown in the presence of nitrogenous compounds such as urea, cyanides, etc. have resistivities much below 10^{12} ohm-cm. Apparently the donor level is rather far removed from the conduction band with nitrogen substitution;[38] from a chemical point of view this would correspond to the high stability of the CN^- ion.

C. Diffusion of Impurities into Diamonds

The diamond lattice is not a close-packed one, and, as Fig. 1 shows, the holes in it are just about large enough, on the average, to contain atoms the size of carbon, about 1.55 A in effective diameter. Of course no real diamond is a perfect single crystal, and one would expect real diamonds to contain imperfections such as vacancies, impurities, or lines or planes of dislocations along which foreign atoms might diffuse relatively rapidly, particularly if the foreign atoms were small. It is observed that the densities of most relatively pure diamonds are about 0.5–1% lower than the theoretical density calculated from the lattice parameters determined by X-ray diffraction. (Diamonds containing metallic inclusions may of course have higher densities.)

It was found that boron diffused into natural or synthetic diamond crystals relatively easily at high pressures and temperatures.[34] For example, a ten-minute exposure at 60 kb and 1700°C to a 50:50 mixture of graphite and B_4C changed the average resistance at 25°C of 0.3 mm synthetic diamond crystals from about 10^{10} ohm to about 10^4 ohm. Natural diamonds were similarly affected. Such crystals were *p*-type semiconductors with activation energies for conduction of about 0.02 to 0.05 eV.

By examining broken crystals, Bovenkerk found that the highly conducting region was confined to a thin skin on the crystal surface. When such conducting diamonds were re-exposed to high pressures and temperatures in the presence of graphite alone, the average conductivity usually fell and the conduction activation energy tended to increase, implying that the high surface concentration of boron had been reduced by diffusion into both diamond and graphite.

After being exposed to boron-rich mixtures at high temperatures and pressures for times of the order of an hour, the diamond surfaces tended to change color toward blue or blue-gray. The crystals were not usually uniformly colored, which suggests that the permeability to boron is not the same everywhere in a particular crystal. Electron-diffraction patterns of the surfaces of boron-treated diamonds showed reflections not allowed for plain diamond but which would arise from a diamond lattice in whose holes other atoms (presumably boron) were regularly present.

The higher the treatment temperature with boron, the higher is the resulting conductivity of the crystals, provided that the pressure is high enough for the diamonds not to turn into graphite while being exposed to boron. If one assumes that the resulting conductivity of similar starting crystals is proportional to their boron contents, then a plot of the logarithm of the average crystal resistance vs. the reciprocal of the treatment temperature (for the same treatment time) should furnish a measure of the activation energy involved in the diffusion process. Such a plot indicates that this activation energy is about 80 kcal/mole, a value comparable with chemical bonding energies. Presumably the entry of boron into diamond crystals involves the breaking of chemical bonds among boron and carbon atoms.

It was noticed that more boron entered the crystals at higher pressures (65 kb) than at lower pressures (8–35 kb) at approximately the same temperature, about 1600°C. The crystals treated at lower pressures often showed evidence of partial graphitization such as black spots. Pressure may enhance the rate of entry of boron for two reasons: the rate of graphitization is reduced at higher pressures so that boron may enter the diamond surface faster than the surface is being changed to graphite, and the partial pressure of fugacity of boron is higher, relative to pure diamond, the higher the pressure. The diamond lattice is relatively stiff, so that external pressures of 60 kb do not compress it much or restrict diffusion inside it.

VI. CONCLUDING REMARKS

The discussions in this paper have been rather more qualitative than quantitative, principally because the field of endeavor is relatively new. Quantitative measurements at high pressures and temperatures are still difficult and may still often be subject to errors of 10% or more. Furthermore, it is always desirable to have a qualitative understanding before attempting to make quantitative measurements; probably future work, which will make use of ever-improving techniques, will more precisely define the phenomena observed. It is hoped that this present treatment has at least exposed to view some of the many factors which must be taken into account in the study of the transformation of graphite into diamond or in similar transformations.

Obviously in the presentation of a subject as complex as this one the author needs to draw upon work other than his own, and he is indebted to his colleagues not only for that work of theirs which he has described or referred to, but also for many valuable discussions with them. He particularly wishes to thank A. J. Nerad, H. M. Strong, H. P. Bovenkerk, E. H. Hull, F. P. Bundy, and P. Cannon for their contributions to this paper.

References

1. Bundy, F. P., Hall, H. T., Strong, H. M., and Wentorf, R. H., Jr., *Nature* **176**, 51 (1955).
2. Franklin, R. E., *Acta Cryst.* **4**, 253 (1951).
3. Bridgman, P. W., *J. Appl. Phys.* **12**, 461 (1941).
4. Rossini, F. D., and Jessup, R. S., *J. Res. Natl. Bur. Std.* **21**, 491 (1938).
5. Berman, R., and Simon, F., *Z. Elektrochem.* **59**, 333 (1955).
6. Bundy, F. P., Bovenkerk, H. P., Strong, H. M., and Wentorf, R. H., Jr., *J. Chem. Phys.* **35**, 383 (1961).
7. Stishov, S. M., and Popova, S. V., *Geokhimiya* **10**, 837 (1961).
8. Wentorf, R. H., Jr., *J. Chem. Phys.* **34**, 809 (1961).
9. Wentorf, R. H., Jr., *J. Geophys. Res.* **67**, 3648 (1962).
10. Bundy, F. P., and Wentorf, R. H., Jr., *J. Chem. Phys.* **38**, 1144 (1963).
11. Alder, B. J., and Christian, R. H., *Phys. Rev. Letters* **7**, 367 (1961).
12. DeCarli, P. S., and Jamieson, J. C., *Science* **133**, 1821 (1961).
13. Bundy, F. P., *J. Chem. Phys.* **38**, 631 (1963); *Science* **137**, 1057 (1962).
14. Wentorf, R. H., Jr. Unpublished work.
15. Bovenkerk, H. P., Bundy, F. P., Hall, H. T., Strong, H. M., and Wentorf, R. H., Jr., *Nature* **184**, 1094 (1959).
16. Liander H., and Lundblad, E., *Arkiv Kemi*, **16** (9), 139 (1960).
17. Hall, H. T., *Rev. Sci. Instr.* **31**, 125 (1960).
18. Wentorf, R. H., Jr., and Bovenkerk, H. P., *Astrophys. J.* **134**, 995 (1961).
19. Grenville-Wells, H. J., and Lonsdale, K., *Nature* **181**, 758 (1958).
20. Bever, M. J., and Floe, C. F., *Trans. AIME* **166**, 128 (1946).
21. Wentorf, R. H., Jr. Paper presented at symposium "The Physics and Chemistry of High Pressures," London, 1962; published by Society of Chemical Industry, London, 1963, pp. 185–190.
22. Sanbongi, K., and Ohtani, M., *Sci. Rep. Res. Inst. Tohoku Univ.* **A5** (3), 263 (1953).
23. Cannon, P., *J. Am. Chem. Soc.* **84**, 4253 (1962).
24. Bovenkerk, H. P., *Progress in Very High Pressure Research*, John Wiley, New York, 1961, pp. 58–69.
25. Tolansky, S., *Proc. Roy. Soc. London* **A263**, 31 (1961).
26. Muan, A., and Hahn, W. C., Jr., *J. Phys. Chem.* **63**, 1826 (1959).
27. Ostwald, W., *Z. Physik. Chem.* **22**, 289 (1897).
28. Grisdale, R. O., in *The Art and Science of Growing Crystals*, John Wiley, New York, 1963, pp. 163–173.
29. Bridgman, P. W., *Sci. Am.* **193**, November, 1955.
30. Eversole, W. G., U.S. Patent 3030188.

31. Takasu, S., Private communication.
32. Custers, J. F. H., *Physica* **18**, 489 (1952).
33. Austin, I. G., and Wolfe, R., *Proc. Phys. Soc. London* **B69**, 329 (1956).
34. Drake Bell, M., and Leivo, W. J., *Phys. Rev.* **111**, 1227 (1958).
35. Wentorf, R. H., Jr., and Bovenkerk, H. P., *J. Chem. Phys.* **36**, 1987 (1962).
36. Kaiser, W., and Bond, W. L., *Phys. Rev.* **115**, 857 (1959).
37. Huggins, C. M., and Cannon, P., *Nature* **194**, 829 (1962).
38. Urlau, R. R., Logie, H. J., and Nabarro, F. R. N., *Proc. Phys. Soc. London* **78**, 256 (1961).

AUTHOR INDEX

The names of authors and joint authors of chapters in the present book, and the page numbers at which these chapters begin, are printed in **bold type.** Page numbers of bibliographical references as listed at the ends of the chapters are in *italics* and citations in the text are in ordinary type.

SUBJECT INDEX

The letters ff indicate that the stated subject is mentioned again on one or both of the next two following pages. Page numbers in **bold type** denote the beginning of a chapter about the entry, where a list of the chapter subdivisions may be found. These subdivision heads are not all re-indexed here.